Horse
Management

Horse Management

SECOND EDITION

Edited by

JOHN HICKMAN

Emeritus Reader in Animal Surgery
University of Cambridge, Cambridge, UK

1987

ACADEMIC PRESS

Harcourt Brace Jovanovich, Publishers

London   San Diego   New York
Boston   Sydney   Tokyo   Toronto

ACADEMIC PRESS LIMITED
24–28 Oval Road, London NW1 7DX

United States Edition published by
ACADEMIC PRESS, INC.
San Diego, CA 92101

British Library Cataloguing in Publication Data

Horse management. — 2nd ed.
1. Horses
I. Hickman, John
636.1′083   SF285.3

ISBN case: 0-12-347218-0

Phototypeset by
Dobbie Typesetting Service, Plymouth, Devon

Printed in Great Britain by
Galliard (Printers) Ltd., Great Yarmouth

CONTRIBUTORS

A. F. CLARKE

Department of Animal Husbandry
Langford House
Langford
Bristol BS18 7DU, UK

P. J. DAVISON

1 King's Court
London W6 0RN, UK

F. L. C. DAWSON

The White House
Hadstock
Near Linton
Cambridgeshire CB1 6PF, UK

D. R. ELLIS

Reynolds House
166 High Street
Newmarket
Suffolk CB8 9AH, UK

J. HICKMAN

The White Lodge
57 High Street
Haslingfield
Cambridgeshire CB3 7JP, UK

M. HUMPHREY

174 Links Way
Croxley Green
Hertfordshire WD3 3RN, UK

D. W. SAINSBURY

Department of Clinical Veterinary Medicine
Madingley Road
Cambridge CB3 0ES, UK

D. H. SNOW

Animal Health Trust
Research Station
P O Box 5
Snailwell Road
Newmarket
Suffolk CB3 7DW, UK

D. R. WISE   Department of Clinical Veterinary Medicine
                              Madingley Road
                              Cambridge CB3 0ES, UK

I. M. WRIGHT   Department of Clinical Veterinary Medicine
                              Madingley Road
                              Cambridge CB3 0ES, UK

PREFACE

Good horse management is essential for the well-being and welfare of horses. Many injuries and diseases are due to a lack of knowledge and attention to detail, and can be prevented by the practice of good management which ensures that horses are kept fit and healthy.

In this second edition chapters have been introduced dealing with important subjects such as Care of the Mare and Foal, Stable Environment in Relation to the Control of Respiratory Diseases and Basic Training, from handling the foal through to introduction to jumping.

The original chapters dealing with Equine Reproduction, Housing the Horse and Exercise and Training have been revised and brought up to date. Food and Feeding has been replaced by Nutrition of Horses which presents a more practical aspect of the subject. The chapter on The Foot and Shoeing has been rewritten and Stable Management, which contained much covered in other chapters of the book, has been replaced by a chapter entitled Equipment for the Care and Protection of Horses.

This edition reflects the recent advances in horse management. Some aspects are developing rapidly and therefore of topical interest; others have become established practices, or have stood the test of time but still require critical evaluation. Inevitably, controversial new ideas and views are expressed in the book which it is hoped will stimulate thought and further studies.

Without knowledge of the basic scientific problems underlying the many aspects of horse management it is difficult to understand the problems involved. For this reason every effort has been made to keep each chapter concise and logical, while providing a balanced approach between the scientific aspects and practical applications of the subject.

This volume is primarily written for veterinary and agricultural students, but should also serve as a useful reference book for veterinary surgeons — and all who keep and breed horses.

J. Hickman

CONTENTS

1   Equine Reproduction

F. L. M. DAWSON

INTRODUCTION

Although a number of reputable books has been written specifically on the subject of equine reproduction, it is hoped that this short contribution may interest the general reader, as including some information not readily available elsewhere, and points based on practical experience.

In many tropical and sub-tropical countries the source of agricultural power is, increasingly, the horse, mule, and ass, replacing much bovine and human labour (FAO, 1966, 1976, 1984). (Between 1976 and 1983 world numbers rose by a million head.) The literature contains few data on horse-breeding operations in these low latitudes where the requirement for work stock is increasing. Most of the reproductive data has been accumulated in the temperate zones, where the horse is strictly a seasonal breeder. Both sexes' breeding activity is controlled by the impact of daylight on the sense organs: activity increases and decreases in close relationship with daylength. The extent to which the knowledge of reproduction now current is applicable to breeding operations in low latitudes can be evaluated only in the light of future experience.

REPRODUCTIVE FUNCTION IN THE MALE

Applied anatomy

In general the anatomy of the stallion's genital tract conforms to the basic mammalian pattern. The gross anatomy of the reproductive system is described by Sisson and Grossman (1938); more concise accounts are given by Evans *et al.* (1977) and Rossdale (1981).

A most important anatomical feature, not always recognised, is the prolongation of the urethra at the tip of the penis. This free tube, up to an inch long, called the urethral process is surrounded by a circular fossa lined

Horse Management 2nd edition
ISBN: 0-12-347218-0 case

by thin skin rather than mucous membrane; its upper part is termed the urethral sinus. It is important to be able readily to gain access to the urethral fossa and sinus for purposes both of diagnostic sampling and treatment, as these are predilection sites for micro-organisms which may interfere with fertility. The fossa is formed in the substance of the glans penis and is protected by the inner of the two preputial sheaths.

The unstriped retractor penis muscles are of practical importance, because their resistance has to be overcome to draw the penis out of its external sheath for washing or sampling. Sampling may be the more important though washing (traditional) is done more frequently. These muscles originate from the external sphincter muscle of the anus, pass along the ventral surface of the penis and split up into bundles as they approach the glans area. At this level the bundles are inserted into the fibrous tunica albuginea which encloses the erectile tissue of the organ.

Physiology of ejaculation

The stallion in action makes up to ten thrusts before ejaculation results, then there may be several ejaculatory surges, detectable as palpation waves palpable in the urethra, and each one signalled by a "flag-wagging" movement of the tail. The procedure is reasonably rapid, but varies a good deal in speed from horse to horse and is always much slower than in the bull.

Techniques have been developed using an artificial vagina (AV), which make possible the collection of a complete ejaculate by fractions, in the order in which these leave the urethra. Bader and Huttenrauch (1966) used a long plastic tube attached to the distal end of the AV with leads to a series of beakers, each beaker received 20 ml volumes of semen in turn during ejaculation. Tischner (1976) designed an AV which enables a succession of funnels to be held directly to the urethral orifice. These methods of study have added a great deal to knowledge of ejaculate composition and the process of ejaculation, previously derived by analysis of the secretions of different parts of the tract.

The ejaculate is in three fractions. The first part, about 20 ml, is watery, including the secretions of the Cowper's and bulbo-urethral glands. The second part, 40–80 ml, is the creamy sperm-rich fraction which contains, too, the secretions of the prostate gland and ampullae. This fraction has a high content of sulphydryl compounds, mainly ergothioneine, produced in the ampullae. In some stallions at certain seasons of the year its appearance is followed by the "gel" fraction, which in appearance resembles melting snowflakes. The final fraction, up to 20 ml, contains citric and lactic acids; it comes, like the gel, from the vesicular glands.

Components of the seminal plasma The function of the accessory secretions is not clearly understood and there is evidence to suggest they may be spermicidal. There is no clear consensus of opinion on what useful function they may serve. Centrifugation to separate and concentrate sperm from the accessory secretions definitely depresses sperm viability, and this effect may be in part the result of exposure to, for example, citric acid, an exposure which can be avoided by fractional collection of the sperm-rich portion.

However, there has also been a suggestion that the concentrations of sulphydryl compounds found naturally in some stallions may reach spermicidal levels in the sperm-rich fraction. The work supporting this suggestion was based on a routine, now outmoded, method of collecting penile and vaginal drip, "dismount samples", from stallions as they left the mare, and analysing these samples as representative of the whole ejaculate (Werthessen *et al.*, 1957).

Investigations by workers in Kentucky found that the higher the concentration of sulphydryl compounds, the quicker the sperm in the sample died, with lower conception rates resulting. The technique used was to equilibrate standard washed sperm from a known fertile horse with a sample of the seminal plasma under study, using the standard animal's own seminal plasma as a control. Results appeared to suggest that the fertility of certain horses might be much improved by separating off their plasma from the sperms, and replacing it by a substitute with a lower sulphydryl concentration prior to artificial insemination.

These findings were challenged in Britain by Mann *et al.* (1957), who showed that dismount samples bore no relation at all to the composition of the ejaculate as a whole. Their work suggested that: (i) there was a greater tendency in Kentucky than in Britain for mares to be served when not fully oestrous; (ii) that in the case of such a mare, more of the sperm and sulphydryl-rich fraction might have been included in the dismount sample especially from a mare with a half-closed cervix; and (iii) hence, an artificial correlation between high sulphydryl concentrations and poor fertility might have resulted. Haag (1959) in Kentucky remained unconvinced, and this interesting controversy is still unresolved.

Nishikawa (1959) in Japan stated that equine sperm had been found to survive better in bovine seminal plasma than in that of the donor.

Like those of spermatozoa, the concentrations of the various components of semen plasma vary greatly. For instance, ergothioneine level may vary from 7 mg/100 ml or less, to 18 mg/100 ml or more. As quoted by Rossdale (1981) a possibly typical sample was analysed as follows: ergothioneine 7.6 mg%, citric acid 26 mg%, fructose 15 mg%, phosphorus 17 mg%, lactic acid 12 mg%, and urea 3 mg%.

Puberty

In the horse the onset of male puberty is very gradual. The age of puberty has been defined arbitrarily by Hauer *et al.* (1970) as that age at which an ejaculate contains at least 100 million sperm, 10% of which are motile; an average level of spermatogenesis reached at 16 months of age. A typical ejaculate from an adult should contain not less than 8000 million sperm, with 40% motile (Pickett, 1981). In another experiment (Carson and Thompson, 1979), the colts' semen, at 16 months contained a high proportion of sperms having secondary abnormalities, and the level of testosterone in the blood plasma, of 132 pg/ml, was less than 10% of the adult figure.

Besides testosterone considerable quantities of oestrogenic hormones are found in the urine and blood of the colt and stallion. It is believed that interstitial cells of the testis secrete these oestrogens, hence the increasing concentration of oestrogen in the urine may be valid as a guide to the rate of attainment of puberty (Pigon *et al.*, 1962).

Not one of ten geldings and eight colt foals under a year old ever exceeded a concentration of 4 μg of oestrogen per 100 ml of urine, while the levels in eight adult stallions varied from 1440 to 3000 μg with a mean of 2068 μg. Sixteen colts aged from 12 to 24 months showed very variable levels; thus three colts, each 22 months old, showed levels respectively of 3, 60 and 393 μg, while the level in twelve three-year-olds varied from 200 μg upwards. So it seems clear that by this method of assessment some three-year-olds are very far from adult, considered as potential sires.

Evans (1977) recommended that during the peak of the breeding season, two-year-olds may serve twice, three-year-olds six times, and adults 14 times weekly; he considered that not more than six or seven mares should be served by a two-year-old colt. Early this century, in breeding carthorses on British farms, it was considered that a two-year-old could serve from 10 to 20 mares in a season, whereas 80 were sent to an adult stallion, who would leave about 50 of them in foal.

Covering capacity of stallions

The rate of semen production by the adult stallion may be expected to vary quite widely with the season, nutritional status, and the conditions of his access to mares, but there is also individual variation when these factors are the same for each member of a group.

Bielanski and Wierzbowski (1961) reported a trial in which each of seven stallions was given the opportunity to serve, within 24-hour periods, as often as he wished; this on 14 consecutive occasions occurring at intervals of five days, and at no other time. Three of the seven horses lost interest after only six, ten, and ten occasions respectively, while four were still willing to serve on the 14th occasion. The seven averaged four ejaculates each at the first test, four of them averaged three ejaculates per test, with an individual sperm yield per test varying from 13 to 28 thousand million. One of these horses ejaculated ten times on a single occasion.

Pechnikov (1960) collected 13 ejaculates from one horse in one day; these showed a progressive fall in volume and density but sperm motility was maintained. Eight days later an ejaculate was obtained which showed that recovery from the depletion was complete. However when 32 ejaculates were collected from this horse during one month, both the density and motility became progressively reduced.

In contrast Gebauer *et al.* (1974) were able to collect a daily ejaculate from each of 11 stallions throughout a period of 70 days, and these animals achieved the high average daily sperm production of 8×10^9 (8000 million). Ejaculates collected by Catanzaro (1975) varied in volume between 40 and 320 ml and in density from 30 million to 800 million sperms per ml. With the mean density recorded as 60 million, the average ejaculate volume, for 8×10^9 sperms daily, would be 133 ml. These stallions' ejaculates showed a progressive fall in sperm content when they performed repeatedly on the same day. By the fifth sample ejaculate content had fallen from 8×10^9 to 0.9×10^9 sperms; when ejaculates were collected, respectively six and three times weekly, the total of sperm ejaculated in one week varied very little.

The results obtained by Gebauer *et al.* (1974) corresponded well with the work of Schaefer and Baum (1963). Their daily collections averaged a yield of 8.3×10^9 sperms, and 26×10^9 when collected every fourth day. Collections on alternate days gave the greatest long-term sperm yield and also the best quality, as judged by the persistence of motility after storage in diluent at 5°C, when 27% remained motile after 30 hours.

The conclusion from this work was, that a highly productive stallion could be used most economically by taking three or four ejaculates weekly, as this should allow him to maintain the quality of his ejaculate over the whole season. Obviously such a horse could, thus, impregnate 60 mares in a 30-week season, if he had to serve each one only twice, and they came forward in regular succession when fully in oestrus. Such a situation can rarely occur in actual fact, too many mares enter oestrus at about the same time in the spring.

Especially when covering is strictly controlled by stud staff, difficulties arise in the timing of ovulation and receptivity of the mare, so it often happens that individual mares have to be served not once or twice, but three or four times. The effect of this necessity on the sperm content per ejaculate was investigated for a group of 20 thoroughbred stallions by Boyle *et al.* (1984). In mid-March the group averaged 8.7×10^9 sperms per ejaculate, and from then until the end of May each horse was covering an average of ten times weekly. During this period sperms per ejaculate fell to a mean figure of less than 1×10^9; not surprisingly, some of the stallions had to serve individual mares during several successive cycles. However, by the end of the season, no stallion left fewer than two-thirds of his 40 mares (mean figure) pregnant. Although no stallion of less than 7 years of age was included in this group, the view could be maintained that there was over-use during the peak period and that the final pregnancy percentage could have been much improved by altered management. When a stallion is expected to cover up to 21 times in a single week (Rossdale, 1981) it is easy to appreciate that some mares may return repeatedly to service having received an inadequate semen dose, thus even further increasing the demands on that stallion.

Seasonal variation

Most of the research on the seasonality of the male reproductive function has been carried out in North America between latitudes 35° and 50°.

Spermatogenesis   Pickett *et al.* (1970) conducted a study based on weekly sampling of four entire (uncastrated) quarter horses and four thoroughbred entires; they found the lowest density and percentage of motile sperms in November, with the peak in April and May and a gradual fall thereafter. Horses had achieved maximal libido by March; they were least keen to serve in December. Cornwall *et al.* (1972) studied 13 quarter horses, sampled fortnightly, and obtained a similar result; with minimum sperm densities of 210 million/ml occurring in November or December, which contrasted with a peak density of 402 million as late as September, but total sperm per ejaculate ranged from 6.3×10^9 in December to 16×10^9 in April.

Additional data have accrued from German studs where collection of ejaculates for deep freezing is permitted only when stallions are not required for natural service (Klug *et al.*, 1977a; Blobel and Klug, 1975). Originally January and February samples were obtained but latterly only December sampling has been permitted.

Only 20% of the stallions tested provided satisfactory winter samples, the remainder were disqualified because of poor motility and abnormal sperm structure. The best stallions, if collected from at intervals of eight days, did then achieve sperm concentrations averaging 511 million/ml, 51×10^9 sperms/100 ml ejaculate. One sample, collected and frozen in August, which had been markedly below the May peak figures in density, motility, and normality of sperm structure, was used to inseminate 33 mares during the following spring. Ten mares conceived at the first insemination and 14 others at repeat inseminations, even though 13 of these mares were presented as problem breeders.

Hormone production   Androgen production, presumably in the testis, is responsible for willingness to serve and for male aggressive behaviour. The level of testosterone in circulating blood has been found lowest from October to December, 1.5 ng/ml of plasma, and highest in May at 3 ng/ml, while in between these extremes the rise and fall was gradual (Berndtson *et al.*, 1974). These results agree well with average reaction times observed by the same workers, of 800 seconds in December, falling to 400 seconds in March. The reaction time is the interval between the presentation of a mare to the horse, and his actually mounting her.

These androgen values, obtained with stabled horses, were confirmed in a study by Kirkpatrick *et al.* (1977) in which 34 completely wild stallions, in Montana and Wyoming, each yielded single samples between September and July. Minimum levels of 1.8 ng/ml were recorded in December, with maxima of 3 ng/ml in May. Despite semi-starvation at that season, the winter levels of testosterone were perhaps slightly higher than those found in stabled horses, possibly through unlimited exercise and a wide choice of forage plants.

Fighting between stallions was seen to begin in March, as testosterone levels reached 2.3 ng/ml, coinciding with the start of the natural covering season which ended in mid-July, when levels had fallen to 2.4 ng/ml. The level of oestrogens in stallions' blood also rose, during the peak months of April and May to 130 pg/ml from a figure around 75 pg/ml during other months (Thompson *et al.*, 1978).

Effect of daylength change   Thompson *et al.* (1977) investigated the extent to which changes in daylength influence the seasonal variations in sex steroid levels. Nine stallions responded positively when exposed to artificial light to make a total daylength of 16 hours from October to February. The treatment raised the winter testosterone level to the spring maximum of 3 ng/ml, with a corresponding increase in keenness to serve, but the effect of the additional light was less marked in stimulating spermatogenesis. The effect of light on oestrogen concentrations was in marked contrast; in "winter-lit" stallions the normal spring peak failed and these horses showed a steady level of about 63 pg/ml of oestrogen all the year round.

With natural daylength, I have noticed that the skin of the horse's scrotum often begins to sweat in the stable during the second half of January, which is usually a colder month than December. Presumably, less than one month of increasing daylength has sufficed, markedly to increase the blood supply to the testes, which need to be kept cooler than the rest of the body.

In the cellular structure of the genital tract, the main variation with season concerns the predominance of meiotic figures in the testicular tubules from December to March and their comparative scarcity by June (Ellery, 1971), suggesting that relatively few spermatocytes proceed beyond the stage of meiotic division, primary to secondary spermatocyte, until spring is well advanced.

Conclusions   Although most stallions may be capable of getting mares in foal throughout the winter, this would be at a much reduced monthly rate compared with their potential in spring. A minority of less fertile horses is likely to be infertile in winter. Should such a horse be required to work outside the normal season, exposure to artificial light may occasionally have some practical value. It is remarkable that the collection of winter ejaculates in Germany for use in artificial insemination has given conception rates good enough to enable the research project to continue. It is possible that an international trade in frozen winter-collected semen may build up for use on mares in spring, or in the tropics; thereby enhancing the value of a highly-fertile stallion whose semen freezes well. Fertility may fluctuate between rainy and dry seasons where the daylength is almost constant.

Assessment of male fertility

For many centuries horsebreeding has been practised without any assessment. However, assessment is desirable to determine how many mares to send to a young horse on his first going to stud, and more important, to check on the continued fertility throughout a busy season of a horse expected to cover a large number of mares. Further special cases occur when the artificial insemination of mares is intended, either with a view to using a good horse on as many mares as possible, or in order to use ejaculates collected in winter when mares are not available for natural service.

   The physical health of the horse must first be established. An examination of the scrotum and its contents by palpation should follow: the collection of a testicular biopsy sample is not recommended; palpation *per rectum* of the vesicular glands and associated structures may be rewarding for an operator having sufficient experience to detect disease. But the most important part of the assessment is the collection and examination of a series of semen samples; a single sample is often misleading.

   A useful sample can be obtained by inserting a sheath or condom into the vagina of a mare, but the ease and convenience of collecting an entire ejaculate in an AV, by successive fractions if desired, have made this an even better method. Nishikawa (1959) made unsuccessful attempts to obtain samples from the standing horse, both by massaging the vesicular glands *per rectum* (in the bull a practical method) and by electro-ejaculation.

   In practice it is necessary to have both an AV and some sort of "teaser" for the horse to mount to enable him to make his thrust. For this purpose a quiet mare of suitable size, fully in oestrus, is obviously ideal, but may not be available when a sample is wanted prior to a horse going to stud, or in artificial insemination centres.

Live mounts and "phantoms" The use of a dummy mare or "phantom" has been found quite practicable, independently by several teams of workers operating in different countries. A suitable dummy has legs of steel tubing and a body of steel sheet, is padded and covered with leather; it is mounted on wheels, fitted with brakes, and should be adjustable for height to suit the requirements of individual stallions (Kenney and Cooper, 1974). Stallions required to serve frequently into an artificial vagina soon take a dislike to the necessary oblique angle of their thrust on mounting. This problem can be solved by constructing a dummy having a recess where a real mare's thigh would normally be (Klug *et al.*, 1977b). Over a five-year period 70% of 242 winter season collections were taken by means of this dummy by one research team.

Before they will mount a dummy readily, horses require training. One method is to place a distinctive rug on an oestrous mare. The next time the horse is brought out he is faced with an oestrous mare and the dummy close beside the mare and covered with the distinctive rug. Other methods include blindfolding the horse when he approaches the covering site, and sprinkling the dummy with urine from an oestrous mare. Hafez and Wierzbowski (1961) found that 9% of untrained horses would mount a dummy if blindfolded, if oestrous urine was also used the figure rose to 37%, and these were colts with little experience of oestrous mares. Some stallions are actually percipient enough to associate the sight or smell of an AV with the act of service, and will be eager to mount a dummy on being shown an AV. Trained stallions will, similarly, mount a quiet cow or dioestrous mare instead of a dummy (Schaefer, 1962; Dowsett and Pattie, 1980). In an emergency, 5 mg of oestradiol intramuscularly will produce a four-day symptomatic oestrus in a mare without affecting her ovaries.

There are reports that whatever the stallion mounts, he will produce a better quality of ejaculate if not allowed to mount when he first seems willing, but is led off and walked about before eventually being given the office. Although some investigators discount its value, it has been reported that a stallion provided 11 ejaculates in one day, collected by this "frustration" method.

The artificial vagina More recent types of AV are made from plastic, are much lighter than the original "long" models, and some have been reduced in length to as little as 40 cm (16 in). All are constructed to meet the stallion's need for "something to push against". This may be a tapering inner liner (Catanzaro, 1975) or a fitted circular diaphragm of sponge rubber with a central hole (called the "doughnut" type): alternatively, the blocked distal end of the AV has a small hole for joining it to the attached collecting flask.

Dowsett and Pattie (1980) reported their use of the "doughnut" type to make 216 successful individual collections from 222 horses.

For use the AV's inner liner must be clean, generously lubricated with liquid paraffin, and the interspace between it and the wall filled with water to give a temperature of 40–44°C on the liner. It is usually not difficult to "cap" the horse with an AV as he thrusts, for some models it is recommended that the operator should push against the horse in time with the rhythm of his thrusts. The pulsation waves in the urethra as he ejaculates may be counted by palpation; there are seldom more than nine or ten per ejaculate.

Fractionated collections are ideal, and not difficult to obtain but they do require two people; and usually a third (at the horse's head). Centrifugation tends to depress the fertility of a sample, sometimes quite formidably, but this process is acceptable, though not optimal, for the examination of a non-fractionated "assessment sample".

Examination of semen samples   On collection samples should be inspected; the sperm-rich portion should be creamy, neither brown from blood nor grey from pus, nor watery from a too-low sperm content. The volume may be measured, and a quick estimation of density made, electronically with a Coulter counter or approximately, using a set of calibrated opacity tubes. A thin film of the raw live semen may be examined at once under a low power of the microscope ($\times 50$–100 linear); this will give an approximation of the density and an indication of the sperm motility. If gel is present in an unfractionated sample, it can be filtered off through gauze, and part or all of the residue can for convenience be centrifuged, at not more than 800 g for three to five minutes. Aliquots may be diluted, and longevity tests set up with diluted and undiluted sperm-rich fraction, or with an approximation to this obtained by centrifuge. It is convenient to perform longevity tests at either a standard room temperature of 22°C, or a standard refrigerator temperature of 4°C. With undiluted semen, 10% of the sperm should, in the best samples, still be showing forward movement after six hours at 22°C, though some mares will conceive to samples below this standard.

Zhmurin (1960) described a simple diluent consisting of 7 g of glucose and 0.8 g of egg yolk in 100 ml of water or normal saline. The best samples, thus diluted five to ten times, had 27% of sperms still moving forward after 30 hours in the refrigerator, contrasting with 10% or less in undiluted semen of the same ejaculate (Schaefer and Baum, 1963). The forward movement of sperm in a sample of high density is easier to assess after dilution. However the inclusion of egg yolk as a protective to some extent obscures the sperm from view, while its value as a protective for equine semen is not fully established. Hence an even simpler diluent, without egg yolk, may be preferred, based on pasteurised skim milk, and relying for protective effect on the addition of 1 mg dihydrostreptomycin per ml of milk, and 200 units of polymyxin B sulphate (Hughes and Loy, 1970).

Up to this point, a good deal may have been learnt about the sample without using a high-power microscope. The detailed microscopical examination begins with checks of viability and of the type of movement shown by the sperm, with and without a warm stage at a linear magnification of 400 to 600. An accurate determination of sperm density may be required on a heat-killed sub-sample using the blood-count method. Samples having less than ten million sperm per ml suggest poor fertility.

Morphology and motility of spermatozoa   In evaluating the quality of sperm produced by a horse one must never forget that both the season of the year and the frequency of ejaculation greatly affect the composition of an ejaculate; including, besides volume and density, the morphology and motility of the sperm.

Examination of a single ejaculate is of very limited value, unless it shows a high percentage of good quality sperm from a known regular donor. The equine sperm consists of a head, middle piece, and tail, a peculiarity is that the head measurements show individual variation. The range from horse to horse can vary from less than 6μ to more than 8μ in length, and from 3μ to 5μ in width. For a given animal these measurements may be so consistent and characteristic, that conceivably they could be used to determine which of two horses had served a mare, or whether both had covered her (Dott and Foster, 1981): this can be important when teaser stallions are employed.

In examining spermatozoa microscopically, it is usual to examine live sperm in films, both cooled and on a warm stage, both on collection and after storage; phase-contrast microscopy detects some abnormalities not otherwise apparent, but others show up only in fixed films stained by the Giemsa method or with nigrosin–eosin live-dead staining.

Motility and percentages of abnormal forms. Recent work by van der Holst (1981) has indicated marked characteristic differences in the sperm of different breeds. He found that a consistent production of as little as 30% of normally structured forward-moving sperm seemed to correlate with a satisfactory level of fertility. Dott and Foster (1981) in a survey of 86 thoroughbred stallions showed that though most of them were fertile, 34% of these horses averaged over 50% of abnormal forms in their semen and 46% of horses averaged between 30 and 50% of abnormally-formed sperm, thus leaving only 20% of animals that produced at least 70% of normally-formed sperm. Hendrikse (1966) also found that some highly fertile horses consistently yield around 40% of abnormal forms. Overall, however, his work, based on study of 141 stallions, has shown correlation between the proportion of sperms with

Table 1.1

Percentage of mares pregnant	<66	66–75	>75
Percentage of sperm with normal forward movement	51	53	55
Percentage of sperm with structural abnormality	22	14	16

Data from Hendrikse, 1966.

normal structure and motility characteristic of a particular horse, and the percentage of those mares he had served which became pregnant.

Individual abnormalities. Separated sperm heads and tails may be a genuine primary abnormality (as in many Guernsey bulls) constituting a serious heritable defect, but the heads of equine sperm, normal when ejaculated, often separate after 24 hours storage at 4°C. As in the bull, a defective structure of the acrosome ("knobbed sperm"), detectable by phase-contrast microscopy, has been long recognised in equine sperm. High power electron microscopy has shown no structural differences between sperm abnormality in the bull, which is well correlated with poor fertility, and in the horse, where the connection with fertility is obscure. Of the 86 mostly fertile thoroughbred stallions studied by Dott and Foster, 30% produced some knobbed sperm, averaging 12% of the total ejaculated. In the Friesian breed of horse, knobbed sperm seem to be normal, and most of the animals studied (consistently producing up to 60% of knobbed sperm) have proved to be quite fertile.

Little is known about the heritability of abnormal sperm structure and its effect on fertility. David (1981) reported that the production of sperm with two heads on one tail is transmissible from sire to son; a bent-tail defect found in 20 stallions of one breed was associated with a 50% rate of early foetal death in the mares which they served (van der Holst, 1981). Abnormalities of obscure significance include asymmetry of the head, tightly coiled tails, and tails attached to the head abaxially at an angle. Some observers consider these sperm forms to be normal. However, it may be concluded that a horse producing a large percentage of sperm with one specific structural abnormality is unlikely to prove highly fertile.

Management and utilisation of the entire horse

The philosophical approach to the handling of entires

"One must never relax one's vigilance with a wild animal, or take its good behaviour for granted." (Chipperfield, 1975).

"One should always be on one's guard against an unexpected attack from a stallion; the awe in which stallions are held is a correct and prudent attitude." (Rossdale, 1981).

The first of these quotations was written about circus lions and tigers, although it can equally be applied to stallion handling. The second, specifically referring to the stallion, emphasises his violent nature. Just as violent human behaviour often stems from an inability to cope with the environment, or from imperfect social conditioning, so I believe does wrongly directed (though often expensive) conditioning of the stallion result in the type of unpredictable violent attack mentioned.

The wild horse spends much of his time in active search of food and water. The domesticated horse, with these needs supplied, requires substitute occupation to prevent boredom. If he is not kept occupied with harmless although preferably useful tasks, he will amuse himself to his own detriment with unwanted habits, e.g. coprophagy, crib-biting, masturbation, eating bedding etc.

A well-fed stabled stallion is capable not only of getting his 50 mares in foal, but also of as many daily hours of physical labour as a man. Thus it should be no surprise to find that such a powerful, intelligent, gregarious creature, largely deprived both of occupation and of contact with others of his species, protests against boredom and so occasionally becomes dangerous. Such an animal can be compared with humans who vandalise telephone boxes because society has not taught them to make more constructive use of their energies.

To become useful members of a society, both men and horses must learn to conform to its rules, to avoid "anti-social" behaviour. Just as it may be rarely useful to beat a human to achieve this, so it is not helpful to beat a colt or stallion. A stallion is not only as proud as the person who beats him but is also far stronger, and once he has fought a human being remains "unconditioned". He can then be used only for service, may become dangerous, even kill, and eventually have to be put down.

A practical method for establishing a satisfactory rapport Everard Calthrop (1920) described a simple gentle method by which the most timid human weakling can convince any normal unhandled stallion, that he is the animal's beloved friend, yet must at all times be obeyed. In a small enclosure, well bedded or on soft ground, the horse has his near fore leg strapped up, and while eating he is bridled. Eventually he will "sit", when the handler, kneeling by the off shoulder, uses the bridle to pull the head round leftwards. The horse then has to lie flat, and the handler sits on his head for a time, then releases the leg strap, tells the horse to get up, then rewards and makes much of him. This drill is repeated daily, incorporating the vocal order "lie down". Passing through stages of just holding up the foot, and next merely tapping it with a finger, the handler teaches the horse to lie flat on command and to remain so until the handler permits him to rise. If desired, he may rise with the handler on his back.

In this way I have conditioned four stallions, throroughbred, thoroughbred × Exmoor, dun Norwegian, and New Forest pony. The dun on the first day lay down within 45 minutes, and required only 12 days' schooling; the "Forester" however took two hours the first day and required 20 days. After his "course" the thoroughbred was ridden regularly for some months; he then stood for an hour in harness, lay down in it on command, and was then successfully put to and drew a light waggon. On one occasion a lad of 16 punctually completed a ride of 33 miles on this horse, while in charge of 5 younger boys on mares and geldings. Excluding the most docile stallion of the four, which belonged to a Research Institute, the other three horses were used for a number of years in the Scout movement, for the training of boys aged from 11 to 16. At summer camps the entires were picketed alongside mares on breastlines; heelropes were not used. No-one was hurt, no mares were served, and some dozens of boys rode the horses — two of which were occasionally ridden to hounds by ladies.

Parshutin (1956) claimed that, in entire horses, "the repeated inhibition of sexual reflexes leads to a steady extinguishing of them, and sometimes to complete impotence". David (1981) stated that "young horses out of training are so under discipline that they take no interest in oestrous mares". There was no such inhibition in the behaviour of my horses. If the handler desired them to move, they moved, in any direction irrespective of the presence of mares; if desired to stand, they stood. But lacking a specific contrary order, they would rapidly approach and investigate mares, and were keen to serve if the mares would stand. These horses had of course never been whipped. An inhibition of mating reflexes could well develop, though, should young horses be whipped off mares, while still too immature actively to resent such treatment or to punish those who applied it.

Use of entires for riding and draught   The mediaeval knights rode entire horses in battle, as did the 21st Lancers at Omdurman, 1898. Entires are still effectively used — the Barcelona Mounted Police, for example, used entires in their display at the International Horse Show in 1982.

Thus man has always made use of the entire's natural impulse to defeat his rival in battle, where the aggressive urges of both horse and rider combine against other similar partnerships. This combination may be directed against other forms of "opposition" — the police horse to control restive crowds — the wrangler's horse to combat the natural tendency of a bunch of mares or young stock to scatter and wander. In such situations the sense of comradeship between man and horse is very real, differing subtly and effectively from the rider's rapport with a gelding or mare.

In contrast, the entire can become adaptable and docile. For example, it was standard practice to use his lowering eye and heavy crest to add dignity to funeral processions, where he walked with decorum aside the pole with another stallion.

Exercise requirements of the breeding stallion   Thoroughbred stallions have been ridden and raced, but many entires of other breeds have been schooled neither to saddle nor harness. Not all such horses have been even lunged regularly, while few stud hands *would* lunge a horse for more than 90 minutes daily, so tedious is this for both horse and man.

All stallions should work under saddle or in harness; they may indeed usefully be employed in the cultivations necessary to produce their own feed. Requirement varies with the individual and the breed — but only at the very peak of the breeding season will they require less than two hours of real physical work daily.

Methods for mating   The term 'mating' is taken to mean the arrangements whereby stallions have access to mares.

Extensive systems   One arrangement, similar to that seen in nature, is practised in the management of the moorland ponies of Dartmoor and Exmoor in England. Harems of mares, each under a stallion, forage for themselves throughout the year. Each autumn, all the herds are collected together, and most of the young stock then removed for sale, while the best of the colts and some fillies are returned to the moor to maintain the strength of the breeding herds. This system differs from nature's, only in that there is little fighting between stallions, and the numbers of lower-grade colts, which compete unsuccessfully for available feed, are absent from the range. With this system fertility is high; over 90% of the mares breed each year (David, 1981).

The mares sort themselves out into a hierarchy or "peck order"; they are well known to one another and to the stallion. The process of courtship passes through successive stages, beginning as the mare enters pro-oestrus, and by the time the Graafian follicles in the mare's ovaries approach maximal development, the mares are very eager to be covered. The system is trouble-free, cheap and efficient, with the sole drawback that its product is a small animal.

The "taboon" (drove) system practised in south and south-east Russia (Kalinin, 1947) is very similar; it differs on the credit side, in that a horse of 15–15.2 hands (152–157 cm) is produced: the Don horse is able as a three-year-old to gallop a mile in 1 minute 49 seconds, only 10 seconds slower than the best Russian thoroughbreds. On the debit side, the very cold winters necessitate some feeding of hay, and experience has shown that horses carrying more than 50% of thoroughbred blood cannot withstand these winters in the open.

The stallions are removed from their mares at the end of July and the group of entires winters tranquilly all together on their own separate part of the range. The year's crop of foals over-winters on their dams, and is removed in spring at 10–12 months of age. The stallions then again take charge of their harems each of 18–25 mares, and stay with them from four to five months; by the end of this period 90–95% of the mares are found to be pregnant.

As on the British moorlands, the breeding conditions approach closely those of the wild horse, with individuals forming a close-knit group having established social relationships within it. Again the system is cheap and highly efficient. A feature both of the moors and of the steppes is the width of the range — 50 acres per mare is usual in Russia.

There are few reports on attempts to modify natural mating systems to farm conditions. When horses are used for farm work (there is renewed interest even in North America (Telleen, 1977; Bristol, 1981)), a band of draught mares can be run with a stallion, following sheep or cattle round the pastures, with the schooled entire taking his share of the farm work. In this way the stud may become a highly economic depreciation-free source of traction.

The situation is different when larger numbers of mares are at pasture with a stallion, and those diagnosed pregnant are promptly removed to be replaced by freshly-foaled mares. Established hierarchies are thus disturbed, although the opportunities for successful fertilisation still appear more favourable than when only "hand" mating is practised.

The intensive system

Reasons for the intensive system. The essential requirement for the intensive system is that the stallion lives alone, stabled apart from his mares, whom he encounters only for quite a short time, in the covering shed. This is the system prevalent in western Europe and eastern North America, and is virtually universal for the breeding of thoroughbreds. It is based on the concept that the stallion is of great value, hence his safety is paramount; he has to be safeguarded from jumping out of a paddock or breaking through a fence, and from being kicked by mares. The intensive system does prevent the first, but may invite the second. Also it does much to minimise the risks of injury from malicious human interference — the greatest risk of all. This safeguard, for the most valuable stallions, entirely justifies all the drawbacks of the system, since stables can be guarded far more easily than paddocks. Owners of less valuable stallions may be forced into the expensive, and, for them, relatively inefficient intensive system, simply because their clients, the owners of mares sent to stud, expect this system to prevail.

Oestrus detection using the breeding stallion. The efficiency of the intensive system depends on what dairy farmers call "heat detection". When stranger encounters stranger in the covering shed, apprehension on the mare's part is to be expected, and if she is not fully in oestrus, she is very likely to hurt both the stallion and the stud hands, and not very likely to settle in foal. In a small stud the covering stallion must often himself be used for heat detection. The time-honoured method is to "try him across a gate" with a mare thought to be in oestrus, as a test of her receptivity.

Remember, however, that the bars of a gate constitute an unacceptable risk to the legs of the stallion. A smooth solid wall, 121–136 cm (4–4.5 feet) high preferably with a "rollertop" should be provided where the horse may be accustomed to perform (Evans *et al.*, 1977).

A better method with a well-handled stallion (Miller and Robertson, 1937) is to lead him slowly and quietly through the paddocks where the mares are turned out, when any mare ready for covering will be apt to approach or follow him; an assistant then catches up these mares. Thirdly, veterinary examination of mares including palpation of ovaries through the rectal wall, and vaginal examination by hand or speculum, is an effective aid in heat detection, but may (rarely) lower fertility (Voss *et al.*, 1975).

Oestrus detection using teaser males. In larger studs it is frequently found economic to maintain "teaser males" of low intrinsic value as an aid to heat detection. One method is to keep a pony stallion not tall enough to reach the mares to be served (Evans *et al.*, 1977), to liberate him daily in the paddocks and then watch. When he has identified any mares that appear willing to stand for service, he is caught up using the long rope which trails from his head-collar. The pony eats little, but the method is time-consuming for the men, and has the drawback that the teaser pony does not stay long enough in the paddocks to establish a social relationship with the mares, thus missing some timid mares which may be genuinely oestrous. This difficulty can be overcome by the use of a teaser capable of identifying oestrous mares, but unable to get them in foal, who remains throughout the season in the mare paddocks. A stallion rendered infertile by resection of the epididymal tail (Erbslöh, 1974), or by surgical removal of a part of the ductus deferens, should be totally infertile by six weeks after operation, but remains fully potent. One disadvantage, since he can actually cover mares, is the risk of his infecting them with a venereal disease, should he become a carrier.

This risk can be reduced by the use of a teaser interested in oestrous mares but incapable of service. The preparation of a gelding as a teaser has been achieved easily (e.g. Swift, 1972) by the implantation, subcutaneously in the neck, of 200 mg of the hormone testosterone in tablet form. After only seven days the gelding became interested in mares, and proved effective, as a teaser incapable of service, until the implant was removed at the end of the breeding season five months later.

The covering procedure. Whatever the method used, it is usual to investigate a band of mares for receptive individuals daily, or on alternate days, once they have arrived at the stud where the stallion is kept. When found to be receptive, they are brought up to the covering shed, where it is quite common, even usual, to apply restraints to them in case they should kick the stallion. It is routine procedure to apply a twitch, put on a bridle, bandage the tail, wash the vulva, and hold up a foreleg until the stallion has mounted. More troublesome mares may be hobbled, blind-folded, or even sedated with tranquillisers (Evans *et al.*, 1977; Rossdale, 1981). If a mare's foal should accompany her into the covering shed, in theory she should be less upset.

Once considered receptive, mares are served, usually on alternate days, until oestrus is over. Teasing continues, and the whole process of teasing and service may go on until the mare is pregnant, or through to the end of the season.

Possible reasons for low fertility levels. Large-breed stallions on hand service are allowed 40–80 mares each season, exceptionally up to 130 (Wallace and Scott-Watson, 1923), and on average they leave only 60% in foal (David, 1981).

This poor level of fertility might be associated with the unnatural conditions under which the mare may be mated. If she has to be twitched, etc., she is presumably under psychological stress. In the human subject, spasm of the Fallopian tube musculature, from psychological stress, is recognised as a cause of infertility (Sandler, 1960). A similar cause might well result in, *inter alia*, spasm of the equine oviduct. The artificial conditions of hand service might also affect the quality of the stallion's ejaculate; e.g. for the horse to bite the mare he mounts is normal, but in some studs he is muzzled. Breeding operations will be made easier if a good rapport is developed, as described, between the stallion and his handler. Well worked, the horse will be less inclined to savage or alarm stranger mares and he may safely be led through the paddocks for the attraction of oestrous mares, thus gradually preparing both him and his mares for the act of covering.

High fertility with the intensive method may depend on a gentle approach by a well-roused stallion and on a fully oestrous, unfrightened mare, in little need of physical restraint.

A note on bacteriological sampling of stallions. Sampling by swabbing the urethral fossa is frequently mandatory. Recently access to the fossa has often been obtained by means of sedation with a tranquilliser, which can be very expensive, and carries a considerable risk that priapism may be induced. This puts at risk the horse's value for breeding (Bolz, 1970; Gerring, 1981)—see below. With correct handling, a horse will allow his "yard to be drawn" without sedation, risk, or restraint (Miller and Robertson, 1937, p. 193).

Pathology of the male genitalia

Non-infectious abnormalities

Cryptorchidism. The failure of testicular descent to the scrotum from the abdominal situation occurs frequently in horses but is hardly relevant to the subject of reproduction, since the undescended testicle is incapable of spermatogenesis. Normally both testes are present in the scrotum by the time the foal is born, but they may be retracted again subsequently. Occasionally both testicles may be retained in the abdomen until the colt is three years old; there is some evidence that in ponies particularly, one testicle may frequently descend early while the other is retained and descends only after the age of three years (Cox *et al.*, 1979).

If it is intended to breed from a colt in which testicles cannot be palpated in the scrotum, the presence of abdominal testicles can be confirmed by an assay of testosterone concentration in peripheral blood plasma, sampled 30–90 minutes after the intravenous injection of human or equine chorionic gonadotrophin—a dose of 6000 i.u. of HCG is suggested. Levels of over 400 pg/ml indicate the presence of abdominal testicles while in their absence the testosterone level does not exceed 50 pg/ml (Cox *et al.*, 1973; Cox, 1984; Cox and Redhead, 1985). In subjects over three years of age a high blood concentration of oestrone sulphate will indicate the presence of abdominal testes; it can be determined in a 10 ml sample of clotted blood obtained without previous HCG priming (Cox, 1984).

Reports have been published claiming that the descent of abdominal testicles has sometimes resulted from the injection of anterior pituitary-like hormones (Adams and Hector, 1942). One cause of cryptorchidism is genetic abnormality; such cases will never breed. Diagnosis is made by the preparation of a karyotype from the nuclei of white blood corpuscles: some such animals have a fully female karyotype with masculine body formation (Dunn *et al.*, 1979).

Priapism. Since the introduction, as tranquillisers, of such drugs as the phenothiazine derivatives, they have been widely used to assist withdrawal of the penis for examination and sampling. In some cases priapism has unfortunately resulted. The penis remains extruded, becoming engorged with blood, and it may eventually become necessary to amputate the organ (Pearson and Weaver, 1978). Breeding stallions should therefore be trained to allow the penis to be withdrawn manually for sampling. A safe, but costly, alternative is to bring about extrusion with the relaxant xylazine.

Prolapse and paralysis of the penis, a rare condition in breeding stallions, has to be distinguished from priapism, as it results from exhaustion or debility and spontaneous recovery ensues as normal vigour or health is regained. Simmons *et al.* (1985) however, described an all-but-incredible case of long-term neglect in Britain, resulting in complete penile paralysis in three entires (and four geldings), of which only two made a partial recovery.

Tumours. Seminomata occur rarely, affecting both testicles almost never. They may cause pain during, and hence impair, the covering act, but treatment by excision of the affected testicle is likely to restore some degree of normal function.

Abnormalities of semen. The commonest abnormality as a low density of normal motile sperm; its commonest cause is over-use, especially of immature animals. In aged stallions, following atrophy and fibrosis of the testicles and epididymes, total azoospermia may ensue. In some younger animals azoospermia may result from a genetic defect of the chromosome complement. One four-year-old stallion was allowed to serve 98 mares in one year, none of which conceived (Draca and Davidovic, 1962). Over-use was first suspected, but serial sampling during a rest period consistently revealed a total absence of sperm from the ejaculate. Presumably a genetic defect was responsible—the semen resembled that of an entire mule.

Doubtless much important information concerning breed-specific semen defects (examples are mentioned above) remains "classified" by the breeders and is unavailable for collation.

Miscellaneous conditions. Penile rupture can occur during the act of ejaculation, when the corpora cavernosa are subjected to pressure and sometimes tear. Small tears may pass unnoticed at first, but organisation of the clotted blood produces a haematoma which may require surgical removal (Bignozzi *et al.*, 1978). Tears over 1 cm long can result in the discharge of a stream of blood during covering, with a need for immediate sedation, restraint, and suturing with chromic catgut. In one case, the stallion resumed stud duty after four weeks rest (Pascoe, 1971).

Genital infection There are several specific equine infections transmissible at coitus; some of the organisms responsible seem to be primary pathogens, while others occur widely among normal horses and cause disease only in the presence of a predisposing "X-factor"—secondary or facultative pathogens (Lovell, 1943).

Contagious equine metritis (CEM). The role of CEM (*Haemophilus* or *Taylorella equigenitalis* infection) in the stallion has become better clarified since 1982. When an adult stallion becomes infected he harbours the bacterium (CEM organism, CEMO) particularly on the urethral process and sinus and transmits infection with around 90% efficiency to the mares covered. Diagnosis is based on the recovery of CEMO in culture from swabs, most readily from swabs exposed in the urethral sinus. Since the development of media capable of isolating streptomycin-sensitive strains of CEMO from heavily contaminated samples (Timoney *et al.*, 1982; Atherton, 1983), the method has been rendered markedly more efficient, but it still seems likely that some swabs submitted for culture have not been adequately exposed. However, because Timoney and Powell (1982) over 5 years recovered CEMO from 16 colts (and two fillies) of up to 4 years old which had never bred, the fact must be faced that foalhood or congenital infection sometimes occurs. It may fail to be diagnosed even with expertly-taken swabs and the best culture media, though a finding by Ward *et al.* (1984), that plates may require up to 13 days' incubation to show positive should help. A reluctant recourse to test-mating has given diagnoses when confidence in the culture of swabs had been lost. Serological diagnosis is unlikely to be helpful, though the passive haemagglutination test which will detect some infected mares missed by the other three blood tests (and vice versa) has seldom been applied in entire males.

Fortunately, treatment has never yet failed to clear a stallion of infection. Some of the most valuable thoroughbred stallions in the world have become infected two or three times each, but as a result of prompt treatment have lost little of a covering season. For details see Crowhurst *et al.* (1978).

Klebsiella. Klebsiella infection in horses involves a large number of capsule type variants of the organism. Certainly capsule types 5 and 1 are primary pathogens, while most of the others (including capsule type 7 which is most frequently recovered from stallions) are probably harmless (Platt *et al.*, 1976).

Certain variants can be very virulent. One stallion infected with a strain of capsule type 5 did not respond to treatment, infected a number of mares most of which became infertile, and himself developed infection of the renal pelvis, bladder, and prostate gland (Crouch *et al.*, 1972.)

Dourine (Trypanosoma equiperdum infection). This is a fatal infection with a primary pathogen venereally transmitted. Its prevalence has been greatly reduced, largely by means of artificial insemination, in the course of the last 35 years, but the disease has recently been reported (1978–81) in southern Africa, Italy and Russia. It was widespread up to the 1960s in most of the states bordering the Mediterranean Sea. Infection was introduced into North America in 1882 and eradicated by 1920; it has never been known to occur in Great Britain.

An early symptom is exaggerated libido, followed by oedema and ulceration of the external genitalia and sometimes by hyperaesthesia. There is a mucopurulent discharge from the urethra and a diagnosis is made by identifying the protozoan in samples. In later stages a general paralysis precedes death. Symptomless infection of the male is a feature of dourine; such animals spread the disease, but its presence can be diagnosed accurately by a complement fixation test. Two negative complement fixation tests on samples of blood collected at an interval of one month provide good evidence of freedom from infection. Treatment is not satisfactory.

Equine coital exanthema ("horse-pox"). The disease occurs commonly throughout the world—a description was published in 1796. Pustular or vesicular lesions develop on the external genitalia within hours after coitus with an affected animal, but heal spontaneously within ten days. There is little evidence of any adverse effect on fertility, or of the development of any degree of immunity. Nevertheless, neutralising antibody is raised in the blood serum. During 1964–68 the disease was shown to be due to a herpes virus, closely related to, but serologically distinct from, Equine Herpes Virus Type 1 which causes virus abortion. Non-venereal transmission of horse-pox between studs by human vectors is also known.

Streptococcus zooepidemicus infection. Disease produced by this organism is a good example of activity by a facultative or secondary pathogen. In one study, the Lancefield Group C streptococcus was isolated from the semen of 26 out of 39 apparently normal stallions. Such findings have resulted in the routine introduction of antibiotic mixtures into the uterus, after service by carrier stallions, a practice which seems to have limited the spread, and which has produced reasonable fertility in the mares covered (Simpson *et al.*, 1975; Weiss *et al.*, 1975). But in some stallions the presence of infection causes disease. One such horse yielded a low density of pus cells in the ejaculate, along with the streptococci, continuously over two years. Of 19 "clean" oestrous mares served naturally by him, nine remained normal, but ten developed a light endometritis which was resolved naturally within ten days. Presumably in these ten mares a natural X-factor was operative which made them susceptible to infection. In twelve other mares, an artificial X-factor was then superposed, by inseminating them during dioestrus, when all twelve developed a more severe and persistent endometritis (Spincemaille and Vandeplassche, 1964).

Other infected stallions, shedding streptococci in the semen, have shown a progressive testicular atrophy, and abscess formation in the epididymis. The vesicular glands may be similarly affected, but if no other of the genitalia show lesions, excision of these glands may restore normal breeding function (Deegen *et al.*, 1979).

Since 1970, the intra-uterine prophylactic use of antibiotic mixtures in thoroughbred breeding has greatly reduced the "population" of *Streptococcus zooepidemicus* in the genitalia, but this "vacuum" has been filled by proliferation of other facultative pathogens such as *Pseudomonas*, certain haemolytic strains of *Escherichia coli* and by a variety of capsule types of *Klebsiella*. For *Pseudomonas* infection, treatment with gentamicin (4.4 mg/kg of body weight, given twice daily for eight days) has sometimes proved effective both in stallions and mares (Hamm, 1978: Pedersoli *et al.*, 1985). However, a more logical control of the effects of facultative pathogens would be achieved by identifying and eliminating the X-factors.

REPRODUCTIVE FUNCTION IN THE FEMALE

Applied anatomy

Simpson and Eaton-Evans (1978) were the first to obtain evidence that the CEM organism could frequently be recovered in culture from the clitoral sinuses in the dorsal aspect of the glans clitoridis, even when it was absent from all other sites of the female tract. Some mares have 7 clitoral sinuses (Turnbull and Shreeve, 1984), more usual are one (larger) central and two lateral sinuses. Even in 13-hand ponies all sinuses can be swabbed with a 3–4 mm diameter standard wire-mounted bacteriological swab. The sinuses are usually filled with a smegma rich in bacteria; their relationships with the glans and clitoral fossa have been described and illustrated by Powell *et al.* (1978).

The equine cervical canal is wider and less convoluted than that of the cow, allowing a catheter easy passage, the equine vagina also is wider and more elastic than the cow's; hence, it may readily be distended to a diameter of 10–13 cm. In consequence, a manual examination is quite possible in nulliparous subjects, a small hand can easily be inserted into the vagina of a pony filly of 12.5 hands (125 cm) and aged three years or less.

The conformation of the equine vulva shows much individual variation. Of practical importance is the angle of the labia in relation to the horizontal plane. Labia at right angles to the floor are preferred to those with a cranio-dorsal to caudo-ventral slope—a conformation which is characteristic of many thoroughbred mares, and which may predispose them to vaginal windsucking (pneumovagina) and hence, endometritis.

Puberty

There is less interest in the age at which the filly rather than the colt attains puberty. Unlike colts, pubescent fillies do not have to be kept separated from other horses, and fillies are more often required for riding or draught before they are expected to breed. Thus may be explained the paucity of data on the age of puberty in the filly.

Wesson and Ginther (1981) in an abattoir study on Shetland and Welsh ponies found that ovarian cycles began between 12 and 15 months of age, presumably in the late spring or summer of the year following birth. Earlier, a similar pubertal age had been attributed to the equine species as a whole by Evans *et al.* (1977). It is however probable that the age of puberty varies greatly with climate, latitude, and availability of feed. In Senegal (latitude 15°N) the onset of cyclic activity in the local breed of large ponies, 14–14.2 hands (140–145 cm), was found consistently to occur at the early age of 7–8 months, and cycles continued throughout the year (Redon and Fayolle, 1957).

Whereas most mares are expected to work first and breed later, there may be an economic advantage in getting a foal from a draught filly before she is old enough for regular work on the land. In Britain it was an accepted practice for heavy-draught fillies, weighing up to one ton, to be served as two-year-olds to foal at ages of under 36 months (Wallace and Scott-Watson, 1923).

The ovarian cycle — seasonal variation

Features of the cycle   Throughout the winter most empty mares fail to show oestrus and pass into anoestrus. Many show a long period of anoestrus before they first ovulate for the season. This first oestrus is prolonged, lasting 10–30 days, and may or may not be accompanied by ovulation. Two experiments in the north temperate zone have shown, during January and February, ovulation in 20–25% of the mares studied. This is probably near a maximum figure, however.

During the early spring months, in the northern hemisphere February to April, periods of behavioural oestrus become more conspicuous and therefore apparently longer. Their lengths are quite variable, thus in one experiment they averaged nine days during these months, as against only four days in spring and summer (Quinlan *et al.*, 1951); in another trial they averaged 15 days during April, falling to seven in June (Ginther, 1974). A more important difference from the winter situation is that a small proportion of mares, some 20–30% does consistently achieve a natural ovulation during these months. In one investigation by Ginther (1974) of 14 mares, one ovulated in December, none in January, February or March, three in April and 13 in May. Van Leeuwen (1981) studied a group of mares from March 16 onwards which averaged 48 days to first ovulation — this took place on the "notional date" of May 4.

The very low level of natural fertility during these months conflicts with their breeders' wish to have thoroughbred mares foal as early as possible in the year. Hitherto the start of the breeding season in mid-February has led to a great deal of wasted effort, and enormous expense in labour costs, but recently, methods of advancing the onset of the "ovulation season" have been developed, and these should constitute means to a significant reduction in costs, for breeders of some show horses as well as thoroughbreds. For most other types of horse, May is the best month to start breeding.

Induction of ovulation in early spring

Artificial lighting. Artificially increasing "daylength" to 16 hours, starting on any day between November 15 and December 15, has been shown to produce ovulation up to 85 days earlier in the year than occurred in control groups (Merkt and von Lepel, 1969; Oxender and Noden, 1975; Kooistra and Ginther, 1975).

More recently it has been shown that at the beginning of the season, concentrations of luteinizing hormone (LH) in plasma are increased 14 times by this means (Vandeplassche *et al.*, 1971), and those of follicle stimulating hormone (FSH) 20–30 times — an effect which is dependent on the pineal body's production of melatonin (Sharp, 1981; Colquhoun *et al.*, 1987). In the absence of treatment, during the succession of prolonged anovulatory oestrous periods in early spring, the level of FSH in plasma is high at first, but shows a sharp drop about 20 days before the first ovulation.

Progestagens. Even without artificial light, treatment with a progestagen can convert the natural prolonged anovulatory heat to a five-day ovulatory heat during the early spring. In one trial using proligestone, all of 14 mares treated on March 16 had ovulated by April 2 (van Leeuwen, 1981). In another trial, Squires *et al.* (1979) combined the use of artificial light from November 1 with a 12-day course of allyl trenbolone in late January or early February. All of 17 mares ovulated during a seven-day oestrus period which began within three days of progestagen withdrawal; 15 of them became pregnant after covering then or during the ensuing 6–7 weeks. The optimum daily dose of allyl trenbolone has been calculated at 0.176 mg/kg of body weight (Webel, 1981). From 1983, this method has been quite extensively practised by breeders. Its suppliers have invested large sums in sales promotion of the hormone, which is most easily given by mouth daily.

GnRH analogue. Allen *et al.* (1985) have reported superior results from February on with a proprietary analogue administered by subcutaneous implant of $60\,\mu$g for a period of 28 days.

The normal cycle

Features of the cycle   In the north temperate zone the normal cycle is shown by most normal empty mares from mid-April to mid-September. Many mares may continue longer into the autumn with ovulatory cycles, depending, *inter alia* on the quality of the feed supply. A number of reports conclude that the modal length of the normal cycle is 23–24 days, with the corpus luteum (CL) functioning for about 14 days; the length of the follicular phase being more variable, and "standing oestrus" lasting seldom less than four or more than seven days.

During early oestrus both ovaries contain several follicles, one or two attain a diameter of 5–7 cm. The largest follicle becomes less tense to palpation shortly before it ovulates. Ovulation takes place only at the ovary's ovulation fossa, to which the fimbrial funnel of the oviduct is closely apposed.

An experienced veterinary surgeon can hope to diagnose ovulation by a rectal examination of the ovary on every second day the mare is in oestrus. Single ovulations are usual and tend to occur three to four days after the mare will first stand to the horse; within 24 hours a blood clot replaces the ruptured follicle. On palpation this clot closely resembles a ripe follicle, but as the follicle luteinizes it is possible to palpate the firmer tissue of the CL, though not for more than three days, as its final situation is well within the ovary. At the stage of maximum development the CL may weigh 10 or 11 g with a diameter of 3.5 cm. The presence of large follicles in both ovaries throughout the oestrous period complicates diagnosis, as does the occurrence of two follicles ovulating at the same or different times during one oestrous period.

The uterus of the mare is turgid, feeling compact and highly tonic on palpation when progesterone is at a high level in the circulation, while during oestrus it is flaccid — the reverse of what is felt in the cow. Typical oestrous behaviour is best shown by mares acclimatised to their environment, including adjustment to their ranking in the herd hierarchy, although normal oestrus is often shown by mares pastured alone or with a goat or donkey. Expression of behavioural oestrus is often impaired when mares have been recently sent away to stud, to find there a strange pasture, strange mares for company, and an over-excited stallion. In typical oestrus the mare will approach the stallion (or a gelding), and may even position herself between his forelegs. The tail will then be held to one side, there is jetting of urine and "winking" of the clitoris: if no male is present she will show increased interest in the distant sound of hoofbeats or a whinny.

In the vagina and cervix, the oestrous period is characterised by oedema, vascular congestion and the production of large quantities of stringy mucus from the open cervix. During the luteal phase the cervix is tightly closed and dry; its external os is projected horizontally into the vagina and the vaginal mucous membrane is pale. Many reports on levels of steroid and gonadotrophic hormones in the blood and in follicles during the oestrous cycle have been published; there is some diversity in the results, though it may be stated that progesterone concentrations in peripheral plasma seldom exceed one ng/ml during oestrus, rising to 6–15 ng/ml during dioestrus. During oestrus the concentration of oestradiol present in the largest follicle may exceed 251 ng/ml of follicular fluid. At other times, quite large follicles may be virtually oestrogen-free.

Acceleration of ovulation and curtailment of oestrous periods   With labour-intensive "hand mating", prolongation beyond five days of an oestrous period, plus the additional coverings needed on alternate days till the period ends, adds greatly to the expense of breeding operations, especially in the early part of the season. Since 1939, worthwhile results have been obtained from the use of injected hormone preparations to cause ovulation on about the fourth day of stallion-acceptance. The first preparation tried was human chorionic gonadotrophin (HCG) at doses of 2000–3000 i.u. given intravenously or intramuscularly, usually on the second or third day of standing oestrus. In a controlled trial restricted to oestrous periods beginning after May 1, Burwash *et al.* (1974) induced in 36 mares oestrous periods with a mean length of five days, and 2.5 services during each period; untreated controls averaged eight days and 3.6 services. Fertility was similar for both groups at this first oestrus, but with these mares, bred over three successive oestrous periods, 94% of the treated group became pregnant, contrasting with only 81% of control mares. Other reports do not stress this differential fertility but do agree that the method of treatment is effective for its purpose, both in hastening ovulation and reducing oestrus length. In a trial by Davison (1947) HCG was not given until, on serial rectal palpation, the largest follicles were found to be growing less tense. Treatment of 23 mares was followed in 19 of them by ovulation within 48 hours, and when a single artificial insemination was done 24 hours after the injection of HCG, five out of nine mares became pregnant.

The hormone in HCG, a protein, is similar to though not identical with LH, which on release from the anterior pituitary gland triggers natural ovulation. Loy and Hughes (1966) administered 2000 i.u. of actual LH on the second day of acceptance; 32 of 36 mares treated ovulated within the next 48 hours, on the fourth day of the period, while on average the control animals ovulated only on the seventh day; fertility was similar in both groups.

Both HCG and LH are large polypeptide molecules and as such are antigenic. Results have sometimes suggested, though scarcely proven, that their use in successive oestrous periods may become self-defeating, the mares developing neutralising antibody. Synthetic gonadotrophic releaser hormone (GnRH), with a much smaller molecule (decapeptide) does not have this possible drawback. GnRH causes the release of LH into the circulation; an experiment by Foster *et al.* (1979) in which 0.5 mg was injected at differing stages of the ovarian cycle, showed that the responses in terms of LH release was greatest late in oestrus. Nevertheless, when 2 mg of GnRH was given on the third day of acceptance, both ovulation and the end of standing heat occurred three days earlier than in control mares (Kreider *et al.*, 1976). Humke and Beaupoil (1979) in trials using buserelin, a more potent analogue of the natural GnRH, showed 40 µg to be the appropriate dose; when given on any one of the first four days of an oestrous period, 86% of subjects ovulated within 48 hours, with a conception rate of 68%. Only 48% of untreated controls ovulated equally early and only 28% conceived then. There is a suspicion, though little statistically valid evidence, that the use of any of these hormones may increase the proportion of twin ovulation and conceptions, which are highly undesirable.

Diagnosis and management of multiple ovulation

Prevalence of multiple ovulation. There is a measure of general agreement in the literature that the ovulation of two, rarely more, follicles occurs in a significant number of individual oestrous (heat) periods. The proportion varies with the breed, the age of the mare, and the month of the breeding season, with a tendency to higher proportions late in the season, and in the youngest and oldest mares. Interest in multiple ovulation is accentuated in horse breeding, because fraternal twin pregnancy is a disaster, precluded if only a single follicle ovulates during the service heat.

In studies by Henry (1981) and Newcombe (1981) regarding breed variations, there was agreement that in light breeds, including thoroughbreds, the all-age, whole season average figure, for periods showing multiple ovulation, was 20–25% of the total of oestrous periods; there was also agreement that in ponies the figure was lower — one of 11% resulting from that study more broadly based on abattoir material, whereas, with assessment based mainly on rectal palpation, ovulation of only one follicle characterised over 99% of ponies' oestrous periods. These two studies differed concerning the prevalence in mares of heavy breeds, giving 32% and 10% respectively. Further it was found that 60% of multiple ovulations were almost synchronous, while 40% were separated by from 1 to 4 days. Palmer (1981) using the ultrasonic scanner, mainly in thoroughbreds, reported 40% of periods with multiple ovulations, though Sanderson (1984) recorded a figure of not over 20%.

Sanderson reported, too, that only 5.5% of recognised twin pregnancies followed covering during 603 oestrous periods which presented with twin ripe follicles, though on smaller numbers, Simpson (1982) had found an ensuing 20% of twin pregnancies.

Quantitative investigations on prevalence of advanced twin pregnancy among different breeds are few. Szymanski (1976) reported in thoroughbreds a figure of 3.8% of twin pregnancies relative to total pregnancies, in step with the high figures mentioned above for multiple ovulations; these were pregnancies recognised only at abortion or full-term birth. The corresponding figures for twin pregnancies in Anglo-Arabs, pure Arabs and heavy horses were only 0.7, 0.5 and 0.2% relative to total pregnancies. Overall, of 143 twin pregnancies, 100 were aborted and 43 went to full term. Szymanski suggested that line breeding from winners might account for the high figures for twinning found among thoroughbreds.

Control of breeding by rectal palpation. During the Second World War, the revival of food production in Britain awoke both veterinarians and farmers to the great value attaching to veterinary breeding control in dairy cattle, based largely on rectal palpation. Partly as a result, a comparable expansion in veterinary breeding control in the British equine bloodstock industry occurred from 1945, and eventually in horses world-wide.

Aid which could thus be given to stud managers included the identification of ripe Graafian follicles, not only by their size (of up to 7 cm), but also by the lowering of tension within them as ovulation time approached. This help was useful particularly in early spring, when many oestrous periods are prolonged and others are not associated with ovulation; conserving the limited covering powers of stallions by restricting service, to the more appropriate stages of the least abnormal oestrous periods. By means of rectal palpation, it was possible to advise when mares might be worth covering although showing no overt behavioural oestrus—a "silent heat" which may be no less fertile than the norm. Most useful of all, pregnancy could be diagnosed, often before the 25th day after covering and in good time for commencement of the necessary extra feeding.

The passage of years has amply confirmed the value of these aids to breeding. Some veterinarians also claimed that their advice could often prevent twin pregnancy, by withholding service when multiple ovulations were likely close together during a period, or by delaying it 24 hours when these occurred at an interval, so as to minimise the chances of fertilisation for the first ovum shed.

The conditions peculiar to the breeding of thoroughbreds, as distinct from horses of other breeds, greatly enhance the desirability of veterinary control. For example: the pressure for foals to be born early in the year, so that they will be well grown when required for two-year-old racing. (In the United Kingdom the flat racing season starts in March.)

Conversely, it is more difficult to justify veterinary breeding control in breeds which need not foal until May and show little tendency to twin pregnancy. The availability of the hormones HCG and GnRH for inducing ovulation even further reduces, in non-thoroughbreds, the need for veterinary control of ovulation. Nevertheless, the value, both of early pregnancy diagnosis and of at least one rectal examination near to the intended time of covering is, in any breed of horse, difficult to contest. Veterinary breeding control is most effective when the veterinarian is closely involved with the stud management's decisions.

Recent research has tended to discredit the concept that rectal palpation based decisions are effective in preventing twin pregnancy. Collectively, Simpson (1982) and Pascoe (1983) diagnosed only a single ovulation at the covering heat preceding 38 of 129 twin pregnancies—errors probably due to the assessment as one, of two contiguous follicles; some follicles may contain two oocytes. Since 1980 the development and application of the ultrasonic scanner has made possible an accurate diagnosis of twin pregnancy at 18–20 days after covering or earlier—in ample time for the termination of such a pregnancy before 30 days, or for an attempt at converting it to a single-foal pregnancy. If synchronous dual ovulation has been diagnosed at the covering oestrus, there is no need to waste time by withholding service if a scanner is available to diagnose any resulting twin pregnancies in time for treatment. The scanner will frequently detect two early pregnancies occupying the same uterine horn, which is rarely possible by rectal pregnancy diagnosis. Thus, Pascoe diagnosed 29 twin-pregnant mares, not only as having ovulated singly, but also as carrying single pregnancies 25–30 days after service.

Although use of the scanner has, beyond all doubt, greatly increased the efficiency and earliness with which twin pregnancy can be diagnosed, even the scanner is fallible! Sanderson and Allen (1983) reported diagnosis by scanner of 35 bilateral twin pregnancies at 18 days and of 32 twin pregnancies which were contiguous in the same uterine horn. Of these 32, 16 were confirmed as time passed, but 16 were revealed as erroneous, due to abnormally placed endometrial cysts juxtaposed to single embryos. The scanner had also diagnosed as single-pregnant, 11 (of 1364) mares which in fact aborted twins.

Owners of either mares or cows tend to be apprehensive about unfortunate results from rectal palpation of the genital tract. Such results have, indeed been known, but they are very much the exception. In a controlled trial by Voss *et al.* (1975), of 59 mares palpated daily before and during the covering heat, only 11 (19%) became pregnant, contrasting with 17 of 30 (57%) control mares which were not palpated. But here each palpated mare was palpated for lengthy periods by two people, one of them under instruction, so the conditions were quite unlike those of normal professional practice.

Diagnosis of multiple ovulation by the ultrasonic scanner. The scanner was developed for external use in human medicine. It has been adapted and applied in the mare in the first instance by Palmer (1981) and by Driancourt (1979) in France. The equipment comprises a screen, and electronic units carried on a trolley, with a probe (attached by a long lead), which can be inserted into the mare's rectum. To determine ovarian contents, the inserted probe is placed as close as possible to the mare's ovary. Used in differing positions, it will show "simulated transverse sections" of the entire organ on the screen. Follicles of diameter as small as 2 cm show as very black areas, because their fluid content does not give rise to an echo, while the young corpus luteum is highly echogenic. Checks on genitalia obtained post-mortem show that the scanner will give fairly accurate measurements of follicles.

Since the ovary is not subjected to manual pressure, the method is comparatively safe; serial scans show clearly the development of a follicle to ovulation, or whether two or more large follicles are present: a corpus luteum may show at as few as 12 hours after ovulation. The scanner cannot reveal the reduction in follicular tension, palpable shortly before ovulation, but in other respects regarding determination of ovarian contents it is much superior to rectal palpation.

Those already competent in equine gynaecology soon learn to use the scanner, but it is an expensive instrument. Advertisers are reluctant to disclose prices, but a scanner costs about as much as a fair-sized, comfortable, four-wheel-drive motor vehicle. Despite the cost, it is certain that a scanner is necessary where effective control of twin pregnancy is important; its application is fast coming to be regarded as essential in all public studs. Its use on the uterine contents, including the diagnosis of pregnancy, is more important than its use to monitor ovarian contents, but clearly, each of these uses complements the other.

Resumption of the cycle after foaling

Length of the gestation period. The equine gestation period is not much less than a year; efficient reproduction in nature, outside the tropics, would seem to demand that foals should be born in April and May (southern hemisphere in October and November), for the good spring grass and for a daylength favouring prompt resumption of cycle. There is in fact a need for rather a rapid resumption of cycle if a mare is not to be overtaken by the calendar.

Studies by Hintz *et al.* (1979) and Ropiha *et al.* (1966) on the duration of pregnancy agree not only on a mean gestation of 335–340 days, but also that a large proportion of gestation lengths diverge widely from the mean. Thus Ropiha *et al.* reported that 90 mares carried their foals for 47 weeks, 170 mares for 48 weeks, and 130 mares for 49 weeks, which is about 329, 336, and 343 days respectively, leaving 132 mares, or 25% of the total, which carried for less than 329 or more than 343 days. Authenticated gestations (ending in normal foaling) of 371–387 days have been recorded by Onstad and Wormstrand (1972).

Behavioural oestrus and the first ovulation. It is clear that when a gestation exceeds 49 weeks, resumption of the cycle must be extra rapid if the mare is to keep in step with the calendar—she must have conceived again three weeks after foaling: but if she has carried for only 330 days she has 35 days in hand. Thus at least 50% of mares do not require to conceive at the "foal heat", but can afford to await the second cyclic ovulation after giving birth. In a trial conducted by Matthews *et al.* (1967), onset of the foal heat occurred as in the following table.

Table 1.2. Interval to onset of behavioural oestrus

Days after foaling	6	7	8	9
Number of mares	54	97	94	48

Of a total of 428 mares, 293 figured in the table, a further 77 showing onset between the 9th and 18th days, and a final group of 59 (14%), in which onset occurred at around 27 days, suggesting a preceding "silent" ovulation 7 days after foaling.

Du Plessis (1964) and Burkhardt (1948) mention that in some mares, the first ovulation after foaling is in fact not accompanied by behavioural oestrus, possibly an indication that these mares may not be ready to resume their reproductive function.

Mares vary in the quantity of milk yielded, and there is evidence that milk production does affect the resumption of ovarian activity and also the subsequent fertility. A number of foaling mares show no oestrus until lactation has ceased, while others after showing a foal heat, lapse either into ovarian quiescence or suboestrus (silent ovulation). Whether advantage should be taken of the opportunity to cover such mares at this foal heat is uncertain; the balance of evidence may be thought to suggest that it should not.

The foaling–service interval. Data collected during 14 years from the German thoroughbred industry show a consistent pattern of foetal resorption prevalence, related to lactation and the foaling-covering interval. Foetal resorption occurred mainly 25–35 days after conception, when milk yield reaches its peak, and was diagnosed by rectal palpation before 25 days, repeated after 35 days. Over the 14 years, the prevalence of foetal resorption in lactating mares varied from 12 to 24%, while in dry mares only from 2–6% resorbed (Merkt and Guenzel, 1979; Merkt, 1966, 1968). A more detailed survey of 735 lactating and 319 dry mares showed that resorption affected 17% of lactating mares which had conceived at the foal heat, 11% of those conceiving during the next oestrus after the foal heat, and only 7% of those conceiving later than this. Only 2% of non-lactating mares resorbed. Ten years later the situation was little changed, overall 17% of 2278 lactating mares resorbed, and it was noted, that of lactating mares which carried their foals to full term, 25% had been covered at the foal heat, 39% at the next following oestrus, and 36% later still. Of lactating mares which resorbed, 39% had been covered during the foal heat, 36% at the next following, and only 23% at later periods; of 2400 dry and maiden mares, only 5% resorbed. A difference in conception rates, evaluated on the pregnancies of 537 mares, 254 dry and 283 suckling was also apparent (Merkt and Guenzel, 1979).

Feo (1980) demonstrated that, of 22 mares which were covered at the foal heat but conceived three weeks later, 19 became pregnant in the uterine horn opposite to that recently pregnant. This result would suggest that decreased fertility soon after foaling may be due to incomplete regeneration of the endometrium. This work needs repeating on larger numbers, since some migration of embryos from one uterine horn to another does take place.

Table 1.3

Percentage conceiving	Dry mares	Lactating mares
At foal heat/first service oestrus	62	37
At next following period	26	26
At later periods	12	37
Total	100	100

Data from Merkt and Guenzel (1979).

Pregnancy

Physiology of pregnancy

Corpora lutea. The initial corpus luteum (CL) developing from the follicle whence the pregnancy originated, is reinforced in the mare by a succession of secondary corpora; there are waves of follicular growth around the 20th and 40th days of pregnancy (van Rensburg and van Niekerk, 1969). Some of these follicles ovulate and luteinize, and others luteinize without ovulating, to give rise to the secondary corpora. The corpus luteum-dominated phase of pregnancy ends between the 160th and 180th days of gestation (Squires *et al.*, 1974), when all the CL regress together and cease functioning. As late as the 90th day, the primary CL has been found to be still rich in progesterone. There is experimental proof that progesterone supplied by the corpora lutea sustains the development of a pregnancy until well into its sixth month (e.g. W. E. Allen, 1977).

Gonadotrophin production. A gonadotrophin which has a follicle-stimulating effect is present in the serum of pregnant mares, approximately between the 40th and 130th days (Cole and Hart, 1930). This hormone originates from the endometrial cups, ulcer-like outgrowths from the endometrium of the pregnant horn, which begin to develop from the 35th day. However, these cells are not of maternal origin, but are derived from specialised trophoblast cells of the chorionic girdle (Allen and Moor, 1972). Allen (1984a) suggested that the girdle cell "invasion" elicits maternal reaction which is at first aggressive (leucocytic destruction of young cup tissue) but later protective of the pregnancy; because there is evidence of gonadotrophin involvement in growth of the accessory CL (e.g. Urwin and Allen, 1983) Volsen *et al.* (1985) did demonstrate a marked difference in antigenic phenotype between girdle and cup cells.

If the conceptus should be removed from the uterus after the 40th day of gestation, either surgically or as a result of embryonic death, the endometrial cups continue to function, secreting gonadotrophin for several more weeks (Mitchell and Allen, 1975). Tests for pregnancy based on the presence in serum of gonadotrophin, will in such cases give false positive results. Among equids, gonadotrophin production is almost specific to the horse; its concentration in pregnant donkeys and zebra is respectively 12% and 1% of that found in horses: when a mare is carrying a conceptus sired by a jackass, her production of serum gonadotrophin (pmsg or eCG) is greatly reduced.

Steroid hormone concentrations. During the first half of gestation, progestagen (mainly progesterone) concentrations in blood plasma peak at 8–15 ng/ml between days 6 and 14 after ovulation, and steadily decline to 4–6 ng/ml at days 30–35. With maturation of the first accessory CL a pronounced secondary rise occurs, maintaining levels of 8–25 ng/ml until the regression of all corpora from the end of the fifth month. During the second half of gestation, progestagen levels remain below 4 ng/ml, except that from the 300th day or later, a steady rise begins; concentrations at term reach 20–40 ng/ml.

The concentration of oestrogen in plasma remains very low during the first 100 days of pregnancy, rises steadily to a peak around the 300th day, and then declines rapidly up to foaling (Nett *et al.*, 1973). The mare is an unusual mammal in that large quantities of the ring B unsaturated oestrogens, equilin and equilenin, are excreted in the urine during the second half of pregnancy, and that levels of plasma progestagen rise while oestrogen levels fall during the last weeks of gestation.

Evans *et al.* (1977) and Rossdale (1981) have published good concise accounts of the young embryo's morphological development, and that of its membranes.

Diagnosis of pregnancy

Hormone assay tests. Laboratory diagnosis based on the recognition of gonadotrophin in the serum of mares 40–130 days pregnant was at first, and indeed for 30 years, effected by a bioassay using female mice or rats. The first account of an immunological test was published by Wide and Wide (1963); their method greatly reduced the requirement of laboratory animals and the time needed to obtain a result from at least 24 to 2.5 h. Currently gonadotrophin tests are widely performed; kits are available, and an enzyme-linked immunosorbent assay has cut the time needed to 2 h, with an accurate estimate of the serum concentration (wide variations), following the original quantitative assay by Allen (1969). Although 7 ml samples were routinely requested and are desirable, tests can be done on only 1 or 2 ml of serum. Samples must be cooled immediately, and then decanted off the clot. Between 1982 and 1986 charges fell for pmsg estimations. This test is certainly useful in mares which resent rectal examination, with owners who will not allow adequate restraint!

Tests for pregnancy, based on the detection of oestrogen in the urine or blood serum after the 90th day from covering, may still be useful on occasion. A sample of urine need not exceed a volume of 15 ml, may be sent by post, and stores well at 4°C for up to 8 weeks (Livesey *et al.*, 1983).

Rectal Palpation. The rectal method has great advantages: with a result known immediately, it is applicable very early in gestation, accurate, and quick. A good diagnosis can be made from 17–21 days after covering (van Niekerk, 1965; Bain, 1967; Sindelic, 1972; Lensch, 1961). One drawback is that an early positive diagnosis may be followed by either an iatrogenic or a natural embryonic death. Van Niekerk (1965) reported a high rate of embryonic death in poorly-fed mares between day 25 and 31, which is the period of a change from chorio-vitelline to chorio-allantoic placentation. Hence it may be wise to forego rectal examination of some mares between 25 and 31 days after covering.

Most mares will tolerate rectal examination. A minority may require that a foreleg be held up by a confident and determined assistant, and some mares may also need a twitch applied, but it is far better that the attendant should improve his rapport with such mares, by obedience training as described for the stallion. In mares thus trained rectal examinations can easily be made with the mare standing on all four legs.

In pregnancy the uterine horn, under the influence of progesterone, shows a high degree of muscular tone, in all except some mares, served at the foal heat. It is therefore practicable to encircle the horn with the finger and thumb, thus systematically to examine for the subspherical swelling due to the presence of the conceptus. Both horns must be examined in case there should be a conceptus in each. Before 20 days the bulge, usually sited ventrally near the bifurcation, has a diameter of about 2.5 cm; at 20–25 days the swelling becomes more fluctuant, "like a bantam's egg" (van Niekerk, 1965); and at 28 days it may measure 5 cm. Some 10% of pregnant mares are, at 20 days, in oestrus as well as pregnant, with follicles of up to 5 cm in diameter present in the ovaries.

In later stages of pregnancy the differences from the empty uterus are often obvious, but in awkward, very large, or fat mares, it may be useful to examine for increased tension caused by the weight of the conceptus, affecting the mesovaria and broad ligaments. From five months of gestation a "snap" diagnosis may be possible by finding the characteristic fremitus or thrill in the enlarged uterine artery.

Use of the ultrasonic scanner. Besides its use for the detection of multiple ovulation the scanner, used on the uterus for early examination of the conceptus, constitutes a great advance on past methods (Driancourt and Palmer, 1979; Palmer, 1981). A conceptus may be definitely identified as a dark body by the 14th day; by day 16 its diameter can be measured, on average 2 cm. The straightforward use of the scanner to diagnose single pregnancies in some thousands of farm mares raised the proportion of covered mares pregnant from 60 to 74% for the season (Palmer and Chevalier, 1983). A result almost as good could have been obtained by rectal palpation. Hence either the current veterinary fees were too high, or the owners were unwilling to pay reasonable fees; they benefited from a subsidised experiment. Ricketts (1983) found a single scan far preferable to rectal palpation for diagnosis of single pregnancies in foal-heat-served mares, in which the uterus remains atypically flaccid.

The main value of the scanner method is for the early diagnosis of thoroughbred twin pregnancy, including that of a proportion of unilateral cases. As mentioned above (page 30) Sanderson and Allen (1983) reported success in diagnosis of 35 bilateral twin pregnancies. They correctly diagnosed 16 unilateral twinnings, and missed 11 of these, while making 16 false positive diagnoses; thus in their material there occurred 35 bilateral and 27 unilateral twin pregnancies. Palmer and Chevalier (1983) also reported, after double scanning, a majority of bilateral twins, 61%; with 28% unilateral, and 11% which appeared to change from one type to the other between scans (migration of conceptus).

Without a scanner, Pascoe correctly diagnosed 76 bilateral twin pregnancies by rectal palpation at around 25 days, missing 29 which later foaled or aborted twins and which had, presumably, been unilateral. Evidence exists that more unilateral than bilateral twin pregnancies fail early in gestation, so that Pascoe's results are not inconsistent with Sanderson's.

There is general agreement that diagnosis of unilateral twin pregnancy by scanner is not easy. A second scan following uterine massage per rectum may show two vesicles which had previously scanned as only one. Palmer and Chevalier agreed with Sanderson and Allen on the difficulty caused by endometrial cysts, common during the first 30 days post-partum. In Palmer's material only 60% of twin pregnancies were later confirmed when two vesicles appeared to be shown on the first scan. Ricketts (1983) preferred two scans several days apart, as both conceptus would have grown after the interval, a cyst would not. Greenwood and Simpson (1983) observed that 50% of twin-pregnant mares would by 23 days show two echogenic "white" embryos, in this moiety meeting the difficulty caused by cysts which, like the 16–18-day conceptus are not echogenic and show dark on the screen. Leidl and Kahn (1984) agreed that endometrial cysts caused difficulty, though they considered that these were rarely as big as the 18-day conceptus, and seldom as "beautifully round".

The scanner is useful in estimating the prevalence of early embryonic death, since impending resorption shows as mottling—"snowstorm"—of the dark vesicle. In one trial, 11% of pregnancies present at 23 days did not go to term, 25% of these had already failed by 45 days.

Control of twin pregnancy.   Pascoe (1983) reported that of 130 mares which foaled or aborted a pair of twins, only 14 produced a viable single foal as survivor of a twin pair. Previous to the scanner's advent, the many attempts to convert bilateral twin pregnancies to single ones, by crushing one conceptus *per rectum* (or "squeezing" — the word used for it in France!), almost always failed. The best method of control was to terminate the twin pregnancy totally as soon as diagnosed, using intramuscularly 250 μg of a prostaglandin analogue, fluprostenol or cloprostenol. Pascoe treated thus 12 mares rectally diagnosed as twin-pregnant before the 25th day; all returned promptly, within 7 days, to oestrus and were then normally fertile, thus these mares did not "lose a year" of breeding and the veterinary expenses were cost-effective. With a diagnosis by scanner at 14–18 days, even less time would be lost by this method for control of twin pregnancy.

However, perhaps the most remarkable advance due to use of the scanner is the finding that *if done sufficiently early in gestation*, the conversion of twin to single pregnancies by "squeezing" is often possible. A majority of such cases so treated, and reported both by Palmer and Chevalier, and by Sanderson and Allen in 1983, was successfully converted. Collectively, success in conversion was reported in 72 of 96 twin-pregnant mares (74%). In some of the residue, both pregnancies were terminated, though others continued twin-pregnant.

Dietary restriction offers an alternative and cheaper approach to conversion. It was found in Germany that two out of each three twin pregnancies diagnosed were in non-lactating mares, which are less likely to be inadequately fed than mares in milk. Therefore a trial was arranged in which 40 mares, diagnosed *per rectum* as twin-pregnant, were immediately, for one month, subjected to dietary restriction. For mares in milk, the concentrate allowance was reduced; dry mares received only hay. Seven mares continued twin-pregnant, eight had no foal at all, but 26 had single foals, of which 24 (60%) proved viable (Klug and Merkt, 1981). This percentage contrasts with Pascoe's 11% control figure for viable foals, when twin pregnancies were left untreated.

Care of the pregnant mare   A failure to exercise pregnant mares, beyond what exercise they will take when turned out in a paddock, is bad practice. Mares not exercised may become too fat, increasing the risk of a difficult foaling; the hard work of foaling is best performed by a mare physically fit as regards both the heart and abdominal muscles. Hind-leg oedema can develop from lack of exercise; though not a serious condition, owners sometimes find it alarming. Mares may be kept in moderate work up to the ninth month, and then in light work until they foal.

Nutrition of the pregnant mare obviously ought to receive careful attention, but is frequently deficient. The taking of a metabolic profile, based on analysis of blood samples, may be well worth its cost; it should indicate any deficiencies in time for them to be made good. Its timing should be related to the season and to any changes in the feeding regime, rather than to the stage of gestation. As term approaches, it is important to prevent any tendency to intestinal impaction with straw or excessive quantities of hay. This is most easily controlled by allowing a run out at grass, even if only during daylight. Hence, some grass paddocks should be held in reserve for winter grazing, even though most of the winter feed input comes from concentrates and hay. The risk of strongylosis (redworm infestation) is often overlooked, especially in small studs where the pastures are not grazed in rotation by sheep or cattle. In all studs, control measures should be practised, with worm egg counts at regular intervals and anthelmintic treatment given as necessary. To ensure the immediate acceptance by the mare of her foal, tradition lays down that the udder and adjacent parts should be gently handled daily by the attendant from the time that hypertrophy becomes evident — especially in the mare pregnant for the first time.

If a mare has been "stitched" (episioplasty) for vaginal windsucking, the sutures must be removed before foaling. Therefore, a routine digital examination should be carried out on the lower commissure of the vulva of all mares sent to stud. If sutures are then detected, they should be removed, either during the last weeks of pregnancy, or at exactly the right moment at the end of the first stage of labour, which is preferable.

Induction of parturition   To wait up to 50 days for a mare to foal must be tedious; it has been suggested that foals grow rapidly *in utero*, gaining up to 1 kg in weight for each day after 11 months gestation, thus increasing foaling risks.

The most satisfactory method of induction is by the intravenous or intramuscular injection of oxytocin. Doses of from 5 to 150 i.u. have been used; a mean of 62 i.u. was used in one experiment, Rossdale *et al.* (1976) recommended dosage at 0.26 to 0.64 i.u. per kg of body weight. Foaling begins 20–50 minutes after injection, placental delivery should take place normally. It is believed that a mare showing no prepartal development of the udder is not yet ready for induction, but a closed cervix and inelastic vagina need not deter (Klug and von Lepel, 1974; Lang, 1975).

Ousey and Rossdale (1983) and Ousey *et al.* (1984) have established a range of preparturient normal values for the concentrations of sodium, potassium, and calcium ions in the "pre-milk" estimated over a period. When the graph lines for sodium and potassium levels crossed, natural foaling occurred in about three days, so that induction at crossing time is appropriate; a dose of only 5 i.u. was found sufficient in pony mares already near full term.

In cattle, injection of 20 mg of dexamethasone is followed by calving two days later. In mares, 70 mg on each of four consecutive days has no effect on *pregnancy*, though this dose, given for other purposes, may well prove adequate (Burns, 1973). There is agreement that a dose of 100 mg on each of four consecutive days is sufficient to induce parturition in the mare, and results may be satisfactory, but too few have been published for a definite recommendation to be made. This method has resulted in good foals when injections were begun as early as the 321st day after ovulation (First and Alm, 1976).

Injection of the prostaglandin analogue fluprostenol has also resulted in a satisfactory series of inductions. In cattle such induction is usually followed by placental retention, but so far, with mares, this has not been a feature. In all of nine mares, doses varying from 250 to 1250 mg resulted in parturition, after intervals of 45 to 200 minutes, with placental delivery two hours later (Rossdale et al., 1976). The risk to the life of the foal remains only partially evaluated. Advocacy of induction, to be carried out as a routine for all foalings on a stud, has reached at least the discussion stage (Purvis, 1978).

Parturition

It is quite rare for equine parturition to take other than a normal course; but if the process should be held up, action has to be taken very swiftly if the mare is to have a reasonable chance of survival: a fortiori the life of the foal is even more at risk. The keeping of mares about to foal under continuous close observation is an economic proposition with valuable stock, and when foaling is occurring continuously over a five-month season. In other circumstances, the advent of safe, effective, and cheap methods of induction has released breeders from a dilemma; in the past "ordinary" mares had to take their chance, and sometimes died, though they might well have survived had assistance been available at the appropriate time.

If mares are not to be watched, they are better foaling in the open, as there is less risk of the newborn foal being overlaid. The only suitable bedding for a foaling box is straw, plenty of it; there should be provision for watching the mare unobtrusively without entering the box (video monitoring is best) and details of a "foaling alarm" have been published (Ruckebusch et al., 1978).

Signs of impending parturition   Some mares shown no signs at all, such animals have even successfully delivered unsuspected foals at equestrian events during the winter months.

Usually the udder begins to hypertrophy at any time between 14 and 7 days before the mare foals; her back may begin to hollow and the pelvic ligaments perceptibly to relax at the same time or a little later. The teats may fill up with "pre-milk" 6–4 days before foaling, and the appearance on them of scales of a waxy material may occur from 4 to 2 days in advance, or else, not at all. If milk drips from the teats, foaling may occur within 24 hours, though not always. Cox (1968) found no change in the mare's body temperature to indicate impending parturition, contrary to findings in the cow. However, daily histological study of vaginal smears has been reported as rewarding. Using Shorr's haematoxylin stain after ether ethanol fixation, the proportion of red-stained to blue-stained cells proved critical. The "blue" cells, comprising 23–30% of the total on a given day, had all disappeared from the film taken 24 hours later, and then foaling usually ensued within 12 hours and after 52 hours at most (Bader *et al.*, 1978).

The act of foaling   As a rule mares foal at night; in one survey 86% of foalings occurred between 7 p.m. and 7 a.m., 40% within two hours either side of midnight (Rossdale and Short, 1967). The first stage of labour is very variable in length, and its typical signs may sometimes occur as a false alarm, a week or more before foaling. If outside, the mare first leaves the herd. Stabled, she walks, or even trots, round the box; rolls, begins to sweat, and looks at her abdomen. There may be intermissions with normal behaviour. The second stage of labour, parturition proper, begins with the discharge of foetal fluids, sometimes inconspicuous. When mares are under observation, the time of fluid discharge should be recorded, as, unless delivered, the foal from that time has from one to three hours to live inside the mare.

Mares normally pass through the second stage lying down, and the foal should be born from 10 to 20 minutes after the fluids have been discharged. Within five minutes of their discharge, one or both of the foetal forelegs, covered with the amnion, should appear; if not, it is traditional that the attendant should at once make a vaginal examination to see whether the vagina is in fact empty, in which case the veterinary surgeon is called to the mare with the least possible delay. The foal is usually born within the amnion; a lively foal breaks its way out at once, but should there be any delay, the attendant must open the sac. On completion of parturition, the mare rises to her feet; in doing so she breaks the umbilical cord, though just occasionally it may be necessary to sever the cord by hand, scrape it through or ligate it 5 cm from the foal's navel, a traditional practice.

Dystokia   Dystokia means a failure of the normal process of foaling. There is already dystokia if the foetal forelegs are not present in the vagina within 5 minutes of the discharge of fluids. Obviously, failing a vaginal examination, there is dystokia if birth has not ensued by 20 minutes after the escape of fluids has been noted. Normally one foot appears first past the vulva, closely followed by the other and then by the head. If the head appears without the feet, or one foreleg appears as far as the knee without the other, a malpresentation is probable and assistance must be given at once: similarly assitance is needed if both forelegs appear without any sign of the head. Should the mare alternate standing with lying during her second stage, this deviant pattern of behaviour may indicate that delivery is held up; or nothing, beyond that the mare resents being watched, as many do (Rossdale, 1981). The correction of dystokia may require a quite exceptional degree both of strength and skill on the part of the attending veterinary surgeon.

Placental separation "Cleansing", the third stage of labour, generally occurs spontaneously, soon after the foal is born, often within 15 minutes; no concern need be felt until two hours have elapsed, but it has been claimed that after 12 hours retention there is already danger of its serious consequences, metroperitonitis with laminitis, developing some 24 hours after foaling. Nevertheless, spontaneous cleansing delayed until 8–24 hours after the foal's birth is not uncommon, and need not be attended by any harmful consequences (Vandeplassche *et al.*, 1971).

Placental retention is associated with the few twin pregnancies which reach an advanced stage; there is also an association with uterine inertia following assisted delivery of animals in poor condition. Failure to allow the chorio-allantois to rupture spontaneously, i.e. premature rupture by hand, has been claimed as a cause of retention (Khalil-Agha, 1956). Manual removal of the retained placenta is sometimes possible. The usual treatment is by the intravenous injection of oxytocin, by means of a slow drip over 25–60 minutes in 1–2 litres of normal saline solution. Doses of from 50 to 120 i.u. have been used successfully, the placenta often separating about 1–2 hours after injection has begun. If the placenta resists all attempts at removal, and has to be allowed to decompose *in situ*, systemic antibiotic support may prevent puerperal septicaemia (Mason, 1971). No very harmful consequences need be expected, if the mare should happen to eat a placenta which has separated spontaneously.

Infertility

Non-infectious infertility

Genetic infertility. Chandley *et al.* (1975); Hughes and Trommershausen-Smith (1977) and Blue *et al.* (1978) among others, have shown that abnormalities of the chromosome complement are a significant cause of infertility in the mare. The commonest divergence from the normal female sex chromosome complement XX, is an absent chromosome (XO), followed in order of frequency by mosaics (XO/XX) and by cases of testicular feminisation (XY). In the human subject the XO abnormality is characterised by low stature and by multiple somatic abnormalities (Turner's syndrome); most such conceptions abort, but 1 in 2700 live female births is affected. "XO mares" have shown no convincing evidence of small size or somatic abnormality; they have small inactive gonads and are completely infertile, though some XO mares have accepted service.

Mares showing a mosaic sex chromosome pattern (64 XX/63 XO or 64 XX/64 XY) need not be totally infertile. Halnan (1985) reported that of 10 such mares, thoroughbreds diagnosed by him, five had each produced one foal, and had subsequently been vainly served for up to 5 seasons. Diagnosis of genetic abnormality requires the preparation of chromosome spreads from white blood corpuscles.

Nutritional infertility. Too little attention has been paid to defective nutrition in relation to poor fertility; in the natural season, with mating in May, the fertility on spring grass is likely to be good. When mares are mated in winter to bring early foals, nutritional infertility is a real risk. Oats and hay are deficient in the amino-acids arginine, lysine, and tryptophane; when these deficiencies are supplied,the proportion of mares showing ovarian activity may be greatly increased (Bengtsson and Knudsen, 1963). Oats and hay are also low in carotene and vitamin E, present in fresh grass. There is evidence that carotene intake must be adequate if fertility is to be normal. Cereals are low in calcium, and in practice a failure to balance the excess of phosphorus, which results in an incorrect Ca/P ratio, is common, and is a contributory cause of ovarian dysfunction (Sippel, 1969; Wolter, 1972). Dietary supplements for horses, specially prepared to correct these deficiencies, are available.

In a detailed study of nutrition on ten studs, Kronfield (1979) revealed a tendency to over-feed young stock beyond their capacity to respond in terms of growth rate; while lactating mares and mares in late pregnancy were underfed, the latter receiving no concentrates. A deficient input of energy is a proven common cause of infertility in cattle but comparable equine studies are lacking.

Ovarian dysfunction. A failure of the ovarian cycle, which is especially common in July, may be related to nutritional status. Typically there is a stasis in the luteal phase, with a persistent corpus luteum, resulting in a "prolonged dioestrus" which may last for months (Stabenfeldt and Hughes, 1981). Diagnosis can be reached only from serial estimation of progestagen level in plasma, which should show continuous high levels, above 5 ng/ml. Treatment by the intra-muscular injection of 250 g of fluprostenol is effective in restoring cyclic activity.

The occurrence in the mare of true cystic ovarian disease, analogous to the disorder affecting cows, has often been questioned. The large size of the normal ripe follicle, and the occurrence in early spring of long oestrous periods without ovulation have caused confusion. None the less, true cystic ovarian disease does affect the mare, and it has been diagnosed both histologically and by serial clinical examinations (Zwick, 1968; Baier *et al.*, 1972). Admittedly one of its causes is the injudicious injection of oestrogenic hormones. There may be a good response to treatment by injection of 500 mg of progesterone, repeated every five days.

Ovarian tumours are not uncommon; Meagher *et al.* (1978) have published a description of 78 cases of granulosa cell tumour. Such cases may show continuous oestrus, or else masculine behaviour. Often there is a return to a normal cycle after surgical removal of the tumour.

Vaginal windsucking (pneumovagina). This develops in a proportion of mares after foaling, especially in those with a predisposing conformation of the vulva. In one sample of 200 thoroughbred mares sent to public studs, 30 had been subjected to episioplasty (Caslick's operation). If pneumovagina is not corrected by surgery, a damaging non-specific endometritis frequently results. Modifications of Caslick's operation have been described (Brown and Coffman, 1972; Berthelon, 1973; Shires and Johnson, 1974; Pouret, 1982).

Infertility due to infection

Non-specific endometritis. Secondary or facultative infection plays an important part in infertility of the mare, producing endometritis, often a mixed infection; formerly, *Streptococcus zooepidemicus* played a dominant role, but although itself new largely eliminated by means of streptomycin, the "X-factor" still remains which permitted the organism to become pathogenic, and in such animals the streptococci are often replaced by streptomycin-resistant genera, *Klebsiella, Pseudomonas, Escherichia coli*. The antibiotic gentamicin is effective against these genera, (Amtsberg and Krabisch, 1975: Pedersoli *et al.*, 1985) but it is expensive. X-factors have not been well identified, despite obvious suggestions. However Liu *et al.* (1985) found that, while mares resistant to these endometrites had uterine neutrophiles effective against the infecting agent, and susceptible mares "with an X-factor" had uterine neutrophiles incapable in this respect, peripheral blood leucocytes from susceptible mares were fully effective. Bouters and Vandeplassche (1985) confirmed these differences, and went on to treat (cheaply, and with excellent results) twelve mares affected with pyometra, using intra-uterine infusion of the mare's own blood, heparinised, 100–150 ml on each of five successive days. The treatment and the prognosis of non-specific endometritis have been advanced considerably by the results of two research projects based on uterine biopsy (Kenney, 1978; Doig *et al.*, 1981).

Contagious equine metritis (CEM). This important emergent disease was hardly known until 1977. A good source of information on many aspects remains the symposium on it published in *Equine Vet. J.* (**10**, No. 3, 1978). Some more recent findings are briefly noted above (pp. 20–21) and the present writer has summarised these developments rather more fully elsewhere (Dawson, 1979–84 and 1986–87).

Recovery of the organism (*Haemophilus* or *Taylorella equigenitalis*) in culture, usually from the clitoral sinuses, constitutes the definite diagnosis. To achieve it has often proved difficult and expensive — the collection of a single set of samples may cost the owner £60. The value of serological diagnosis, which costs little, has been under-rated because the disease has been sought for mainly in thoroughbreds—a definite bacteriological diagnosis can be afforded and is particularly useful in the special conditions attaching to this breed. Should the disease ever become widespread among work stock in developing countries, there would be no practical alternative to testing serum. In a French context, Pitre *et al.* (1979) stated that "serological tests form a safeguard against defective culturing, and are the only tests cheap enough to be performed extensively in non-thoroughbred horses".

There are four available tests, the agglutination, anti-globulin, complement fixation and passive haemagglutination tests (Benson *et al.*, 1978; Croxton-Smith *et al.*, 1978; Fernie *et al.*, 1979). An infected mare may react in any one of these tests, hence a diagnostic titre in an infected animal is more probable if paired samples are subejcted to all four tests. This method should identify all mares infected during an outbreak; most carrier mares will show reactions, though not all have done so.

Equine herpes virus 1 (pneumonitis) infection. EHV1 infection results both in disease of the respiratory tract and in abortion. Two strains or subtypes have been identified, subtype 1 (strain RAC/H) which has been isolated mainly from abortions, and subtype 2 (strain MD) from the respiratory tract.

Besides investigating the possibilities of a bivalent vaccine to produce immunity, recent research has considered the immunity produced by a killed vaccine derived from subtype 1, "the abortion strain"; in the United Kingdom this strain has been recovered also from cases of respiratory disease, although in Britain most isolates from pneumonitis and abortions have been of subtype 2.

The killed vaccine evaluated entered the British market in 1981 from the USA, where Bryans and Allen (1982) reported in large-scale field trials, abortion rates of 7 and 2 per thousand, respectively in control and vaccinated groups. But in a careful experiment Burrows *et al.* (1984), challenging vaccinated ponies with the homologous strain, subtype 1 virus, found the vaccinates not protected against abortion, though the vaccine did mitigate the respiratory effects of challenge. Contrariwise, Mumford and Bates (1984), challenging similar vaccinates with subtype 2 virus, showed a good cross-immunity in 20 of the 30 subtype 1-vaccinated ponies.

Thus, on present evidence, the killed vaccine may be worth using. The procedure recommended after a diagnosis of EHV1 infection has been summarised (*Vet. Rec.* **114**, 56, 1984). Edington *et al.* (1985) demonstrated that EHV1 infection could, in apparently recovered animals, enter a stage of latency; viraemia could recur following imposition of a "simulated stress".

Artificial insemination

Insemination with unprocessed semen   The artifical insemination (AI) of the horse has a long history; yet few horse breeders in the West regard the practice as other than a threat, since their basic aim is to restrict the use of the best stallions, and thus to exploit the scarcity value of their progeny. Even within the limitations thus imposed, the efficiency of breeding could, from two aspects, be improved by the use of AI.

Firstly, in disease control: due to the impact of CEM upon the thoroughbred breeding industries of England, Ireland, France, and the USA in 1977–78, attention was drawn to the salient fact that AI could prevent the infection of stallions from carrier mares. The requirement was merely to collect the whole ejaculate, using perhaps a phantom as mount, and, within minutes, inseminate it all into the mare to be covered; a very slight departure from accepted practice, and one that would also give some protection to an uninfected mare. Despite the activity of pressure groups, and as a result of the partial control of CEM by other means, even this form of AI is not yet generally available for thoroughbreds. However, AI of thoroughbreds is commonplace in Russia.

Secondly, a single ejaculate may well contain sufficient sperms to impregnate 100 mares. By sub-division of an ejaculate, within-minutes AI on an individual stud would make certain of adequate impregnation, at peak periods, of all mares sent to a stallion; without affecting, other than favourably, the economics of the existing system. There need be no increase in a horse's quota of mares, only an improvement in the percentage rendered pregnant. Since no horse would need to ejaculate more often than once daily, only good quality ejaculates would need to be used. To some extent this form of AI is in fact currently practised in the breeding of trotters on both sides of the Atlantic.

In China it appears that equine AI for a proportion of mares used in agriculture and transport, has been mandatory, in order to exploit the genetic potential of the best sires. Cheng (1961) published a claim that in the course of two years, 101 ejaculates were collected from one horse between May and July, and used to inseminate 6000 mares of which 4000 conceived; other sires closely approached this achievement.

A recent study by Bristol (1981), on the potential of AI using unprocessed semen, applied techniques of synchronisation to concentrate the breeding season into a nine-day period. Advantage was taken of the most suitable time of year and each cyclic mare received, between May 16 and May 31, two injections of a prostaglandin preparation at an interval of 13 days. Foaling mares received implants of a synthetic progestagen, presumably removed on May 31. The "nine-day breeding season" began on June 3. A long row of mares in a "chute" was teased by a stallion moving along outside it; mares appearing to be in oestrus were inseminated daily until they ceased to be receptive. On average, a mare had thus to be inseminated about 2.5 times. Of about 250 mares, 72% became pregnant from insemination between June 3 and June 12. The primary purpose of this exercise was the recovery for sale of oestrogen from the pregnant mares' urine, but latterly the cart foals bred met a strong demand.

Methods for collection and dilution, and for short-term storage of semen, have been described above. An insemination dose of 100 million sperms may suffice; a larger dose, if available, may be preferred. A semi-rigid insemination catheter of plastic, as used for the cow, may be employed; others prefer a more flexible type, with a stilette, removed after passage into the uterine body. The semen may be placed in an attached large syringe, or a "concertina" bottle of the type used for bovine intro-uterine irrigation.

Development of the application of frozen storage   Equine semen is less easily frozen than that of the bull. Early attempts were unrewarding until the pellet-freezing technique of Nagase (1967) made a breakthrough, for reasons still not well understood. Insemination of 51 mares with pellet-frozen semen resulted in 24 conceiving; equilibration in 1.75% glycerol for 5–10 h preceded the dropping of 0.1–0.2 ml of semen, diluted 2–4 times, on to a block of solid carbon dioxide to produce the frozen pellet. The method was taken up by a determined team of German research workers who modified it by eliminating the use of the centrifuge. At first they inseminated 100 pellets of the sperm-rich fraction into each mare, after thawing in sterile milk at 40°C. The first two live births were recorded in 1967 by Krause. The German workers were handicapped by the restriction to winter collections of semen, and by the obligation to inseminate mares of doubtful initial fertility. Despite collection in winter, it proved possible to reduce the insemination dose to 16 pellets, to average 96 doses from each ejaculate (range 52–193), and to achieve a foaling rate of near 50% on about 400 mares over a four-year period (Klug, 1976). Suitable ejaculates contained 75% of sperms motile, when fresh, which fell to not less than 40% on thawing. In 1981 pellet-freezing was abandoned for a better technique, and the project continued with a much wider range of sires available as a result of the change (Klug, 1981).

Tischner (1981) reported an improved method for assessing the fertilising potential of frozen-thawed samples of semen. He attempted the recovery after AI of eight-day embryos by uterine flushing via the cervix. Of 52 stallions tested by this method, five were selected as outstanding; 27 of the mares (out of 46) inseminated from among these five sires foaled. Tischner also (1976) referred to repeatable results in Russia, from pellet-freezing using normally fertile females in the normal breeding season. Of 211 mares, 122 (57%)

became pregnant by insemination during a single oestrus period, to doses of 600–700 million thawed sperms. However, the embryonic death rate was high, 15–20%, to give a foaling rate due to AI during one period, of little over 40%.

Embryo transfer

A review by Douglas (1979) recalled that good pregnancy rates had for some years been obtained in recipient "surrogate dams". Donors are AI'd, and flushed for embryos (as in the cow) at 8–9 days after ovulation (as determined by progesterone assay). Ideally, recipients ovulate two days later than donors; embryos are transferred surgically in dorsal recumbency. Allen (1984b) reported 72% pregnancy rates in recipients, and mentioned that, in the USA, 3000 commercial transfers had been made within 9 months. Applications of embryo transfer include production in recipients of several foals each year from a good-performing mare donor, and uninterrupted breeding from mares with damaged uteri or oviducts. Mares and fillies in production for the horse butcher may usefully serve also as recipients. Foal sizes tend to follow that of the recipient dam; thus 15-hand Exmoor ponies might, for instance, be bred.

Embryo transfer is currently in use to improve the reproductive rate of endangered wild species. Thus two reports (Anon., 1984; Summers *et al.*, 1985) together report six advanced pregnancies or foalings, from the transfer of embryos of the Prjevalsky wild horse to 14 pony recipients.

The artificial production of identical twins has proved possible by the bisection of a normal embryo and the transfer of each half into a separate recipient. Starting with 15 embryos, Willadsen *et al.* (1980) produced one pair of full-term identical pony twins from two surrogate dams. Allen and Pashen (1984) produced two pairs of such twins from 19 embryos processed.

REFERENCES

Adams, S. H. and Hector, A. (1942). A case of double cryptorchidism treated successfully with Antuitrin S. *Vet. Rec.* **54**, 6.

Allen, W. E. (1977). Effect of prostaglandin analogue on progesterone-treated pony mares during early pregnancy. *Equine Vet. J.* **9**, 92.

Allen, W. R. (1969). A quantitative immunological assay for pregnant mare serum gonadotrophin. *J. Endocrinol.* **43**, 581.

Allen, W. R. (1984a). Influence of foetal genotype on equine endometrial cup development: transfer of inter-species embryos between horses and donkeys. *Vlaams diergen. Tijdschr.* **53**, 253.

Allen, W. R. (1984b). Equine embryo transfer; methods, results, and potential application in horse breeding. British Equine Veterinary Association Congress, Bath.

Allen, W. R. and Moor, R. M. (1972). Origin of equine endometrial cups; production of PMSG by foetal trophoblast cells. *J. Reprod. Fert.* **29**, 313.

Allen, W. R. and Pashen, R. L. (1984). Monozygotic horse twins. *J. Reprod. Fert.* **71**, 607.

Allen, W. R., Sanderson, M. W. and Jackson, P. S. (1985). Induction of ovulation in anoestrous mares with a slow-release implant of GnRH analogue. Society for the Study of Fertility, Winter Meeting, Abstract no. 38.

Amtsberg, G. and Krabisch, P. (1975). Results of bacterial examinations of swabs taken from mares during the period 1969–73. *Dtsch. Tieraerztl. Wschr.* **83**, 110.

Anon. (1984). Prjevalski foals born to surrogate mares. *Vet. Rec.* **115**, 27 and *Veterinary Times* **14**, no. 9, 20.

Atherton, J. G. (1983). Evaluation of selective supplements used in media for isolation of the C.E.M. organism. *Vet. Rec.* **113**, 299.

Bader, H. and Huttenrauch, O. (1966). Fractionated ejaculates from stallions. *Dtsch. Tieraerztl. Wschr.* **73**, 547.

Bader, H., Genn, H., Klug, E., Martin, J. and Hummler, V. (1978). Vaginal cytological studies in the horse. *Dtsch. Tieraerztl. Wschr.* **85**, 226.

Baier, W., Berchtold, M. and Brummer, H. (1972). Treatment of ovarian dysfunction in mares. *Wien. Tieraerztl. Mschr.* **59**, 13.

Bain, A. M. (1967). Manual pregnancy diagnosis in thoroughbred mares. *N.Z. Vet. J.* **15**, 227.

Bengtsson, G. and Knudsen, O. (1963). Feed and ovarian activity of trotting mares in training. *Cornell Vet.* **53**, 404.

Benson, J. A., Dawson, F. L. M., Edwards, P. T., Durrant, D. S. and Powell, D. G. (1978). Serological response in mares affected by contagious equine metritis, 1977. *Vet. Rec.* **102**, 277.

Berndtson, W., Pickett, B. W. and Nett, T. (1974). Seasonal changes in the testosterone concentration of peripheral plasma in the stallion. *J. Reprod. Fert.* **39**, 115.

Berthelon, M. (1973). Curative interventions for excessive permeability of the mare genital tract. *Rev. Méd. vét.* **124**, 489.

Bielanski, W. and Wierzbowski, S. (1961). Depletion test in stallions. *4th. Int. Cong. Anim. Rep. The Hague* **2**, 279.

Bignozzi, L., Busetti, R. and Gnudi, M. (1978). Surgical treatment of haematoma of the penis in stallions. *Clinica Vet.* **101**, 531.

Blobel, K. and Klug, E. (1975). Artificial insemination in horses under practice conditions. *Berl. Muench. Tieraerztl. Wschr.* **88**, 465.

Blue, M. G., Bruere, A. N. and Dawes, H. F. (1978). The significance of the XO syndrome in infertility of the mare. *N.Z. Vet. J.* **26**, 137.

Bolz, W. (1970). Prolapse and paralysis of the penis in the horse after the administration of neuroleptics. *Vet. Med. Rev. (Leverkusen)* **4**, 255.

Bouters, R. and Vandeplassche, M. (1985). Report of the (Belgian) committee for the control of sterility. *Vlaams diergen. Tijdschr.* **54**, 118.

Boyle, W. S., Cound, J. and Allen, W. R. (1984). A survey of semen characteristics of thoroughbred stallions. British Equine Veterinary Association Congress, Bath.

Bristol, F. (1981). Synchronization of ovulation for timed insemination of mares. British Equine Veterinary Association Congress, Cambridge.

Brown, T. and Coffman, J. R. (1972). A modified technique for episioplasty in the mare. *Vet. Med. Small Anim. Clin.* **66**, 110.

Bryans, J. T. and Allen, P. A. (1982). *Dev. Biol. Stand.* **52**, 493. (cited Burrows *et al.*).

Burkhardt, J. A. (1948). Some clinical problems of horse breeding. *Vet. Rec.* **60**, 243.
Burns, E. J. (1973). Clinical safety of dexamethasone in mares during pregnancy. *Equine Vet. J.* **5**, 91.
Burrows, R., Goodridge, D. and Denyer, M. S. (1984). Trials of an inactivated EHV 1 vaccine: challenge with a subtype 1 virus. *Vet. Rec.* **114**, 369.
Burwash, L. D., Pickett, B. W., Voss, J. L. and Back, D. G. (1974). Relation of duration of oestrus to pregnancy rate in normally-cycling, non-lactating mares. *J. Am. Vet. Med. Assoc.* **165**, 714.
Calthrop, E. R. (1920). The Horse as Comrade and Friend. 243 pp. Hutchinson and Co., London.
Carson, R. L. and Thompson, F. N. (1979). Effects of an anabolic steroid on the reproductive tract in colts. *J. Eq. Med. Surg.* **2**, 221.
Catanzaro, T. E. (1975). Collection of stallion semen. *Vet. Med. Small Anim. Clin.* **70**, 333.
Chandley, A. C., Fletcher, J., Rossdale, P. D., Peace, C. K., Ricketts, S. W., McEnery, R. J., Thorne, J. P., Short, R. V. and Allen, W. R. (1975). Chromosome abnormalities as a cause of infertility in mares. *J. Reprod. Fert.* (suppl. 23), 377.
Cheng, P. L. (1961). Artificial insemination of horses in China, and investigations on increasing conception rate of mares and breeding efficiency of stallions. *4th. Int. Cong. Anim. Rep. The Hague* **1**, 143.
Cheng, P. L. (1963). *Sci Abstr. biol. Sci. China* **4**, 185.
Chipperfield, J. (1975). My Wild Life. MacMillan, London.
Cole, H. H. and Hart, G. H. (1930). The potency of blood serum of mares in progressive stages of pregnancy, in affecting the sexual maturity of the immature rat. *Am. J. Physiol.* **93**, 57.
Colquhoun, K. M., Eckersall, P. D., Renton, J. P. and Douglas, T. A. (1987). Control of breeding in the mare. *Equine Vet. J.* **19**, 138.
Cornwell, J., Guthrie, I., Spillman, T., McCraine, S., Hauer, E. and Vincent, C. (1972). Seasonal variation in stallion semen. *J. Anim. Sci.* **34**, 353.
Cox, J. E. (1968). Rectal temperature as an indication of approaching parturition in the mare. *Equine Vet. J.* **1**, 174.
Cox, J. E. (1971). Urine tests for pregnancy in the mare. *Vet. Rec.* **89**, 606.
Cox, J. E. (1984). Cryptorchid test for horses. *Vet. Rec.* **114**, 127.
Cox, J. E. and Redhead, P. (1985). A comparison of HCG and PMSG effects on steroid production by the equine testis. Society for the Study of Fertility, Winter Meeting, Abstract no. 22.
Cox, J. E., Williams, J., Rowe, P. and Smith, J. (1973). Testosterone in normal, castrated, and cryptorchid male horses. *Equine Vet. J.* **5**, 85.
Cox, J. E., Edwards, G. B. and Neal, P. A. (1979). An analysis of 500 cases of equine cryptorchidism. *Equine Vet. J.* **11**, 113.
Crouch, J. R., Atherton, J. and Platt, H. (1972). Venereal transmission of *Klebsiella aerogenes* on a thoroughbred stud from a persistently-infected stallion. *Vet. Rec.* **90**, 21.
Crowhurst, R. C., Simpson, D., Greenwood, H. and Ellis, D. R. (1978). Contagious equine metritis — treatment of the infected clitoris in the pregnant mare. *Vet. Rec.* **102**, 91.
Croxton-Smith, P., Benson, J. A., Dawson, F. L. M. and Powell, D. G. (1978). A complement fixation test for antibody to the contagious equine metritis organism. *Vet. Rec.* **103**, 275.

David, J. S. E. (1981). Examination of stallions in relation to fertility. British Equine Veterinary Association Congress, Cambridge.

Davison, W. F. (1947). The control of ovulation in the mare, with reference to insemination with stored sperm. *J. Agric. Sci., Camb.* **37**, 287.

Dawson, F. L. M. (1979–84 and 1986–87). Reproduction and Infertility. In: *The Veterinary Annual*. Scientecnica, Bristol (John Wright Ltd).

Deegen, E., Klug, E., Lieske, R., Freytag, K., Martin, J. and Guenzel, A. R. (1979). Surgical treatment of chronic purulent seminal vesiculitis in the stallion. *Dtsch. Tieraerztl. Waschr.* **86**, 140.

Doig, P. A., McKnight, J. D. and Miller, R. B. (1981). The use of endometrial biopsy in the infertile mare. *Can. Vet. J.* **22**, 72.

Dott, H. M. and Foster, G. (1981). Microscopic examination of semen: limits of diagnosis and prognosis. British Equine Veterinary Association Congress, Cambridge.

Douglas, R. H. (1979). A review of equine embryo transfer. *Theriogenology* **11**, 33.

Dowsett, K. F. and Pattie, W. A. (1980). Collection of semen from stallions at stud. *Aust. Vet. J.* **56**, 373.

Draca, P. and Davidovic, A. (1962). Contribution on azoospermia in the stallion. *Vet. Glasn.* **16**, 1140.

Driancourt, M. A. and Palmer, E. (1979). Use of sonography for ovary observation and very early pregnancy diagnosis in the mare. Society for the Study of Fertility, Third Franco-British Meeting, Abstracts p. 34.

Dunn, H., Vaughan, J. and McEntee, K. (1979). Bilaterally cryptorchid stallion with female karyotype. *Cornell Vet.* **64**, 265.

du Plessis, J. L. (1964). Observations and data in thoroughbred breeding. *J. S. Afr. Vet. Med. Assoc.* **35**, 215.

Edington, N., Bridges, C. and Huckle, A. (1985). Reactivation of EHV 1 by corticosteroids. *Equine Vet. J.* **17**, 369.

Ellery, J. C. (1971). Spermatogenesis, accessory sex gland history, and the effects of seasonal change in the stallion. *Diss. Abstr. Int.* **32B**, 2485.

Erbslöh, J. K. (1974). Sterlization of stallions by resection of the epididymal tail (Rosenberger's operation). *Dtsch. Tieraerztl. Wschr.* **81**, 385.

Evans, J. W., Borton, A., Hintz, H. F. and van Vleck, L. D. (1977). *The Horse*. W. H. Freeman and Co., San Francisco.

Feo, J. (1980). Contralateral implantation in mares mated during post-partum oestrus. *Vet. Rec.* **106**, 368.

Fernie, D. S., Cayzer, I. and Chalmers, S. R. (1979). A passive haemagglutination test for the detection of antibodies to the contagious equine metritis organism. *Vet. Rec.* **104**, 260.

First, N. and Alm, K. (1976). Dexamethasone-induced parturition in pony mares. *J. Anim. Sci.* **43**, 283.

Food and Agriculture Organization of the United Nations (1966, 1976, 1984). Production Yearbooks and Vital Statistics Summary, **19, 30, 37**.

Foster, J. F., Evans, M. J. and Irvine, C. H. G. (1979). Differential release of LH and FSH in cyclic mares in response to synthetic GnRH. Society for the Study of Fertility, Annual Conference Summaries, p. 36.

Gebauer, M. R., Pickett, B. W., Voss, J. L. and Swierstra, E. E. (1974). Daily sperm output and testicular measurements of stallions. *J. Am. Vet. Med. Assoc.* **165**, 711.

Gerring, R. L. (1981). Priapism and acepromazine in the horse. *Vet. Rec.* **109**, 64.

Ginther, O. (1974). Occurrence of oestrus, anoestrus, dioestrus and ovulation over a twelve-month period in mares. *Am. J. Vet. Res.* **35**, 1173.

Greenwood, R. E. and Simpson, D. J. (1983). The echogenic development of equine conceptus from days 14 to 60 after ovulation. British Equine Veterinary Association Meeting, Coventry.

Haag, F. M. (1959). Tail-end samples evaluation. *J. Am. Vet. Med. Assoc.* **134**, 312.

Hafez, E. S. E. and Wierzbowski, S. (1961). Analysis of copulatory reflexes in the stallion. *4th Int. Cong. Anim. Rep., The Hague* **2**, 176.

Halnan, C. R. (1985). Sex chromosome mosaicism and infertility in mares. *Vet. Rec.* **116**, 542.

Hamm, D. H. (1978). Gentamicin therapy of genital tract infections in stallions. *J. Eq. Med. Surg.* **2**, 243.

Hauer, E. P., Kellgren, H. C., McCraine, S. E. and Vincent, C. K. (1970). Pubertal characteristics of quarter horse stallions. *J. Anim. Sci.* **30**, 321 (Abstract 34).

Hendrikse, J. (1966). The semen of stallions of normal fertility. *Tijdschr. Diergeneeskd.* **91**, 300.

Henry, J. (1981). Twin ovulation rates in mares as determined by post-mortem study. British Equine Veterinary Association Congress, Cambridge, and Dissertation, Gent State University.

Hintz, H. F., Hintz, R. L., Lein, D. H. and van Vleck, L. D. (1979). Lengths of gestation periods in thoroughbred mares, *J. Eq. Med. Surg.* **3**, 289.

Hughes, J. P. and Loy, R. G. (1970). Artificial insemination in the equine subject: a comparison with natural breeding using six stallions. *Cornell Vet.* **60**, 463.

Hughes, J. P. and Trommershausen-Smith, A. (1977). Infertility in the horse, associated with chromosomal abnormalities. *Aust. Vet. J.* **53**, 283.

Humke, R. and Beaupoil, J. (1979). Ovulation induction in the mare with a synthetic RH analogue. *Berl. Muench. Tieraerztl. Wschr.* **92**, 149.

Kalinin, V. (1947). Horse breeding in the Soviet Union. In: *The Book of the Horse* (Vesey-Fitzgerald, B., ed.) 879 pp. Nicholson and Watson, London.

Kenney, R. M. (1978). Cyclic and pathologic states of the mare endometrium as detected by biopsy, with a note on early embryonic death. *J. Am. Vet. Med. Assoc.* **172**, 241.

Kenney, R. M. and Cooper, W. L. (1974). Therapeutic use of a phantom for semen collection from a stallion. *J. Am. Vet. Med. Assoc.* **165**, 706.

Khalil-Agha, K. (1956). Retained placenta in mares. Thesis, Alfort.

Kirkpatrick, J. P., Wiesner, L., Kenney, R. M., Ganjam, V. K. and Turner, J. W. jr. (1977). Seasonal variation in plasma androgens and testosterone in the North American wild horse. *J. Endocrinol.* **72**, 237.

Klug, E. (1976). Artificial insemination in West Germany using fresh and frozen stallion semen. British Equine Veterinary Association Congress, Cambridge.

Klug, E. (1981). Commercial deep freezing of semen: variations between stallions in sperm survival. British Equine Veterinary Association Congress, Cambridge.

Klug, E. and Merkt, H. (1981). Reduction of twin to single pregnancies by dietetic means. British Equine Veterinary Association Congress, Cambridge.

Klug, E. and von Lepel, J. D. (1974). The possibility of initiating normal birth in the mare with oxytocin. *Dtsch. Tieraerztl. Wschr.* **81**, 349.

Klug, E., Guenzel, A. R., Merkt, H. and Krause, D. (1977a). Examinations of stallions for use in artificial insemination with deep-frozen semen. *Dtsch. Tieraerztl. Wschr.* **84**, 236.

Klug, E., Brinkhoff, D., Fluge, A., Scherbarth, R., Essich, G. and Kiehsler, M. (1977b). Routine use of "phantoms" for collection of genital secretions from stallions. *Dtsch. Tieraerztl. Wschr.* **84**, 382.

Kooistra, L. H. and Ginther, O. (1975). Effect of photoperiod on reproductive activity and on hair growth in mares. *Am. J. Vet. Res.* **36**, 1413.

Krause, D. and Grove, D. (1967). Pellet freezing of jackass and stallion semen. *J. Reprod. Fert.* **14**, 139.

Kreider, J. L., Cornwell, J. C. and Godke, H. A. (1976). Effect of GnRH on oestrus, ovulation, and fertility in the mare. *J. Anim. Sci.* **42**, 263.

Kronfield, D. S. (1979). Feeding on horse breeding farms. *J. Eq. Med. Surg.* **3**, 109.

Lang, K. (1975). Experiences in practice with the induction of parturition in the mare with oxytocin. *Tieraerztl. Umsch.* **30**, 341.

Leidl, W. and Kahn, W. (1984). Differentiation of uterine pathology from early equine pregnancy by ultrasound. *Vlaams diergen. Tijdschr.* **53**, 170.

Lensch, J. (1961). Pregnancy diagnosis in mares at 18–30 days. *Zootec. e Vet.* **15**, 186.

Liu, I., Chung, A., Walsh, E., Miller, M. and Lindenberg, P. (1985). Comparison of peripheral blood- and uterine-derived polymorphs from mares resistant and susceptible to chronic streptococcal endometritis. *Am. J. Vet. Res.* **46**, 917.

Livesey, J. H. (1983). Storage of equine urine. *J. Endocrinol.* **98**, 381.

Lovell, R. (1943). The source of *Corynebacterium pyogenes infections. Vet. Rec.* **55**, 99.

Loy, R. G. and Hughes, J. P. (1966). Effects of LH on ovulation, length of oestrus and fertility in the mare. *Cornell Vet.* **56**, 41.

Mann, T., Short, R. V., Weston, A., Archer, R. K. and Miller, W. C. (1957). The "tail end sample" of stallion semen. *J. Agric. Sci. Camb.* **49**, 301.

Mason, T. A. (1971). Retention of the placenta in the mare. *Vet. Rec.* **89**, 546.

Matthews, R. G., Ropiha, R. T. and Butterfield, R. M. (1967). Phenomenon of foal heat in mares. *Aust. Vet. J.* **43**, 579.

Meagher, D. M., Wheat, J. D., Hughes, J. P., Stabenfeldt, G. H. and Harris, B. A. (1978). Granulosa cell tumours in mares — review of 78 cases. *Proc. 23rd. Ann. Conv. Association American Equine Practitioners,* **133**.

Merkt, H. (1966). Foal heat and foetal resorption. *Zuchthyg.* **1**, 102.

Merkt, H. (1968). Embryonal resorption in relation to twin pregnancy in mares. *Berl. Muench. Tieraerztl. Wschr.* **81**, 369.

Merkt, H. and Guenzel, A. R. (1979). A survey of early pregnancy losses in West German thoroughbred mares. *Equine Vet. J.* **11**, 256.

Merkt, H. and von Lepel, J. (1969). Experiments on heat in thoroughbred mares involving changes in daylight hours. *Dtsch. Tieraerztl. Wschr.* **76**, 672.

Miller, W. C. and Robertson, E. D. S. (1937). *Practical Animal Husbandry*, 2nd Edn, 432 pp. Oliver and Boyd, Edinburgh and London.

Mitchell, D. and Allen, W. R. (1975). Observations on reproductive performance in the yearling mare. *J. Reprod. Fert. Suppl.* **23**, 531.

Mumford, J. and Bates, J. (1984). Trials of an inactivated EHV 1 vaccine: challenge with a subtype 2 virus. *Vet. Rec.* **114**, 375.

Nagase, H. (1967). Deep freezing bull, goat, and stallion semen in concentrated pellet form. *Japan Agric. Res. Quart.* **1** (2), 10.

Nett, T., Holtan, D. and Estergreen, V. (1973). Plasma estrogens in pregnant and post-partum mares. *J. Anim. Sci.* **37**, 962.

Newcombe, J. R. (1981). Variation between breeds in twin ovulation and conception rates. British Equine Veterinary Association Congress, Cambridge.

Nishikawa, I. (1959). Studies on reproduction in horses, 340 pp. Japan Racing Association, Tokyo.

Onstad, O. and Wormstrand, A. (1972). Reproduction in mares in Norway. *Nord. Vet. Med.* **24**, 316.

Ousey, J. C. and Rossdale, P. D. (1983). Mammary secretions in the mare as an indication of readiness for birth. Society for the Study of Fertility, Winter Meeting, Abstracts p. 32.

Ousey, J. C., Rossdale, P. D. and Cash, R. S. (1984). Induction and maternal signals of readiness for birth. British Equine Veterinary Association Congress, Bath.

Oxender, W. and Noden, P. (1975). Photoperiod, initiation of oestrus, and ovulation in seasonally anoestrous mares. Proc. 8th Int. Cong. Anim. Rep. Krakow. Abstracts volume.

Palmer, E. (1981). Ultra sound scanning for ovarian examination and early diagnosis of pregnancy. British Equine Veterinary Association Congress, Cambridge.

Palmer, E. and Chevalier, F. (1983.) Results of the wide-scale use of real time scanning for pregnancy diagnosis in thoroughbred and non-thoroughbred mares in France: accuracy and acceptability of the method. British Equine Veterinary Association Meeting, Coventry.

Parshutin, G. V. (1956). The role of the nervous system in the reproduction of farm animals. (Plenary paper). 3rd Int. Cong. Anim. Rep., Cambridge.

Pascoe, R. R. (1971). Rupture of the corpus cavernosum penis of a stallion. *Aust. Vet. J.* **47**, 616.

Pascoe, R. R. (1983). Methods for the treatment of twin pregnancy in the mare. *Equine Vet. J.* **15**, 40.

Pearson, H. and Weaver, B. (1978). Priapism after sedation neuroleptoanalgesia and anaesthesia in the horse. *Equine Vet. J.* **10**, 85.

Pechnikov, P. P. (1960). *Trudy Vsesoyuznyi Nauchno Issledovatel'skii Institut Konevodstva* **23**, 119.

Pedersoli, W., Fazeli, M., Haddad, N., Ravis, W. and Carson, R. (1985). Endometrial and serum gentamicin concentrations in pony mares given repeated intra-uterine infusions. *Am. J. Vet. Res.* **46**, 1025.

Pickett, B. W. (1981). Management factors affecting seminal characteristics in the stallion. British Equine Veterinary Association Congress, Cambridge.

Pickett, B. W., Faulkner, L. C. and Sutherland, T. M. (1970). Effect of month and stallion on seminal characteristics and sexual behaviour. *J. Anim. Sci.* **31**, 713.

Pigon, H., Lunaas, T. and Velle, W. (1962). Urinary oestrogens in stallions. *Proc. 9th Nord. Vet. Cong.* **2**, 491.

Pitre, J., Legendre, M. F. and Voisin, G. (1979). Facts and thoughts on the microbiological and serological diagnosis of contagious equine metritis. *Prat. vét. equine* **11**, 11.

Platt, H., Atherton, J. and Ørskov, I. (1976). *Klebsiella* and *Enterobacter* organisms isolated from horses. *J. Hyg. Camb.* **77**, 401.

Pouret, E. J. (1982). Surgical technique for the correction of pneumo- and urovagina. *Equine Vet. J.* **14**, 249.

Powell, D. G., David, J. S. E. and Frank, C. J. (1978). Contagious equine metritis: the present situation reviewed and a revised code of practice for its control. *Vet. Rec.* **103**, 399.

Purvis, A. (1978). The induction of labour in mares as a routine breeding farm procedure. Proc. 23rd Ann. Conv. Am. Assoc. Equine Practnrs.

Quinlan, J., van Rensburg, S. W. and Steyn, H. (1951). Occurrence and duration of oestrus, and loss of fertility, throughout the year in mares. *Onderstepoort J. Vet. Res.* **25**, 105.

Redon, P. and Fayolle, L. (1957). Puberty and oestrus in tropical mares. *Rev. Elev.* **10**, 257.

Ricketts, S. W. (1983). The echogenic appearance of twin conception, early foetal death, endometrial cysts and ovarian abnormalities. British Equine Veterinary Association Meeting, Coventry.

Ropiha, R. T., Matthews, R. G., Butterfield, R. M., Moss, R. P. and McFadden, W. J. (1966). The duration of pregnancy in thoroughbred mares. *Vet. Rec.* **84**, 552.

Rossdale, P. D. (1981). *Horse Breeding.* David and Charles, Newton Abbot.

Rossdale, P. D. and Jeffcott, L. B. (1975). Problems encountered during induced foaling in pony mares. *Vet. Rec.* **97**, 371.

Rossdale, P. D. and Short, R. V. (1967). The time of foaling of thoroughbred mares. *J. Reprod. Fert.* **13**, 341.

Rossdale, P. D., Jeffcott, L. B. and Allen, W. R. (1976). Foaling induced by synthetic prostaglandin analogue (fluprostenol). *Vet. Rec.* **99**, 26.

Ruckebusch, Y., Bueno, L. and Tainturier, D. (1978). Automatic apparatus for monitoring the foaling of mares (foaling alarm). *Rev. Méd. vét.* **129**, 1639.

Sanderson, M. W. (1984). Reproductive parameters of thoroughbred mares in the U.K. British Equine Veterinary Association Congress, Bath.

Sanderson, M. W. and Allen, W. R. (1983). Analysis of thoroughbred stud records: influence of the scanner on twinning. British Equine Veterinary Association Meeting, Coventry.

Sandler, B. (1960). Emotional stress and infertility. *J. Reprod. Fert.* **1**, 107.

Schaefer, W. (1962). Artificial insemination in horses. *Zuchthyg. Fortpfl. Besam. Haustiere* **6**, 254.

Schaefer, W. and Baum, W. (1963). Semen quality of stallions used for artificial insemination. *Zuchthyg. Fortpfl. Besam. Haustier* **7**, 382.

Sharp. D. C. (1981). Seasonal reproductive patterns and possible mechanisms of action. British Equine Veterinary Association Congress, Cambridge.

Shires, G. and Johnson, J. (1974). Use of silicone rubber implants in mares as a modification of the Caslick procedure. *Vet. Med. Small Anim. Clin.* **69**, 1171.

Simmons, H. A., Cox, J. E., Edwards, G. B., Neal, P. A. and Urquhart, K. A. (1985). Paraphimosis in seven debilitated horses. *Vet. Rec.* **116**, 126.

Simpson, D. (1982). Ultra sound in pregnancy diagnosis in mares. British Equine Veterinary Association Meeting, Newmarket.

Simpson, D. and Eaton-Evans, W. (1978). Sites of contagious equine metritis infection. *Vet. Rec.* **102**, 488.

Simpson, R. B., Burns, S. J. and Snell, J. R. (1975). Microflora in stallion semen and their control with semen extenders. *20th Ann. Conv. Am. Assoc. Equine Practnrs.*, 255.

Sindelic, V. (1972). Experience in clinical diagnosis of early pregnancy in mares. *Vet. Saraj.* **21**, 194.

Sippel, W. (1969). A veterinarian's approach to stud farm nutrition. *Equine Vet. J.* **1**, 203.

Sisson, S. and Grossman, J. (1938). *The Anatomy of the Domestic Animals*, 3rd Edn, 977 pp. W. B. Saunders Co., Philadelphia and London.

Spincemaille, J. and Vandeplassche, M. (1964). *Streptococcus zooepidemicus*: a venereal disease in horses. *5th Int. Cong. Anim. Rep. Trento* **5**, 166.

Squires, E. L., Douglas, R., Steffenhagen, W. and Ginther, O. (1974). Ovarian changes during the estrous cycle, in pregnancy, in mares. *J. Anim. Sci.* **38**, 330.

Squires, E. L., Stevens, W. B., McGlothlin, D. E. and Pickett, B. W. (1979). Effect of an oral progestin on the estrous cycle and fertility of mares. *J. Anim. Sci.* **49**, 729.

Stabenfeldt, G. H. and Hughes, J. P. (1981). Control of luteolysis in the non-pregnant mare. British Equine Veterinary Association Congress, Cambridge.

Summers, P. M., Shepherd, A., Hodges, J. K., Kydd, J., Boyle, M. S. and Allen, W. R. (1985). Embryo transfer as an aid to breeding exotic equids. Society for the Study of Fertility, Annual Conference, Aberdeen, Abstract no. 74.

Swift, P. N. (1972). Use of testosterone in a gelding for detecting oestrus. *Aust. Vet. J.* **48**, 312.

Szymanski, T. A. (1976). Reproduction statistics of Polish state studs. British Equine Veterinary Association Congress, Cambridge.

Telleen, M. (1977). *The Draft Horse Handbook*. Rodale Press, Emmaus, Philadelphia.

Thompson, D. L., Pickett, B. W., Berndtson, W. E., Voss, J. L. and Nett, T. M. (1977). Artificial photoperiod, collection interval, seminal characteristics, sexual behaviour, and concentrations (in stallions). *J. Anim. Sci.* **44**, 656.

Thompson, D. L., Pickett, B. W. and Nett, T. M. (1978). Effect of season and artificial photoperiod on levels of estradiol 17 and estrone in blood serum of stallions. *J. Anim. Sci.* **47**, 184.

Timoney, P. J. and Powell, D. G. (1982). Isolation of contagious equine metritis organism from colts and fillies in the U.K. and Ireland. *Vet. Rec.* **111**, 478.

Timoney, P. J., Shin, S. J. and Jacobson, R. H. (1982). Improved selective media for isolation of the contagious equine metritis organism. *Vet. Rec.* **111**, 107.

Tischner, M. (1976). Preservation of stallion semen in liquid nitrogen. British Equine Veterinary Association Congress, Cambridge.

Tischner, M. (1981). Use of non-surgical embryo recovery to assess fertility of "deep-frozen" stallions. British Equine Veterinary Association Congress, Cambridge.

Turner, A. I. and Shreeve, J. E. (1984). Contagious equine metritis. *State Vet. J.* **38**, 109.

Urwin, V. and Allen, W. R. (1981). Gonadotrophic and steroid hormone profiles in monovular and twinning mares. British Equine Veterinary Association Congress, Cambridge.

Urwin, V. and Allen, W. R. (1983). Chorionic gonadotrophin control of secondary follicular development during early pregnancy in the horse and donkey. Society for the Study of Fertility, Winter Meeting, Abstracts p. 30.

Vandeplassche, M. (1978). Second international symposium on equine reproduction, California, July 1978: a summary. *Vlaams diergen. Tijdschr.* **47**, 422.

Vandeplassche, M., Spincemaille, J. and Bouters, R. (1971). Aetiology, pathogenesis and treatment of retained placenta in the mare. *Equine Vet. J.* **3**, 144.

van der Holst, W. (1981). Variation in stallion semen characteristics between different breeds. British Equine Veterinary Association Congress, Cambridge.

van Leeuwen, W. (1981). Treatment of mares in prolonged oestrus with proligestone. British Equine Veterinary Association Congress, Cambridge.

van Niekerk, C. H. (1965). Early clinical diagnosis of pregnancy in mares. *J. S. Afr. Vet. Med. Assoc.* **36**, 53.

van Rensburg, S. J. and van Niekerk, C. H. (1969). Ovarian function, oestradiol and progesterone in mares. *Onderstepoort J. Vet. Res.* **35**, 301.

Volsen, S. G., Allen, W. R. and Antczak, D. F. (1985). Modulation of antigen expression during development of the endometrial cups. Society for the Study of Fertility, Annual Conference, Aberdeen, Abstract no. 83.

Voss, J. L. Pickett, B. W., Back, D. G. and Burwash, L. D. (1975). Effect of rectal palpation on pregnancy rate of non-lactating normally cycling mares. *J. Anim. Sci.* **41**, 829.

Wallace, R. and Scott-Watson, J. A. (1923). *Farm Livestock of Great Britain*, 5th Edn, 868 pp. Oliver and Boyd, Edinburgh and London.

Ward, J., Hourigan, M., McGuirk, J. and Gogarty, A. (1984). Incubation times for primary isolation of CEM organism. *Vet. Rec.* **114**, 298.

Webel, S. K. (1981). Influence of allyl trenbolone on mares at various times of the year. British Equine Veterinary Association Congress, Cambridge.

Weiss, R., Boehm, K., Merkt, H. and Klug, E. (1975). Investigation of preputial and nasal swabs for *Klebsiella*-like bacteria. *Berl. Muench. Tieraerztl. Wschr.* **88**, 436.

Werthessen, N. T., Marden, W., Haag, F. and Goldzieher, J. W. (1957). The influence of seminal fluid on fertility. *Proc. Soc. Study of Fertility* **8**, 42.

Wesson, J. A. and Ginther, O. (1981). Influence of season and age on reproductive activity in pony mares. *J. Anim. Sci.* **52**, 199.

Wide, L. and Wide, M. (1963). Pregnancy diagnosis in the mare by an immunological method. *Nature (Lond.)* **198**, 1017.

Willadsen, S. M., Pashen, R. L. and Allen, W. R. (1980). Micromanipulation of horse embryos. Society for the Study of Fertility, Annual Conference, Oxford, Summaries p. 32.

Wolter, R. (1972). Feeding and disease in the horse. *Rev. Méd. vét.* **123**, 623.

Zhmurin, L. M. (1960). *Trudy Vsesoyuznyi Nauchno Issledovatel'skii Institut Konevodstva* **23**, 95.

Zwick, V. (1968). Clinical evaluation of ovaries and endometrium of the mare. Dissertation, Munich.

2   Care of the Mare and Foal

D. R. ELLIS

INTRODUCTION

The seasonal polyoestrous behaviour and eleven month gestation period result in most mares foaling during April and May. However, most horses are aged from January 1st and commercial breeders, especially of thoroughbreds, prefer foals to be born as early in the year as possible. This preference gives two main advantages; first, there are more opportunities for the mare to be covered again and become pregnant before the customary end of the stud season (July 15th) and second, the early foal will be better grown, more mature and therefore a more suitable subject for showing or sale as a foal or yearling.

Thoroughbreds, Arabs and their crosses are notable for their rapid growth rate, reaching more than 90% of their mature height and weight in two years. In order to achieve this growth safely, good management of both mare and foal is paramount.

PRE-FOALING MANAGEMENT

Nutrition

During the first eight months of pregnancy the mare requires rations for maintenance only. In the last three months of gestation the foetus doubles its size and consequently the mare's nutritional requirements gradually increase beyond maintenance to a peak, one month after foaling, when lactation is at its greatest. As with any animals, the necessary increments in the diet should be made gradually. Assessment of mares' nutritional needs is difficult as there is great individual variation. A practical "rule of thumb"

Horse Management 2nd edition
ISBN: 0-12-347218-0 case

of ideal bodily condition in mares is that one should be able to feel the ribs easily but only just able to see them. However, overfeeding should always be avoided as it may cause obesity or laminitis.

If grazing is good and the weather reasonable, heavily pregnant mares can live out day and night without being given extra feed until a month or two before foaling. However, if they are due to foal early in the year and it is hoped to have them covered and got in foal during February and March then mares should be stabled at night and fed concentrates and hay during the last three months of pregnancy. This will not only provide for the growing foetus but reduce the energy demands of keeping the mare warm in the winter months and improve her bodily condition up to and following foaling. Hay and concentrates are gradually increased to 0.25–0.5 kg of each per 50 kg body weight per day at foaling. This diet should contain 11% dry matter as protein and a calcium based mineral supplement (30 g of ground limestone per day is adequate).

Stable management

Simple stabling, of adequate size (3.5 m × 3.5 m floor area), well drained and ventilated is ideal for the heavily pregnant mare. Separate loose-boxes are preferable to stalls in barns for groups of pregnant mares as the environment is easier to control and the spread of infectious disease, such as Virus Abortion, may be reduced. The ideal foaling box should be larger (5 m × 5 m floor area) but made octagonal in shape by cutting off the corners. It should have good lighting, ventilation and heating available.

The provision of extra artificial light in the stable in winter from December onwards is a reliable stimulus to bring barren and maiden mares out of anoestrus and into normal oestrus cycles early in the covering season. At least six weeks and optimally three months of increased light should be provided starting in December. This policy should also be applied to foaling mares. A clean 150 watt bulb or strip light, left on until 10.00 p.m. each night or by time switch ensuring a total of 16 hours day and artificial light per day is adequate.

Preventive medicine

If a mare is to foal away from home, at a stallion stud for example, she should be moved there approximately three weeks before her due date in order that she can develop immunity to the microflora of the new surroundings which can be passed on to her foal in the colostrum. To prevent the mare or her newborn foal acquiring infectious disease it is generally safer to foal the mare at her home stud. This is the wisest precaution against the most serious infection, rhinopneumonitis virus abortion. Although it can occur in isolated mares and is unpredictable in its occurrence the abortion strain of the Equine

Herpes Virus Type 1(EHV1) can be shed by carrier mares, foals or yearlings and is more likely to arise when mares are mixed in large stud farms.

A killed vaccine (Pneumabort K: Fort Dodge) is available for use against EHV1 but controlled trials and field experience have shown that it is not completely effective in preventing virus abortion. There is some evidence that vaccinated mares shed less virus following infection than unvaccinated mares, which may reduce spread of the disease. The vaccine is given by intramuscular injection at the 5th, 7th and 9th months of mares' pregnancies. Even if they are vaccinated, pregnant mares should be kept in groups of similar gestational length and separate from older foals, yearlings, animals out of training or other mares arriving from sales or other stud farms.

Vaccination of breeding stock against influenza is advisable as, apart from the typical cough and fever, infected foals may develop a fatal pneumonia. Fortunately the United Kingdom has not had an epidemic of equine influenza for seven years and this is probably the result of more widespread vaccination. Prior to 1979 an epidemic occurred every three or four years. Initial vaccination courses are given to foals or yearlings which should receive annual boosters thereafter. To comply with *Féderation Équestre Internationale* (FEI) and Jockey Club Rules, injections should be given as follows: 1st injection; 2nd injection 21–92 days later; 3rd injection 5–7 months (150–215 days) after the second and annual boosters no more than 12 months after the last injection. The annual booster should be given to pregnant mares a month before foaling so that a high level of antibody is transferred to the foal in the colostrum. Tetanus vaccination can also be maintained simultaneously.

Most modern worm treatments such as thibendazole, pyrantel embonate, fenbendazole and ivermectin are safe to give in therapeutic doses to pregnant mares so a routine monthly or six weekly treatment regime can be maintained all the year round. Pasture management is as important as regular worm treatment and whenever possible pregnant and foaling mares should graze paddocks that have been rested from horses for at least six months. Three or four days prior to being turned onto new pasture the mares, foals or yearlings should be wormed. Biannual (spring and autumn) worm egg counts on all stock are useful for checking on the efficacy of worm control. Removing dung from paddocks is an important part of stud farm management.

CARE OF THE MARE AND FOAL

Foaling and the first days

Foaling should be supervised calmly and allowed to occur as naturally as possible. Untimely or panicky interference can inhibit some mares or cause them to get up and down when they should be recumbent and proceeding through the second stage of labour. A check that presentation of the foal is normal, an episiotomy if there has been Caslick's operation and possibly traction later in the second stage are all that are normally necessary. A few mares foal standing up and in these cases the foal should be supported during delivery.

Once delivery has taken place the mare and foal should be left as quiet as possible. The foal's umbilical cord should be checked to ensure that pulsation has ceased and that it separates at the weak point 3–5 cm external to the abdomen. If this is so, then blood loss is minimal, ligation unnecessary and the umbilical stump is dusted or sprayed with an antibiotic. Some grooms prefer to treat the navel with iodine solution but this is not recommended as it may blister the foal's skin.

Some stud farms weigh the foal after birth but this is not an essential to good management. The mare's placenta should be checked thoroughly following the third stage of labour to ensure that it is all, especially the tips of the horns, present. It is not uncommon for mares to show transient signs of colic, sometimes quite violent, during or following the third stage of labour. If such colic persists veterinary attention is necessary and the commonest cause in older mares is rupture of the uterine artery and haemorrhage into the broad ligament. If the haemorrhage is not contained within the ligament but free into the peritoneal cavity it is rapidly fatal. No treatment can be given beyond analgesia and protection of the foal during the mare's colic.

The normal foal may exhibit sucking behaviour before rising but once it is standing and capable of feeding a careful check must be made that it does suck the teat and swallow milk. Frequently, foals will suck everywhere but the teat and benefit from gentle and patient guidance. First foaling mares may be reluctant to suckle initially as their udder is firmer, tender, and the teat smaller than in the multiparous mare. Again, patient guidance is normally rewarded and even if the mare is very fractious, just restraining her or holding her front leg up may allow the foal to start sucking. It is rare that sedatives or a twitch are necessary in such cases. A few mares, especially first foalers, will be fiercely possessive of their newborn foal and must be approached and handled with care.

Some newborn foals have difficulty in standing, for no veterinary reason, and gentle assistance may be necessary for a few minutes. The state of the bedding is very important at this stage. If there is insufficient the foal will suffer bedsores on its hocks, stifles and elbows in attempting to stand. While the bed must be deep it should also be firm so that the foal does not get obstructed by clumps of long straw.

To prevent meconium impaction it is good management, especially in thoroughbred foals, to give an enema when a few hours old. Warm soapy water from a douche can, liquid paraffin from a syringe or proprietary human enemas in plastic bags can be used. The foal should be restrained standing and the enema is given, without force, via a soft rubber tube. Meconium is usually expelled soon afterwards but if the normal soft yellow dung is not apparent after a few hours and the foal continues to strain then the enema should be repeated.

Foals should be led separate to the mare for the first few days after which a foal head collar can be put on and an attendant leads the foal with his right hand and the mare with the left. Gaining the foal's confidence and teaching it to lead properly and allow quiet handling are important early steps in its education.

Exercise is a valuable aid to the expulsion of any uterine discharge and to uterine involution but it is especially important in these respects and the prevention of laminitis if the third stage of labour is prolonged. Even if the weather is cold, as in the early months of the year, a healthy mare and foal should be allowed some time turned out in a small paddock or yard from the day after foaling. This should be continued and the time and space gradually increased unless the weather is very wet and cold. It has been a custom, but is poor management and not recommended, to confine mares with foals to the stable until the foal heat is over.

If a foaling mare has to be moved to a stallion stud this can be safely done five or six days later provided mare and foal are healthy and the distance is not too great (up to say 200 miles).

The first months

In the United Kingdom it is customary for mares and foals to be stabled at night and run with others in paddocks in the daytime until the foal is two months old or the month of May when, weather permitting, they can live out day and night. However, some studs delay leaving them out until the mare has been tested in foal at six weeks as she needs to be teased regularly and with a foal at foot this can be managed easier from the stable. The mare is tried by the teaser every second day or daily and she is usually separated from the foal for this. However, mares will often behave more naturally for teasing or covering if their foal is held in front of them rather than separated.

If a foaling mare fails to come into season due to lactation anoestrus and does not respond to hormone treatment, turning her out to pasture day and night with her foal may be effective. In this way, especially from April onwards, the mare receives maximum daylight and maximum benefit from the growth of spring grass both of which are key factors in bringing mares out of winter anoestrus and into normal oestrus cycles.

Nutrition   Lactation exerts the greatest energy demands on the mares which at the peak, one month after foaling, may produce 15 litres of milk per day.

At the same time the mare must maintain her bodily condition so that her oestrus behaviour is normal and she can readily become pregnant again. There is great individual variation, from mares which feed their foals well but lose condition themselves however much food they eat, to other mares which stay fat with or without providing well for their foals. The maximum amount a mare will eat is approximately 1.5 kg of food per 50 kg bodyweight per day. Lush spring grass, especially if it contains some clover, would probably meet all her requirements. If the grass is poor the mare can be fed equal weights of hay and concentrate up to 0.75 kg of each per 50 kg bodyweight daily. Dry mature grass would not provide sufficient protein, calcium and phosphate for her needs so these must be the nucleus of any concentrate mix. Grain, such as crushed oats, is the usual source and added linseed or soya bean meal can enhance the protein level. Grain will provide extra phosphorus but a separate calcium supplement, to provide 0.5% dry matter of the ration, must be added.

After twelve weeks of lactation the mare's rations should be reduced, stopping concentrate feed and reducing the hay by half. If the mare is living out day and night supplementary feed should be withdrawn altogether. This lowering of nutritional levels is especially important a week or two before weaning which is best done at 4–5 months after foaling.

From one month old at the latest, foals will already be grazing and eating hay and concentrates from the mare's manger. When they are 3–4 months

Figure 2.1 Creep feed manger in a paddock.

old, particularly if the grazing is beyond its best, they can be given some creep feed in the paddock. If pasture is poor and foals are not thriving creep feeding can start earlier at 5–6 weeks old. Solid mangers are placed in the paddock and surrounded by a square, single rail fence 5 metres square and 1.5 metres high (Fig. 2.1). Foals which gain access to this enclosure must not be fed *ad lib* but started on 0.25–0.5 kg of a concentrate mix containing a mineral supplement per 50 kg body weight daily.

Weaning   The mare must have been prepared by previous reduction of her rations and the foal must be of adequate size, bodily condition and in good health. The first and more traditional method is the abrupt separation of similar aged foals from their dams. The mares and foals are brought into the stable at the usual time in the afternoon but the foals are shut in with a feed and the mares taken to different stables or grazing as far away as possible to prevent them seeing or hearing their foals and vice versa. The foals usually settle within a day or two, particularly if they can see each other. Accidents can occur if the foals become very distressed following weaning so the stables must be checked carefully for safety — no exposed glass windows or jagged edges to doors or mangers etc. When turned out the following day the foals may gallop to exhaustion so it is best that the paddock is small to prevent injury occurring.

The second method of weaning is less traumatic and more suitable for groups living out day and night or with foals of greatly differing ages. One or two mares are taken from the group every few days, removing the quieter and more tolerant mares last from the foals. Often, the weaned foals may seek the company of or suck the remaining mares for a few days.

The weaned mares are best left out at grass without supplementary feed. They should not be milked out as this would only stimulate further milk production. If the udder becomes tense and painful, gentle massage with an emollient oil may ease the discomfort. Only if mastitis occurs should milk be drawn off and intramammary antibiotics given twice daily. Systemic antibiotics should be given if the mare becomes febrile.

MANAGEMENT OF THE WEANED FOAL AND YEARLING

Weaned foals do not need to be separated into colt and filly groups until they are yearlings. It is important that they are caught and handled daily and if time and labour allow, they should be brought into a stable every afternoon for a small feed and have their feet picked out. This discipline makes them much more manageable especially when they become strong, well grown yearlings.

Creep feeding can be continued through either method of weaning, or the foals could be fed from individual mangers on the ground or hooked onto the fence. The pecking order may affect the amounts eaten by individuals and one can only be certain of their intake if they are stabled for feeding. There is no reason why foals should not live out at pasture day and night through the winter providing shelter is available and they are fed concentrates.

The pasture available will vary greatly with the soil, season, seed mix, weather and stocking rate. Therefore it is only possible to quote some basic nutritional requirements for weaned foals and yearlings. Protein, calcium and phosphorus are the three nutrients which are in greater demand for rapid growth of the young horse. A creep feed concentrate may contain up to 18% dry matter as protein but after weaning 16% is adequate which should be gradually reduced to 13% at two years old. Protein quality is important and the diet should contain 0.7% lysine for optimal growth to be maintained. Legumes such as clover or lucerne are rich in this amino acid but if it is added to a concentrate mix then soya bean meal is the best source.

If the protein maintains growth rate and provides the "scaffolding" or matrix of the developing bone, calcium and phosphorus provide the "cement" and its strength. Calcium should be included in the creep ration at 0.85% dry matter reducing this to 0.7% as a yearling and 0.55% as a two year old. The phosphorus level should always be less than the calcium in the diet maintaining a Ca:P ratio of between 1.2 or 3 to 1. Legumes are a rich source of calcium but grains such as oats or maize contain very little calcium and significantly more phosphorus. Therefore, if concentrates are being fed they must contain, or the animals must have available, a calcium rich mineral supplement. Ground limestone is normally the simplest and cheapest source.

The weaned foal can be given 0.5–0.75 kg of a concentrate mix per 50 kg bodyweight per day but if pasture or hay (such as leguminous hay) is rich in protein and growth and bodily condition are maintained then less should be given. For yearlings the level should be reduced to 0.25–0.5 kg per 50 kg bodyweight per day.

Although foals and yearlings seem to grow in height in spurts, especially during the first three months and again at 6–9 months old, their weight gain is normally very even. In the first three months when growth is more intense, thoroughbred foals can gain up to 1.5 kg per day. This rate settles down to 0.5–1 kg per day until two years of age when they are over 90% of their mature weight.

Foals or yearlings which are prepared for sale at auction in December and October respectively will require more concentrate feed and undergo preparatory handling, grooming and walking for up to a month before the sale. The preparation takes longer for yearlings which, in addition, will be lunged for short periods daily. Lunging should not be excessive, 5–10 minutes per day at a steady canter is adequate, in a ring of prepared, well drained sand or wood chips approximately 15 m in diameter. It is easier to teach a yearling to lunge in both directions in a fenced ring but if a paddock is used the surface should be even and soft and the lunge rein kept long to prevent injury.

Later on in the sale preparation, shoes may be applied to the yearling's front feet, but when lunged the shod yearling should have boots applied to the forelegs and not have shoes on the hind feet. The extensive walking of yearlings on a firm surface at auction sales makes the fitting of front shoes essential and of hind shoes optional. Foals can have front shoes fitted but this is rarely done due to the risk of self-inflicted injury.

A similar, but less intense preparation, is usually practised on yearlings before they go into training in the autumn.

Behaviour problems

Certain behaviour problems arise among foals and yearlings which can be alleviated by changes in management. Fighting among yearling colts can result in wounds, blemishes or serious injury. Two typical blemishes are subcutaneous thickening or haematoma formation on the front of the knee cause by kneeling during fights. Repeat injuries make the blemish chronic and slow to disappear. If this occurs, the group should be split up and the particular offender or sufferer kept on its own. Haematomata on the front of the carpus resorb and gradually disappear if they are left alone and not exacerbated by further injury. Drainage is not recommended as healing is very slow and infection and excessive granulation are likely complications.

Box walking may be acquired by colts or fillies after weaning. It can be difficult to stop once established and may reduce the animal's bodily condition. Stabling the animal with another, in open stalls or in stables with intercommunicating windows may help. Obstructions to the animal circling such as bales of straw or tying the animal up may help. Weaving is a similar but less debilitating vice which is rarely seen in foals or yearlings.

Windsucking or crib-biting (in which air is gulped or swallowed while gripping a door or rail with the teeth) can occur in foals or yearlings. Its ill-effects are little apparent in most cases but may include occasional flatulent colic and abnormal wear of the incisor teeth. It rarely reduces the body condition of young horses. It can be acquired by imitation within a group, particularly if the grazing is poor. At most auction sales affected animals are returnable if this vice is not declared by the vendor. Various practical and surgical remedies have been attempted over the years but none is universally successful. Removing the temptation, such as mangers, and fitting a cage above the bottom door, may eliminate the vice in stables. Aversion therapy, using an electrified wire round the top rail or 1 metre inside the rails may succeed in the paddock. Tight throat straps are not practicable in young horses and any remission following surgery is not always permanent.

Coprophagia is not uncommon in foals in their first two months of life. They mostly eat the droppings of their dam and the habit causes no known ill-effects.

The orphan foal

The orphaned foal has two principal requirements which are nutritional and social. Nutritionally the foal under two months of age is best served by a foster mother and is easier to manage. Older than two months it can thrive and grow satisfactorily without fostering but is better behaved and easier to manage if it has a companion mare or pony.

The ideal foster mother is a quiet mare which has reared previous foals and lost her own foal within the previous twenty-four hours. Careful inquiry must be made as to the cause of death of the foal to ensure that there is no risk of the foster mare bringing an infection, such as EHV1 to her adopted foal and its stud farm. The mare should be in good health, especially if she had a difficult foaling, but having lost her foal she should be milked out regularly and frequently to maintain her milk supply. It is helpful to have the skin of the dead foal removed and available for use on the orphan during fostering. The mare is normally, but not necessarily, travelled to the foal and it is wise to let her rest for an hour or so after the journey and become accustomed to her new surroundings. Milking her out is left overdue and a small dose of sedative such as acetyl promazine may facilitate the fostering. The orphan has the skin tied on or it is smeared with the mare's milk immediately prior to introduction. While the mare is held with a bridle the foal, which should be left a little hungry beforehand, is encouraged to suck. This may elicit squealing or kicking from the mare but with careful and quiet handling the foal should be encouraged to persist and later the mare allowed to smell or muzzle the foal. If safe, the foal can be let loose in the stable but the mare should not be released until the attendants are sure she will not bite or kick the foal. The skin should be removed after a few hours and careful supervision must be maintained during the first day or two. Patience is essential as some successful fosterings may take a day or two, but rejection is rare and normally due to insufficient milk supply in the mare. The foster mare and her new foal should be allowed a few days on their own before joining other mares and foals.

The young orphan foal which is not fostered is fed a proprietary equine milk substitute from a bottle or bowl every three or four hours during the first month of life. Most will feed from a bowl but automatic calf feeders can be labour saving and successful. Water, hay and concentrates, containing milk pellets should be available for foals 10–14 days of age or older. Hand reared foals can become very unruly and difficult to handle and may not learn to graze unless they have equine company, preferably an older mare or pony, from two or three weeks of age. When three or four months old the orphan can join groups of weaned foals.

Preventive medicine

The best preventive medicine for the newborn foal is the good passive immunity acquired from the mare's colostrum. Failure of this transfer is not uncommon and places the newborn foal at great risk of suffering infections, particularly bacterial disease. The commonest reason for failure is the mare running milk, and thus losing the colostrum and its rich supply of immuno-globulins, days or even weeks before foaling. Pre-foaling lactation results from premature separation of the placenta and may herald abortion, premature or twin birth. The early separation of the allantois can be caused by placentitis (often bacterial or fungal) or placental insufficiency, as in twin pregnancy, which may lead to the death of a foetus.

A few mares, often first foalers, fail to provide their foal with sufficient immunoglobulin even though they have not run milk. Presumably the mare fails to concentrate the globulins into colostrum immediately before birth. Also, it is important to ensure that the newborn foal does suck and swallow its dam's colostrum as soon as possible after birth. Ideally this should be the first feed and within six hours of birth as immediately after absorption the intestine loses its ability to absorb the large globulin molecules over the ensuing 18 hours.

Detection of the foal's failure to acquire passive immunity can be confirmed by measuring immunoglobulins in its blood plasma or serum. Originally a simple and rapid zinc sulphate turbidity test on the serum, but more recently a foal immunity test (Aglutinade Ab-AgLabs) can be used on plasma or serum at 18–24 hours after full absorption should have occurred. This test can also assess the immunoglobulin content of colostrum, from the mare or a donor. However, visual assessment is just as reliable; the more honey-like the greater the globulin content and vice versa.

It is a crucial part of good foaling management on a stud to have a colostrum bank. Half a pint of good quality colostrum can be collected from free milking mares, after their foal has sucked, and deep frozen in sterile plastic bottles. It can be kept safely for some years but must not be thawed and re-frozen. Before use it is thawed slowly, not in very hot water which will destroy the immunoglobulins, and a half or if available one pint is given as a first feed to the needy foal by bottle or stomach tube. Although this supplement may not give the foal a high plasma immunoglobulin level, experience has shown it to be a valuable preventive measure.

If colostrum was not available for the foal whose dam ran milk or the foal's immunoglobulin level was deficient on a routine test, plasma or serum can be given. Naturally, if the dam did not produce colostrum after birth she will not have high serum levels of immunoglobulins and this would be reflected by a poor response in the foal's plasma following transfusion. However, experience has shown that plasma or serum transfusion, like colostrum supplementation, is valuable. Three to five litres of blood are collected from the mare into acid-citrate-dextrose and following centrifugation or allowing it to stand for 2 hours 1–3 litres of plasma is siphoned off and given slowly, intravenously to the foal. This technique is also valuable in the treatment of some cases of neonatal septicaemia, enteritis or polyarthritis. If neither colostrum nor plasma supplementation is adopted, foals at risk can be given broad spectrum antibiotic cover for at least four days. Routine antibiotic coverage of all foals is not recommended in case resistance develops.

If the mare has been boosted with influenza and tetanus vaccine about a month before foaling and the foal acquires passive immunity from the colostrum, further tetanus protection with antitoxin is unnecessary until vaccination at three months of age. If the passive immunity is poor tetanus antitoxin, 1500 i.u. should be injected when the foal is a few days old. Some studs repeat this injection at 30 days old but as the earliest a foal will show signs of tetanus is 16 days old experience shows that the single dose at a few days old protects against neonatal infection. Antiserum should always be given if an older foal sustains a wound before it is vaccinated with tetanus toxoid. The earlier after three months old tetanus and influenza vaccination is started the better.

Foals should be given worm treatment at 6–8 weeks old and 4–6 weekly intervals thereafter. Ascarids and small strongyles are the principal parasites controlled at this stage so any broad spectrum treatment can be given as a paste or by stomach tube. Piperazine citrate with thibendazole, pyrantel embonate and fenbendazole are commonly used. Ivermectin's larvicidal effect is particularly valuable, especially against strongylus vulgaris in foals and yearlings. Different drugs should be used from time to time to discourage parasites from developing resistance. The grazing of young stock on clean, rested paddocks and regular dropping removal are also important control measures. It is not essential to treat for bots but if it is done then an organophosphorus compound such as metriphonate can be given in November.

Hoof care is a crucial part of management. All young stock should be carefully inspected standing and walking on a firm level surface at least once a month. This will reveal any changes in limb or foot conformation which require corrective hoof trimming. Routine farriery is necessary once a month and may be more often in foals or yearlings needing corrective trimming.

Table 2.1. A classification of neonatal diseases of foals (After Rossdale, 1972)

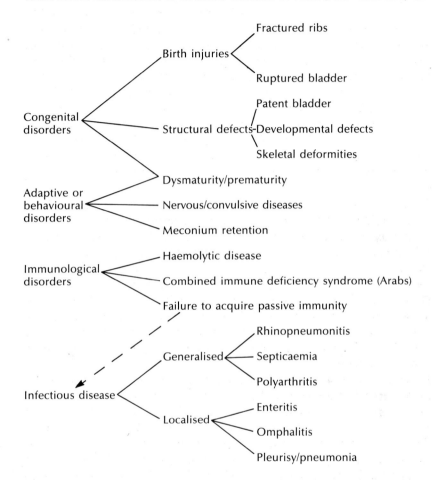

DISEASES OF FOALS

The neonatal period

The first week of a foal's life is the period in which the mortality rate is highest. The neonatal diseases are classified in Table 2.1.

Congenital disorders

Birth injuries   Fracture of the ribs is the commonest injury sustained during birth. Factors which contribute to its aetiology include the keel-like shape of the sternum and the foal's deep chest and large elbows, the size and shape of the mare's pelvic canal and particularly, traction during birth while the mare is standing. In the first few hours after birth the normal rib cage is extremely mobile at the costochondral junction. It is at this site or just dorsal to it that one or more ribs fracture, usually unilaterally and between numbers 2 and 12.

When only two or three ribs are fractured there may be no overt clinical signs and the injury is detected by palpation of oedema, crepitus or a haematoma. When more ribs are fractured asymmetry of the rib cage may be apparent, particularly if inspected with the foal held on its back, the fractured side being concave and the normal side convex. Lameness, difficulty in rising from recumbency, possibly somersaulting forward or inability to stand may be presenting signs. The more serious cases develop localised pleurisy, pericarditis or myocarditis and exhibit weakness, dyspnoea, reluctance or inability to stand and suck and may die soon after birth or after a few days from secondary infection or pathology.

Complications of rib injuries also include abscess formation over the site of fracture, often due to a sequestrum and leading to sinus formation. Rarely, a fractured rib may penetrate the diaphragm and present as a fatal colic with marked dyspnoea.

Rupture of the bladder may be a birth injury or a developmental defect. Some foals are born with a grossly distended urinary bladder, the rupture of which during delivery is quite feasible. However, the rupture invariably occurs on the dorsal surface in the mid-line, sometimes in conjunction with other congenital deformities such as carpal contracture, scoliosis or large umbilical hernia which suggest a congenital weakness or patency.

Clinical signs are not usually apparent until the foal is 40–48 hours old at the earliest but cases have been diagnosed and successfully treated as late as 6 or 7 days of age. Frequent straining is a feature and in most cases it is unproductive but the foal with a small patency or rupture may be able to pass reasonable quantities of urine. The foal becomes weak and listless, gradually going off the suck but characteristically it has a full or distended abdomen in which fluid waves can be seen or palpated during ballottement. The mucous membranes develop a greyish tinge, the pulse rate is not markedly elevated but respiration becomes fast and shallow. The foal's breath smells uraemic but uraemic fits only occur terminally.

Diagnosis is confirmed by paracentesis abdominis, using a large trocar and cannula (0.5 cm in diameter or larger) or if unavailable a large gauge (12G) needle. A small area of skin, 5 cm antero-lateral to the navel is shaved, cleaned and anaesthetised. With the foal held in lateral recumbency the trocar and cannula are introduced into the abdomen and all urine (up to 2 gallons can be present) is drained. The trocar or an obturator must be re-inserted before withdrawing the cannula to prevent dragging omentum through the skin wound. It is very rare to do harm trocarising a foal at this site. However, if the clinician needs encouragement to trocarise, a blood urea level above 8 mmol/1 would provide it, remembering that less marked increases in blood urea occur in dehydrated foals.

Following confirmation of diagnosis and drainage of urine from the abdomen a few hours should elapse to allow the foal to restablise after the loss of heat and fluid and hopefully, decrease its uraemia, before surgical repair. Rarely a pervious urachus may leak into the abdomen causing identical signs and requiring surgical repair in foals up to one month old.

Table 2.2. A classification of congenital disorders in foals (After Platt, 1979)

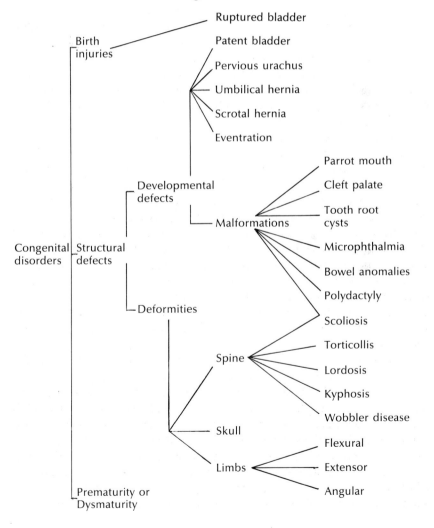

Structural Defects   These are classified in Table 2.2 and patent bladder and pervious urachus may be considered as developmental defects within this group. Pervious urachus is characterised by the dribbling or passing of urine through the navel. It is not uncommon and usually warrants local antisepsis and systemic antibiosis to prevent secondary infections. They mostly heal spontaneously and do not require surgery.

Umbilical hernia is common but not always apparent until the external cord stump has detached at a few weeks of age. Small herniae disappear with growth and cases with larger sacs are treated surgically, clamped or blistered when the foal is a few months old. Complications such as incarceration of an umbilical hernia are rare.

Congenital scrotal hernia is rare but may occur in conjunction with other congenital defects of the limbs or spine. It invariably reduces spontaneously and without complication by a few weeks of age.

Congenital eventration of small intestine through an incomplete but normally small umbilical ring is almost invariably accompanied by other congenital defects. Therefore the foal must be examined thoroughly before reduction and repair as in most cases destruction on humane or economic grounds may be necessary.

Parrot mouth or prognathism is the commonest malformation encountered in foals. It is characterised by malocclusion of the tables of the incisor teeth, the upper being anterior to the lower. Corresponding malocclusion of the molar arcade is also often, but not invariably, present, which in later life may require regular dental attention. Prognathism is believed to be inherited but a definite diagnosis cannot be made in slight cases until they are a few months old. Some will improve with growth but spontaneous correction of a marked case is unlikely and there is no recognised treatment.

Cleft palate is a rare but serious malformation which can involve only the soft palate or both hard and soft palates to differing degrees. Characteristically milk comes down the nose during feeding and diagnosis is not difficult except in cases involving the soft palate only which may require endoscopic examination with the foal sedated or anaesthetised. A persistent right aortic arch or patent ductus arteriosis constricting the oesophagus may also cause milk regurgitation. Rarely a foal may discharge milk down the nostrils during its early days without a palatal defect. This is presumably an inability to swallow large volumes of milk which it gradually overcomes.

The foal with a cleft palate does not thrive and inhalation pneumonia is a common complication. Surgical repair is not recommended.

Tooth root cysts are a rare malformation and can involve incisor, mandibular or maxillary roots. In the latter the sinuses may also be implicated. They rarely prevent the foal sucking and thriving in its early life but may require radiographic and surgical exploration and treatment.

Microphthalmia may be uni- or bilateral and can accompany other deformities such as anencephaly. Enucleation of the single affected eye is unnecessary but the resulting hollow orbit may appear preferable to the small button-like eye. Surgery can be undertaken after weaning.

Anomalies of bowel development mostly involve the large bowel and, like atresia coli and anal agenesis, can be very difficult to correct surgically. An exception is pyloric stenosis which is characterised by colic after sucking or feeding, confirmed by passive drainage of milk from the distended stomach via a naso-gastric tube. A laparotomy and longitudinal incision of the pyloric muscularis effects a cure.

Polydactyly or extra phalanges usually originate from the fetlock or cannon and require ligation or surgical excision depending on their size.

Scoliosis or lateral curvature of the thoraco-lumbar spine may be a curvature of normally formed vertebrae or in the more severely curved a malformation. Some kyphosis may accompany it. Curvature is usually maximal at posterior thoracic level. It is invariably accompanied by other deformities and severe or complicated cases should be destroyed. Simple, mild cases can race and breed satisfactorily and the deformity becomes less obvious with growth. Torticollis is a rare cause of dystocia, which unless the foal is small, usually results in delivering it dead, either by embryotomy or caesarian section. Lordosis is rarely seen in newborn foals. It usually becomes apparent at a few months of age. Likewise Wobbler Disease does not present clinical signs until foals are a few months old even though lesions have been recognised in foeti. Typically the foal shows hind limb ataxia with loss of proprioception and possibly some front limb involvement. The compression of the spinal cord more commonly occurs at third and fourth cervical level in foals and lower in the neck in yearlings. The sudden severe case must be differentiated from spinal fracture. If confirmed such a case warrants destruction on humanitarian grounds.

Lateral curvature of the skull, from cranium to incisor teeth is uncommon, varies in severity and may be accompanied by other deformities. Slight curvatures of the nasal part of the skull have no ill-effects, providing nasal passages are not occluded or there is no dental malocclusion, and they become less obvious with growth. Severe lateral curvature of the anterior maxilla and premaxilla is a serious deformity which warrants humane destruction. The incisor arcades do not meet and one nasal passage is occluded. In spite of this deformity, afflicted foals can suck but grazing is difficult. Complicated surgery is not recommended.

The commonest congenital deformities involve the distal limbs which exhibit abnormal flexion, extension or angulation. Initially, almost all the cases have an imbalance of tendon or ligament tension around joints with no bony deformity. However, if the limb deformity is not corrected during the first month or two of life the abnormal loading of the body on the deformed limb may result in bone, joint or growth plate damage.

The aetiology of congenital limb deformities is uncertain but it is likely that malpositioning, the size of the mare in relation to the foal, lack of free foetal movement in utero and pressure moulding (corresponding pressure lesions can be seen on the allantois from some cases) are factors. There is no firm evidence that heredity predisposes.

All cases should be assessed generally and carefully, with the foal standing, if possible, on a firm level surface. It must be remembered that most cases will correct spontaneously or with careful conservative management.

Flexural deformities (hyperflexion or contractures) of the newborn can involve interphalangeal, fetlock or carpal joints, singly or in combination. The hock is rarely affected. Mild cases, in which the foal can stand or walk without persistently knuckling over, usually correct with exercise. Flexural deformity of the fetlock (Fig. 2.2) or phalanges may require the careful

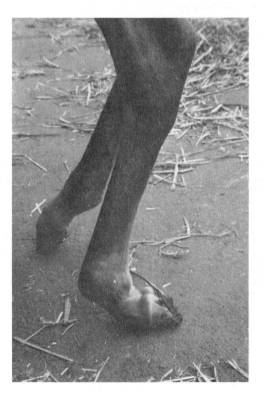

Figure 2.2. Congenital flexural deformity of the left hind fetlock which will respond to splinting.

application of shaped fibreglass splints onto the front of or onto the front and back of a well bandaged lower leg for two days at a time until normal posture can be maintained. Front or hind fetlocks may be uni- or bilaterally involved but the carpus is usually affected bilaterally. Splinting the knee is unsatisfactory and if support is necessary, as it can be following rupture of the common digital extensor tendon on the front of the knee, the application of a fibreglass cast from foot to elbow for 2–3 weeks is successful. If correct supportive measures are employed, surgery is hardly ever indicated for congenital flexural deformities. Severe flexural deformity of the carpus with a limit of extension of approximately 90° does not straighten even after section of tendons above and below the carpus. Such cases may cause dystocia, be accompanied by other deformities, such as scoliosis or eventration and require humane destruction.

Flexural deformities of fetlock joints, especially in hind legs, may also have some angulation of the joint which may require corrective surgery later.

Extensor deformities (hyperextension) can involve phalangeal, fetlock, carpal or tarsal joints, are invariably bilateral and sometimes quadrilateral. Due to tendon, muscle and ligament weakness almost all correct with exercise within the first 2–3 weeks of life (Figs 2.3(a) & (b)). An exception is hyperextension of the distal interphalangeal joints in hind legs which may take some months to correct. The bulbs of the heels may become abraded and benefit from bandage protection. Generally, bandage support of extensor deformities is unnecessary. Some cases may have accompanying angular deformities which mostly correct simultaneously.

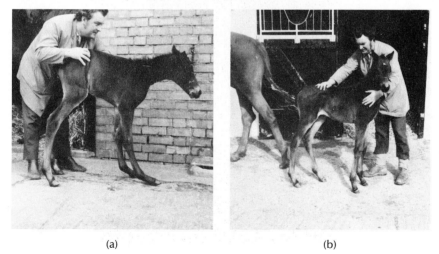

(a)   (b)

Figure 2.3(a). Day old foal with hyperextended front legs. (b). Same foal as in (a) at one week old having been allowed exercise, but given no treatment.

Angular deformities occur in fetlocks, knees or hocks, singly or in combination. Commonest is carpal valgus which is often bilateral and accompanied by outward rotation or angulation at the fetlock. Cases of congenital carpal varus are rare and invariably have a valgal deformity of the contralateral knee. Even quite marked carpal angulations can correct spontaneously or with restriction of exercise and corrective hoof trimming (rasping down the unworn—lateral in valgus—side of the hoof) during the first two months of life.

Angulations of the fetlock may be helped by the application of a splint during the first few days of life if there is accompanying flexural deformity. Most valgal angulations will improve with corrective hoof trimming and age. The most serious fetlock angulation is varal in a hind leg which is usually unilateral and can be complicated by angulation of the cannon bone.

Tarsal angulations are usually bilateral and can be valgal or valgal and varal. The latter case may look severe in the newborn foal which has a tilted pelvis and walks with its limbs rotating uncontrollably (Fig. 2.4). However, provided there are no malformations such as a dislocated tendon or curved cannon bone, these cases improve rapidly with exercise and can be normal

Figure 2.4. A one-week-old foal with valgus and varus deformity of its hind legs. With exercise it became normal by one month of age.

within a month. The severely varal hock is angled at tarso-metatarsal level and may have a curby or sickle-shaped conformation which persists into adulthood but does not necessarily prevent them racing. The majority of valgal angulations of the hock correct spontaneously.

Surgical treatment of angular limb deformities must be undertaken at the correct age as it depends on residual growth at the growth plate correcting the deformity. Growth ceases early at the distal cannon (by 6–8 weeks of age) and the metacarpal and metatarsal growth plates close at 5–6 months old. Therefore, surgery for correction of fetlock angulations must be performed by about one month old. The ideal time for surgery of the carpus or tarsus is by three months old to ensure rapid correction and minimal blemish.

There are two principal surgical techniques. The traditional method involves retardation of growth on one side of the epiphysis by temporary transepiphyseal bridging using staples or screws and wire across the growth plate. Thus, the side of the growth plate compressed by the angular deformity (the lateral side in carpal valgus) grows and straightens the leg while the implants temporarily prevent growth on the opposite side. This process usually takes 6–8 weeks and the implants must be removed when the leg is straight.

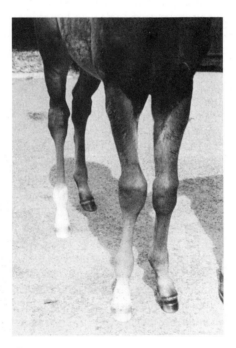

Figure 2.5. Ten-week-old foal with carpal valgus of the right foreleg which corrected following periosteal transection and stripping.

The second and more modern technique relies on hemicircumferential transection and stripping of the periosteum on the metaphysis close to the growth plate stimulating extra growth on that side of the leg (Fig. 2.5).

Persistent congenital deformities of limbs may result in bone pathology or remodelling. Extensor and flexural deformities rarely cause such changes but they can be seen in chronic carpal flexure (over at the knee) in which the carpal bones may become wedge shaped and develop secondary osteoarthritis. Chronic flexure of the distal interphalangeal joint is more commonly acquired as a growth disorder in foals approximately two months old which can result in a narrow hoof with a dropped sole and rotated pedal bone. Hocks with a weak, curby conformation may acquire crush fractures of the tarsal bones leading to osteoarthritis, especially in premature foals.

The commonest pathology resulting from chronic angular limb deformities is compression or a Type 5 injury to the growth plate. The abnormal loading on the angulated leg compresses the epiphysis into a wedge shape and as a result growth ceases on the narrowed side of the growth plate which closes prematurely. These changes are usually initiated in the fetlock in the first month of life and in the knee or hock between one and two months of age when the angulation may suddenly worsen. Rest and surgery are urgently indicated. Premature or dysmature foals with hypoplastic bone and deficient ossification usually suffer this and more severe pathology such as crush fractures of carpal or tarsal bones as early as one or two weeks of age; if such changes occur the prognosis is poor.

Adaptive and behavioural disorders

Dysmature and premature foals   These are foals which are born undersized and underweight (less than 46 kg or 110 lbs for the thoroughbred which on average is 50 kg or 100 lbs approximately). Premature foals are born before 320 days of gestation but only if they are no more than four weeks early (i.e. no less than 300 days) can they be expected to survive. The commonest cause of premature birth is placental insufficiency which can be infectious in origin (bacterial or fungal) or non-infectious (twin pregnancy). It is often characterised by premature separation of the placenta in late pregnancy which causes the mare to run milk for days or weeks before birth. Placental insufficiency can also result in the dysmature foal which is also smaller and lighter than normal, but born at or up to two months after term.

Premature and dysmature foals can show other clinical features in addition to undersize. Their skin may be short haired and silky, the forehead domed and prominent and the distal limbs slack and hyperextended, especially at the fetlock. The oral mucous membrane can have a cherry red colour and the foal may chew persistently. The case which makes encouraging progress during the first 24 hours, possibly standing and sucking may subsequently deteriorate, collapse and die. The small foal which survives will be prone to infectious disease, especially if its dam ran milk prior to foaling and colostrum was not supplemented. Also, in relation to its small size, it will be overfed by the mare, become very overweight and may acquire severe angular limb deformities during the first month. These deformities will be due to the hypoplastic and incompletely ossified bones, particularly in the carpus or tarsus, collapsing under the body weight and stress of exercise.

Whether to persist with a premature or dysmature foal can be a difficult veterinary and management problem. In general, the very small sick foal should be destroyed but, in borderline cases, it is best to watch progress before deciding, remembering that the small premature or dysmature foal will not grow to full size as an adult.

Nervous or convulsive disorders These are not common but have been seen in most breeds of foal. They used to be described by their general appearance as dummies, wanderers and barkers, in ascending order of severity. More recently, they have been titled the Neonatal Maladjustment Syndrome. In the majority of cases the nervous disorder occurs soon after a normal or easy birth. The meconium is expelled normally and some foals may suck and behave normally for a few hours before going off the suck and convulsing.

The common feature and mildest sign is an absence of suck reflex (dummy foal). The foal ignores or bites a finger or artificial teat inserted into its mouth, and will appear blind and not try to find the mare's udder. Following feeding and nursing, maybe for a few days, such cases develop a suck reflex and behave normally. The more serious case has mild clonic convulsions, standing or recumbent, is blind and will continually walk round the stable, ignoring the mare and walking blindly into her or the walls (wanderers). These foals will sleep if laid down carefully and may require some days of feeding and nursing before returning to normal. The most severe convulsive disorder is characterised by inability to stand or right itself onto its brisket, hyperpnoea, clonic convulsions into extensor rigidity and a few foals make loud expiratory noises like a yapping dog (barkers). Such cases may die during a convulsion and require urgent, intensive care if they are to recover.

Treatment of these foals is directed at keeping them adequately fed, warm and free from convulsions until they can feed themselves. It is very rare for a case never to develop a suck reflex. Nursing is crucial and for the recumbent foal includes maintenance of a high ambient temperature in the stable (75–80°F) and a normal body temperature. The limbs should be bandaged from foot to hock or knee, the body covered by a jersey or blanket and the foal should lie on a deep soft bed covered with a blanket. If in lateral recumbency it should be turned every ½–1 hour. Entropion may occur and should be treated by inserting two or three vertical mattress nylon sutures to evert the lower eyelid. The mare must be milked out fully and regularly (every 1½–2 hours) to maintain her milk supply and provide for the foal. The warm milk is fed gradually by stomach tube every 2–4 hours with the foal held in sternal and not lateral recumbency if it is unable to stand; 400 ml per feed is adequate for a thoroughbred foal and no more than 0.5 litre should be given at a time. The mildly convulsing foal may improve after feeding but great care should be exercised with the collapsed or comatose foal which may die during feeding. The stomach tube can be inserted for each feed or a narrow bore tube (7–10 mm in diameter) can be left *in situ* and stitched to the lateral canthus of the nostril.

A permanent tube is best inserted as far as the thoracic oesophagus and should be stopped or corked to prevent ingestion of air and subsequent colic. Constipation is a common side effect of prolonged artificial feeding and may require treatment with liquid paraffin in the feed.

Vascular catheters may be introduced to ease sampling and administration of fluids and drugs. Antibiotic cover should be maintained and anticonvulsant treatment is a crucial aspect of the severe case. The most reliable drug is primidone which is given as a suspension with milk by stomach tube. However, it may take up to 20 minutes to have its effect which is a long time for a convulsing foal and its attendants. Intravenous treatments which have quietened convulsions within a few minutes include detomidine or acetyl promazine, both of which lower the blood pressure markedly, diazepam or pentobarbitone. Following intravenous sedation, maintenance doses of primidone can be given as necessary.

While most convulsive foals will recover, to lead a normal life following intensive empirical treatment, nursing is crucial and secondary problems such as infection can be a serious complication.

Meconium retention   This is the commonest cause of colic in the neonatal foal. It invariably responds to prompt and careful conservative treatment. There is impaction of firm, dark brown meconium in the rectum, mostly palpable, just anterior to the pelvic canal but sometimes higher up in the small colon. With ingestion of milk and failure to pass the meconium, gas accumulates in the large bowel causing a tympanitic colic with persistent straining. Signs normally occur 20–24 hours after birth but, in the early stages, foals continue to suck in between bouts of colic in which they roll and "scratch their ears" with their hind feet. Temperature and mucous membrane colour remain normal but pulse and respiration may be elevated during pain or if there is tympany.

Treatment is directed at relieving pain and softening the meconium and encouraging its passage. Sedation such as acetyl promazine or analgesia such as flunixin meglumine can be given. Liquid paraffin is given by stomach tube (50–90 ml) or as an enema (or warm soapy water) via a soft round-ended rubber tube. Some clinicians remove meconium with a specially made meconium spoon but great care and expertise is necessary. This procedure and enemas must be administered with the foal standing. Passage of gas, meconium and soft yellow dung are signs that recovery is imminent but intermittent colic or straining may persist for a few more hours. The most protracted meconium cases usually clear up with conservative treatment in 24–36 hours.

Meconium retention must be differentiated from the rare malformation of the lower bowel, agenesis coli characterised by a blind colon or rectum. More serious causes of colic, such as volvulus or intussusception are rare in the neonatal period and exhibit total anorexia, unremitting pain and deteriorating vital signs. Patent or ruptured bladder does not give rise to colic and is rarely evident as early (40–48 hours old) as meconium retention (20–24 hours old).

Immunological disorders

Haemolytic disease   Haemolytic disease or isoimmune erythrolysis arises from the foal ingesting colostrum containing antibodies to its own erythrocytes. The mare is sensitised to certain antigens in her foal's blood so she should have had a previous pregnancy for the disease to occur. The foal would inherit these antigens from the stallion, which would cross the placental barrier during pregnancy to cause the isoimmunisation, producing haemagglutins and haemolysins in the colostrum.

The affected foal will be normal for a period after birth and depending on the severity of haemolysis show clinical signs from 12 hours to 5 days old. Very mild cases may show no obvious signs and will only be detected during a routine examination, such as for insurance purposes. Generally, the earlier the signs appear the more severe the anaemia and the more guarded the prognosis. Lethargy, yawning, hyperpnoea with exercise and less vigorous and less frequent sucking are early signs. The mucous membranes are pale and jaundiced and the pulse rapid and thready. Haemoglobinuria gives the urine a characteristic port wine colour. The severely anaemic case is hyperpnoeic at rest and may die of exhaustion trying to follow its dam at pasture. The first case a mare has may be milder than subsequent ones. However, further haemolytic foals are likely even if she goes to a different stallion.

Diagnosis is confirmed by the Direct Sensitisation Test and characteristic haematological changes. Total red cell count, haemoglobin and packed cell volume are all reduced markedly and if their respective values fall below $2 \times 10^6/mm^3$, 4 g% and 10% the prognosis is poor and immediate treatment essential. Less severe cases can recover without treatment even if their haematology falls to $3 \times 10^6/\text{mm}^3$, $5\,\text{g}\%$ and 14% respectively. Such cases would show clinical signs at a later stage, possibly 4 or 5 days old and require regular monitoring and stable rest. However, the clinical condition of the foal does not always correspond with the haematological parameters. A marked elevation of total white cell count (over $15,000/\text{mm}^3$) or plasma bilirubin over $20\,\text{mg}\%$ usually indicate a poor prognosis. Differential diagnosis of haemolytic disease is not difficult as other causes of jaundice in the neonate, such as Actinobacillus equuli septicaemia and Equine Herpes Virus Type 1 abortion strain do not cause anaemia.

The haemolytic foal can be treated by exchange transfusion of whole matched blood or infusion of washed red blood cells from its dam. The former is more difficult as a suitable donor has to be found and exchange transfusion is more complicated, requiring simultaneous withdrawal and infusion of blood. The simpler technique involves collection of 3–4 litres of the mare's blood into acid citrate dextrose which is submitted to a laboratory for centrifugation and washing. This yields 1–2 litres of washed erythrocytes which is then given intravenously to the foal. Antibiotic cover is customary and the foal should be kept stabled until it is three weeks old to prevent exhaustion. There is no point in withholding the mare's milk from a clinical case as the colostral antibodies which have done the damage will be absent from the milk 24 hours after birth.

Prevention depends on withholding the isoimmunised mare's colostrum from the susceptible foal. Rising antibody titres to the certain antigens known to be involved can be detected from serum samples taken from the mare approximately four weeks and one week before she is due to foal. If this rise occurs, the foal should be muzzled after birth, fed donor colostrum and artificial milk, usually by bottle, for the first 24 hours during which its dam must be regularly and fully milked out. A mare which has produced a haemolytic foal should always undergo pre-foaling antibody tests. Sometimes these tests can produce an equivocal result so a direct Coombs test should be performed between the foal's blood (cord blood will suffice) and the colostrum before the foal has sucked. Rarely, this test is negative when the pre-foaling tests on the mare are positive.

Combined immunodeficiency syndrome   Occurs in Arab foals and is rare. It is characterised by a profound lymphopaenia and complete and fatal susceptibility to infection.

Failure to acquire passive immunity   This is the commonest factor predisposing foals to neonatal infections. The causes and remedies have been discussed in foal management.

Infectious diseases

There are a small number of infections which can be acquired in utero and result in the foal being born ill or in a collapsed state. The most serious of these is virus abortion or Equine Herpes Virus Type 1. This virus mostly causes disease of the upper respiratory tract in young stock but occasionally the abortion strain of Type 1 causes abortion in mares, mostly between 7 months and term, although rare cases have been diagnosed as early as four months of gestation. An infected foal can be born at term but in a collapsed or comatose state. Such foals are jaundiced and usually die quickly, within 24 hours, in spite of treatment. Early diagnosis is crucial to preventing spread of the virus. The clinical signs described are strongly indicative and supported by a total white blood cell count of less than $1000/mm^3$ in the foal. Virological techniques may take a few days to confirm but virus can be cultured from naso-pharyngeal swabs or the buffy coat of lithium heparinised blood of foals.

Diagnosis is confirmed at post mortem. The gross features include pleural and peritoneal effusion, minute foci of liver necrosis, oedema of the lungs and necrosis of the thymus gland. Histologically the necrotic foci in the liver exhibit eosinophilic inclusion bodies in the nuclei of surrounding parenchymal cells. A fluorescent antibody technique and culture of the virus from body tissue, usually thymus, liver, lung and adrenal, are conclusive.

Immediately a mare has lost her foal due to EHV1 she should be isolated for at least one month well away from other horses, particularly pregnant mares, and confined to the stable. Her previous stable should have the bedding burnt and be thoroughly cleaned and disinfected. Any mares which have possibly or definitely had contact with the infected case should also be isolated individually or as a group away from other pregnant mares. Those mares which are exposed to infection from a primary case will almost certainly produce a live, healthy foal if it is born to term within 10 days. However, of those exposed to EHV1 which have more than 10 days to wait before foaling, 60% or more may lose their foal. Live foals born within an outbreak of virus abortion or its ataxic form may be asymptomatic carriers of the virus within their upper respiratory tract for up to a month. Sick foals may develop pneumonia. In one outbreak of the ataxic form of rhinopneumonitis, signs in foals 2–19 weeks old included fever, mucopurulent nasal discharge, ataxia of hindlimbs, atony of the bladder and penile prolapse, pneumonia, diarrhoea and hypopyon with iritis. No foals died.

Strict hygiene and isolation procedures are crucial in the control or prevention of an outbreak of virus abortion, which, it must be remembered, can occur in mares vaccinated against EHV1.

Foals born of mares with bacterial placentitis may suffer a perinatal septicaemia which is clinically similar to rhinopneumonitis but from which they can recover with intensive treatment. The venereal pathogen *Klebsiella aerogenes* Capsule Type 1 has been known to cause such disease, unlike *Haemophilus equigenitalis* (CEM). Foals born prematurely due to fungal placentitis often suffer secondary bacterial infections when 24–48 hours old.

Infections acquired after birth show signs from 24 hours of age at the earliest. If the onset is gradual the foal may be febrile but the peracute and terminal case is usually hypothermic. In contrast to the nervous or convulsive disorders, infectious diseases are characterised by deterioration of the foal's condition. The foal goes off suck, becomes dull and weak. There may be lameness with single or multiple joint infection, enteritis leading to dehydration and terminally, convulsions. Diagnosis is confirmed by haematological changes, principally a neutrophilia, but peracute cases may have a leucopaenia as opposed to leucocytosis in the less acute illness. Plasma fibrinogen is elevated and the immunoglobulin level may be lower than normal, reflecting the susceptibility to infections. A blood culture may yield the causative organism and those which commonly arise in neonatal infections include *Actinobacillus equuli*, which may cause jaundice, *Escherichia coli*, staphylococci or streptococci or *Salmonella typhimurium*.

Intense treatment has to be initiated early in the disease process to be successful. Naturally it has to be empirical in most cases and is based on high doses of broad spectrum bactericidal antibiotics given two to four times daily. Nursing, rest, fluid therapy and plasma administration are also important.

The principal localised infection is joint-ill or septic polyarthritis. Any young foal which goes lame without an obvious cause should be suspected of having and empirically treated for septic arthritis. In most cases the joint infection will soon be obvious but the lameness may vary. Increased lameness after getting up or forced flexion of the affected joint are characteristic. There may be joint effusion or painful swelling of the epiphysis close to the joint. The foal is dull and febrile and haematology and raised plasma fibrinogen confirm infection. Blood culture may yield the causative organism in severe, febrile cases but culture of synovia from infected joints is usually unproductive. It is more informative to centrifuge the synovia, gram stain a smear of the sediment and look for organisms microscopically.

Treatment should be urgent and vigorous. High levels of systemic antibiotics must be maintained for one or more weeks. If practicable, drainage and flushing of infected joints is very worthwhile, particularly if the infection is localised in the synovial membrane.

The prognosis is poor if subchondral bone or the growth plate are infected or osteomyelitis develops. It is also poor if many joints rapidly become infected or if *Salmonella* or *Corynebacterium equi* are implicated. Experience has shown that the prognosis is worse for femoro-tibial joint infection than femoro-patellar, tibiotarsal or carpal joints.

Failure to acquire passive immunity is a vital factor in the aetiology of joint-ill. Also, any infective nucleus such as omphalitis or sequestrum formation following a fracture can be a source of secondary infection of joints in the foal. Omphalitis is not always obvious externally and even if present at post mortem of a septicaemic or polyarthritic case it may not be marked or extensive. Navel abscesses can develop externally and are best fomented and allowed to drain naturally. As an abscess has formed it rarely gives rise to secondary joint infection but broad spectrum antibiotics should be given if the foal becomes febrile.

Pneumonia is uncommon as a single entity in the first two weeks of life but can occur in outbreaks of EHV1 or influenza. Bacterial pneumonia may occur in conjunction with septicaemia or polyarthritis. Pneumonia and pleurisy secondary to fractured ribs holds a poor prognosis especially if accompanied by pericarditis.

Enteritis is common in young foals and usually characterised by diarrhoea. Its causes are not always easy to ascertain but include physiological factors as in foal heat diarrhoea, management factors such as the consumption of too much salt from the dam's manger and, most frequently, infections. The diarrhoea can be preceded or accompanied by mild colic. Consequent dehydration and toxaemia vary in degree but may be sudden in onset and severe in the young foal (up to a few days old) with watery diarrhoea. Sunken eyes, tight, unpliable skin, a coated tongue and dullness indicate dehydration. Not all cases go off the suck but if they do so it is more important to give oral or preferably intravenous fluid therapy than to feed milk. Rapid onset of dehydration, weakness and a fast pulse (140 per min) indicate a poor prognosis and the need for intensive nursing and treatment.

The cause of the enteritis is not always easy to determine. Although *E. coli*, *Salmonella* spp, *Campylobacter* or Rotavirus may be incriminated, it has to be remembered that these organisms can also be cultured from the faeces of clinically normal foals. Purity of culture growth and isolation of the same organism within an outbreak will confirm the aetiology. However, especially in the case of *Salmonella* and *Campylobacter*, repeated screening cultures within a stud farm may reveal surprising numbers of healthy carriers.

In addition to electrolytes chlorodyne, kaolin, charcoal or mineral oil can be given singly or in combination once or twice daily by stomach tube. Antibiotics should be included if a specific bacterium is isolated and given systemically if the foal is febrile or showing other systemic signs. Treatment must include rest, maintenance of body temperature and restoration of fluid balance. It is not important, and is sometimes unhelpful, to feed milk to the diarrhoeic foal which is off suck. Fluid therapy is much more important. Meanwhile, the mare should be milked out to alleviate painful distension of the udder and maintain her milk supply for the foal when it resumed feeding.

SOME DISEASES OF OLDER FOALS AND YEARLINGS

Parasitism

Modern anthelminthics and good paddock management can make disease due to strongyles and ascarids uncommon. Parascaris equorum may cause unthriftiness, reduced growth rate with an enlarged abdomen and rough coat. Colic, diarrhoea, intussusception or rupture of the bowel, may result in severe cases. Foals and weanlings are principally affected and should be routinely treated from six weeks of age. Strongyloides has been seen to cause diarrhoea in young (2 weeks to 4 months old) foals. The strongyles are the most significant parasites of young horses. They are normally present as a mixed infection but the large *Strongylus vulgaris* has a longer life cycle and causes more serious loss than the others. Heavy infections are likely to be apparent as anaemia and unthriftiness at between one and three years of age but will only be seen if there is overstocking on infested paddocks without routine treatment. The thrombus formation in the cranial mesenteric artery resulting from migrating *S. vulgaris* larvae may cause intermittent or fatal embolic colic. The diagnosis of strongyle infestation may be suggested by a high faecal worm egg count but is confirmed by raised alpha 2 and beta globulin levels in the serum protein electrophoresis. All animals on a stud farm should be treated every 4-6 weeks with a broad spectrum anthelminthic which is changed twice yearly to discourage resistance. Any stock introduced to the stud should be treated and isolated. Foals and yearlings should always be grazed on paddocks which have been rested from horses for at least six months or as long as possible. The larvicidal drug ivermectin has improved strongyle control enormously and is very effective in the severe clinical case.

Bots and tapeworm are invariably present in young horses but rarely cause clinical disease. Ivermectin and organophosporus compounds given in the autumn remove bot larvae from the stomach. High levels of pyrantel embonate are effective against tapeworms.

Louse infestation is common in foals causing persistent pruritus and bald or raw areas of skin on the head, neck and chest. Dusting the coat with gamma BHC or organophosphorus insecticide at two weekly intervals is effective.

Enteric conditions

Diarrhoea is not uncommon in older foals and can be due to parasitism, or more often, an imbalance of gut microflora when the bulk of the diet is changing from milk to grass at one to three months of age. Occasionally a foal may suffer a persistent diarrhoea for some months but continue to feed and thrive. *Campylobacter jejuni* has been isolated from a few of these cases which normally show only a temporary response to treatment but eventually return to normal.

Gastric or duodenal ulcers are uncommon but have been reported as a serious problem among foals approximately one month old. Clinical signs may include diarrhoea, chronic dull abdominal pain, salivation, grinding of the teeth, intermittent anorexia and unthriftiness. Palpation of the foal's anterior abdomen may cause pain. Gastric reflux can appear at the nose or mouth or be obtained by stomach tube and contain occult blood. Human ulcer treatments, such as Histamine Type 2 antagonists (cimetidine or ranitidine) and antacids have been employed.

Colic is always a potentially serious sign in the foal. Tympanitic colic is not uncommon in the two- to four-month-old foal which is grazing considerable amounts of young grass. It usually responds to antispasmodic or analgesic treatment. However, foals with colic should be treated carefully, as flunixine meglumine in particular can mask the pain so effectively that a necessary laparotomy may be delayed in the case with volvulus or intussusception. The latter conditions usually involve the terminal ileum close to the ileocaecal valve. Unremitting colic, absence of peristalsis and deteriorating vital signs indicate such a crisis for which surgery is urgent. Resection of the small intestine may be necessary if the mesenteric circulation is impaired and end to end or ileum into caecum anastamosis can be performed. Incarceration or strangulation of umbilical hernia is rare in the foal. The former may be reduced manually under general anaesthesia prior to conventional hernia repair but strangulation would require radical surgery.

Impactive colic is rare in the older foal or yearling and if it occurs at pasture is strongly suggestive of early Grass Sickness. This disease is usually fatal and has been seen in yearlings and foals as young as five months.

Corynebacterium equi

This organism causes serious disease in foals between the ages of one and four months which may lead to their death or destruction. The typical manifestation is a suppurative broncho-pneumonia which can be enzootic in some forms. The foal is febrile (often unremittingly), lethargic, has a deep painful cough, is dyspnoeic and exhibits loud moist rales (hence the name rattlers) on auscultation. Due to infection of the abdominal lymph nodes diarrhoea may occur later in the disease. The earlier the diagnosis is made and the more intense the treatment the better the prognosis. The organism is rarely cultured from nasal or rectal swabs but can be found on blood culture, tracheal wash or simply from a plug of bronchial mucus collected in a sterilised stomach tube. Antibiotics should be chosen from *in vitro* sensitivity tests and administered in high doses for at least two to three weeks.

Usually erythromycin (25 mg/kg body weight i.v. twice daily or per os four times daily) is the drug of choice and recent work has shown that its effect is heightened by the concurrent administration of rifampicin (5 mg/kg twice daily). Rest, in a warm well ventilated and dust free stable, is also important.

Corynebacterium can also cause osteomyelitis or septic epiphysitis which are very difficult to cure.

Other respiratory diseases

Rhinopneumonitis or nasal catarrh is ubiquitous in foals and yearlings during the later summer and autumn months. It is usually initiated by Equine Herpes Virus Type 1 or 2 and secondary bacterial infection results in a persistent coryza with enlarged sub-mandibular lymph nodes. Coughing and bronchitis may also occur but affected animals rarely become ill or debilitated. The signs can be suppressed and occasionally eliminated by oral sulphonamide treatment.

Strangles can cause serious illness in foals which may develop abscesses in numerous lymph nodes such as the sub-mandibular, retropharyngeal, parotid or perineal. It is best to withhold antibiotics and allow the abscesses to fulminate naturally but if the foal becomes pneumonic or seriously ill, treatment is necessary. A temporary tracheotomy tube may have to be inserted if throat abscesses cause severe dyspnoea.

Diagnosis is confirmed by culture of *Streptococcus equi* from abscess pus. Failure to culture it from the naso-pharynx does not necessarily indicate that an animal is free from infection.

Preventing the organism spreading to other animals on the farm is important. Clinical cases should be isolated as soon as possible and any contacts which may be incubating the disease should also be isolated for at least one month. Infected cases can be released from isolation one month after the last abscess has healed.

All possibly contaminated areas such as the stable and particularly feed and water mangers must be thoroughly cleaned and disinfected after the infected case has been isolated.

Any developing abscess at any site in a young foal must be treated with caution and the causative organism identified. However, sub-mandibular gland abscessation, not due to *S.equi*, but from which other streptococci or staphylococci are isolated, occurs in foals and yearlings. The inflammation is less intense and the animal remains healthy. Isolation is unnecessary.

Tetanus

Tetanus in the two- to four-week-old foal is rare now that toxoid vaccination of mares or antiserum administration to foals are routine. Susceptible animals should receive a prophylactic dose (1500–3000 i.u.) of tetanus antitoxin if a wound or infective focus (particularly in the foot) arises. Vaccination can be initiated at three months of age. The youngest age at which a foal has developed signs of tetanus is approximately 16 days therefore prophylactic antiserum is best administered at five days old. The fever, stiffness, tetanic convulsions and, in particular, the prolapse of the nictating membrane are characteristic. Young foals can continue to suck in spite of the disease and with intensive nursing and treatment can recover.

ORTHOPAEDIC CONDITIONS

Fractures

Jaw fractures mostly result from kicks but some can occur if the foal grips a railing, gate or door and jerks its head away suddenly. Anterior mandible or premaxillary fractures involving or just caudal to the incisor teeth heal well in the young foal but the older case may require wiring support to maintain fracture alignment. As these fractures are usually compound, secondary infection may be a complication.

Spinal fractures are caused by collisions or falls. Some fractures of the cervical vertebrae are not necessarily serious as the spinal canal is spacious, particularly at the atlas and axis vertebral level and displacement of the floor of the atlas or the dens of the axis can occur without compressing the spinal cord. Fracture of the body of thoracic or lumbar vertebrae usually causes compression of the spinal cord and is either fatal or warrants destruction. Dorsal or transverse process fractures often result in sequestrum and sinus formation.

Pelvic fractures hold a good prognosis provided the acetabulum is not involved, in which case osteoarthritis develops and also provided sequestrum formation does not complicate recovery.

Long bone fracture is a serious injury in horses of any age. If simple, closed and minimally displaced, fracture of the humerus or femur can heal well following prolonged rest. Fractures of the radius or tibia are more likely to be compound and usually require internal fixation. Good surgical technique, prevention of infection and a calm and sensible patient are necessary, but even if these prerequisites are met, results can be disappointing. Foal's bone is very prone to infection and particularly to osteoporosis around implants leading to instability and non-union. Simple uncompounded fractures of cannon bones heal well with external fixation using modern fibreglass casting materials. Fractures of the proximal splint bones result from kicks and destabilise the carpo-metacarpal joint. Following external fixation, healing can allow an athletic career but carpal flexion will remain limited.

Fractures within joints are common. Those of carpal bones often result in avulsion of the fractured fragment and consequent angular deformity. Casting the limb in fibreglass usually allows satisfactory healing. Fractures of the malleoli of the distal tibia or off a trochlear ridge of the tibial tarsal bone may not cause severe lameness and even if removal of a fracture fragment is necessary the prognosis is good. Fracture of the proximal sesamoid bone (Fig. 2.6) is the commonest fracture in young thoroughbred foals up to one month of age. Basal or apical and almost always in forelegs, they are often multiple. If the fractures are not distracted they heal well with rest and bandage support and the prognosis is good. The single distracted fracture responds well to surgical removal of the fragment but if they are multiple the prognosis is poor. Sesamoid fractures are caused by the foal galloping to exhaustion after its dam, especially if they are released in a large paddock following a period of confinement. Running mares with young foals in small nursery paddocks can prevent this injury.

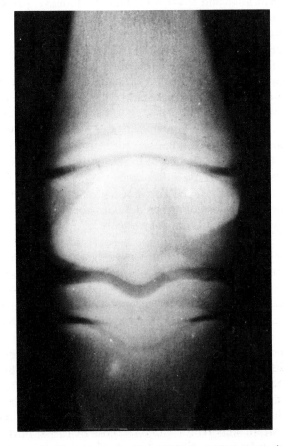

Figure 2.6. Radiograph of a two-week-old foal's fetlock showing a distracted basal fracture of one sesamoid bone.

Fracture of the third phalanx holds a good prognosis in young foals if it is recognised early and the foal is rested in the stable for two months.

Growth plate injuries

Of the six types of growth plate injury classified by Salter and Harris (Fig. 2.7) three are commonly seen in foals. Type 1 or complete separation of growth plates in the leg is seen at the distal cannon or fetlock. They can be relocated easily and heal well within a cast. Separation of the epiphysis at the head of the femur results in chronic osteoarthritis and thus holds a poor prognosis unless fixed internally.

Type 5 or compression injuries are the result of chronic angular limb deformities. The consequent trauma and premature closure of the plate on the compressed side makes angulation more severe or permanent unless timely surgery is undertaken. This pathology most commonly occurs at one to two months of age or earlier, two to three weeks old, in premature or dysmature foals. Similar but less acute injury is seen in older foals and yearlings with epiphysitis.

Type 6 injury or trauma to one side of the growth plate, usually the result of a kick, can cause severe pain and lameness. Although local inflammation develops rapidly, radiographic changes typical of the injury take two to three weeks to appear. There is rarefaction each side of the growth plate and periosteal reaction ultimately forms a bony bridge across the plate. Thus, premature closure of the injured side of the plate can cause an angular deformity.

Due to stronger epiphyseal attachment, Type 2, 3 and 4 growth plate injuries, in which a fracture of epiphyseal or metaphyseal bone crosses the plate, are more common in yearlings than foals.

Growth disorders

Certain orthopaedic disorders occur during the growth of foals and yearlings which may result in a change of limb conformation. The two commonest diseases are acquired flexural deformities and epiphysitis. Their aetiology and pathogenesis are incompletely understood but they affect the lighter boned, faster growing breeds such as thoroughbreds, Arabs, ponies and their crosses.

Acquired flexural deformities or contracted tendons   mainly affect two joints, the distal interphalangeal in front legs of foals and the fetlock joints of older foals and yearlings.

Flexural deformity of the distal interphalangeal joint occurs in well nourished, fast growing foals between the ages of one and six months. It can affect one or both front (very rarely a hind foot) feet and may be accompanied by lameness. Early signs are of an upright anterior hoof wall, concave and with a prominent coronary band. The foal stands and walks

1.   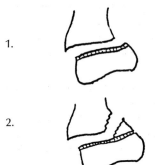   Separation without fracture.

2.   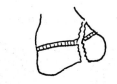   Separation with fracture of a triangular piece of metaphysis.

3.   Transepiphyseal fracture separating at the growth plate.

4.   Transepiphyseal fracture involving triangular piece of metaphysis.

5.   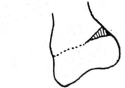   Crush injury to one area of the growth plate resulting in premature fusion of one side.

6.      Periosteal bridge formation as a result of a blow to the periosteum and peri-chondral ring also causing premature fusion of one side of the growth plate.

Figure 2.7. Types of epiphyseal injury (after Salter and Harris, 1963).

on its toe with the heels raised off the ground (Fig. 2.8(a)). The pastern may become upright and most cases acquire a back at the knee conformation. Bilateral cases walk like an animal with laminitis. If the deformity persists, excessive wear at the toe leads to secondary pododermatitis or infections tracking up the white zone. In chronic cases the anterior tip of the pedal bone rotates ventrally and becomes eroded.

The aetiology of this deformity is uncertain. Certain factors such as pain or lameness from another cause and heredity are known to be significant. Overfeeding of concentrates is also thought to contribute. It has been postulated that the deformity results from inability of the inferior check ligament to extend while the cannon bone lengthens but the large proportion of unilateral cases argue against this theory.

Initial treatment should be conservative and comprise stable rest, corrective hoof trimming to lower the heels every two weeks and reduction of feed intake. If a severely affected foal is over three months old it should be weaned and fed 0.5 kg per 50 kg body weight lucerne, clover or good ryegrass hay daily and a handful of concentrate feed containing a calcium supplement. Analgesic or non-steroidal anti-inflammatory treatment may be helpful. If the hoof wall is worn and broken at the toe then a grass tip or shoe can be applied following corrective trimming. The prognosis for restoration of normal conformation is good if the deformity is detected early and treatment

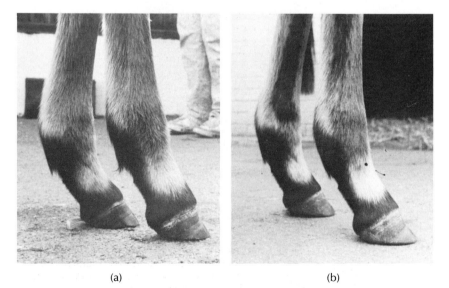

(a)   (b)

Figure 2.8(a) Eight-week-old foal with early flexural deformity of the distal interphalangeal joint of the right foreleg only. (b) The same foal after corrective trimming and two weeks confinement to the stable.

maintained (Figs 2.8(a) and (b)). However, if six to eight weeks of conservative treatment is unsuccessful or the deformity is severe and chronic when presented, then surgery is advisable. If the anterior hoof wall is vertical, section of the inferior check ligament just proximal to its junction with the deep flexor tendon improves the conformation immediately. Post operative daily walking exercise and regularly trimming to lower the heels encourage a rapid return to a normal conformation. The prognosis for racing or other athletic pursuits is good following this surgery. The foal with a very severe deformity, some of which will stand and walk on the anterior wall of the hoof (Fig. 2.9) requires section of the deep flexor tendon just proximal to the bulbs of the heel in order to improve the conformation. Following this surgery extensive thickening of the deep flexor tendon may develop at and above the fetlock and therefore the prognosis for athletic activities is poor.

Preventive measures against this deformity must include regular inspection of all foals and yearlings standing on a firm level surface. The diet must be well balanced and controlled so that concentrates are not overfed and there is a positive calcium to phosphorus ratio. Young stock should not graze young or recently fertilised pasture.

Figure 2.9. Extreme flexural deformity of the distal interphalangeal joint which required section of the deep flexor tendon to restore weight bearing on the sole of the foot. Note that attempts to correct the deformity with an extended toe piece on the shoe were unsuccessful.

Acquired flexural deformity of the fetlock joint occurs in older foals and yearlings. It can involve front or hind legs but is a more serious problem in front fetlocks. Affected animals maintain the foot flat on the ground but the pastern becomes upright and the fetlock joint knuckles forward at rest or when walking. The deformity is invariably bilateral and may lead to a back at the knee conformation. Severe cases may have difficulty in walking.

The inability of the superior check ligament to extend as the radius grows has been thought to cause this deformity. Rapid growth and overfeeding concentrates may contribute to an aetiology which is incompletely understood. Conservative treatment similar to that for acquired flexure of the distal interphalangeal joint should be adopted. Invariably, knuckling of the hind fetlocks responds to this treatment but if deformity of the front fetlocks does not correct within a month, surgery should be considered. Section of the superior check ligament can restore a normal conformation provided the deformity has not been present too long prior to surgery. Changes in the periarticular structure of the fetlock and the back at knee conformation make surgical correction of the longstanding case unsatisfactory. The prognosis for racing or other athletic pursuits is good in cases which make a complete recovery with or without the help of surgery.

Epiphysitis is an inflammation of the growth plate commonly caused by compression. It can result from overfeeding, especially concentrates or lush grass, which produces a rapid growth and increase in body weight, thus placing too great a load on the growth plates at the fetlock or carpus and occasionally tarsus of young horses of light boned breeds. It can also occur in a front leg which bears excess weight due to a chronic lameness in the contralateral leg. It is seen in the fetlock joints of foals between two and six months old (Fig. 2.10) and in the knees, occasionally accompanied by the hocks, of animals six to thirty months of age.

Epiphysitis is characterised by a hot painful swelling, usually on the medial side of the growth plate which, if severe, can cause lameness and lead to an upright conformation and a varal deformity of the affected legs. As, in the normally conformed limb, the majority of body weight is borne down the medial side so compression is greatest on the medial side of the growth plate. This area is thus more painful so the animal with epiphysitis tries to bear the weight on the lateral side of the limb and combined with the Type 5 growth plate injury closing the compressed side of the plate prematurely, develops a bandy or varal deformity of the leg.

Treatment must include rest and reduction of body weight or alleviation of a chronic lameness if that is the cause, and corrective hoof trimming and maintenance of a positive calcium to phosphorus ratio in a minimal diet containing no concentrates and hay given at 0.5 kg per 50 kg body weight daily. Analgesics are not usually necessary. Exercise should not be allowed

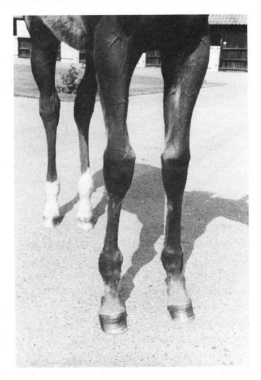

Figure 2.10. Four-month-old foal with epiphysitis of the right fore fetlock joint.

until the growth plate is free from all signs of pain and inflammation. If the conformation has not been altered by the epiphysitis the prognosis for future athletic purposes is good.

FURTHER READING

Hickman, J. (1977). *Farriery*. J. A. Allen, London.

Hintz, H. F. and Schryver, H. F. (1976). Nutrition and bone development in horses. *J. Am. Vet. Med. Ass.* **168**, 36–44.

Lewis, L. D. (1979). Nutrition for the brood mare and growing horse and its role in epiphysitis. *Proc. Am. Ass. Equine Pract.* **25**, 269–296.

Mansmann, R. A. *et al.* (1982). *Equine Medicine and Surgery*, 3rd ed., 2 vols. American Veterinary Publications, Santa Barbara.

Platt, H. (1979). A Survey of Perinatal Mortality and Disorders in the Thoroughbred. Animal Health Trust, Newmarket.

Rossdale, P. D. (1972). Modern concepts of neonatal disease in foals. *Equine Vet. J.* **4**, 117–128.

Rossdale, P. D. and Ricketts, S. W. (1980). *Equine Stud Farm Medicine*, 2nd ed. Bailliere Tindall, London.

3   Housing the Horse

D. W. B. SAINSBURY

INTRODUCTION

There are few more fascinating and important subjects than the housing of
the horse. It may at first thought seem rather a mundane and dead topic,
primarily a question of bricks and mortar with not much to bring it alive.
In fact, the realities are very different. The housing may be the major capital
expense that must be faced by the owner. Buildings on traditional lines have
become extremely expensive so that methods have had to be sought to lessen
these costs. This has led to the increasing use of non-traditional materials
and the adoption of prefabrication, both worthy endeavours but capable of
leading the purchaser into considerable areas of risk. In addition there has
been a trend to the housing of horses under total cover in relatively large
numbers which has created difficulties in environmental control and disease
prevention that are not always easy to solve. Thus we find ourselves dealing
with a dynamic subject where there will be major decisions to be taken and
it is hoped this chapter will help the reader to make the right ones.

Considered only from the point of view of the horse's health, it would
be better to leave the animal outside all the year round. With the development
of a good coat, the provision of adequate food and the acclimatisation from
normal seasonal changes, there is no need for further protection from the
weather other than the use of natural help from windbreaks and trees. The
horse will develop a coat which is long and thick in the cold weather and
by movement and exercise it will generate enough heat to keep warm.

However, there are many reasons why a horse must be housed. A horse must
be kept clean and in good lean condition and its coat has to be clipped short and
must be regularly groomed. It needs to be dry and carefully managed and
handled, and this can only be achieved in good stabling. Nevertheless, the first
essential of housing is to give horses as healthy an environment inside as they

Horse Management 2nd edition
ISBN: 0-12-347218-0 case

would have outside and yet give them adequate protection from the severity of the elements, and a good environment and facilities for the handlers.

A list of the principal essentials for stables are as follows:

1. A reasonably uniform temperature, eliminating as far as possible extremes.
2. A dry atmosphere and freedom from condensation on the surfaces of the building.
3. Generous air movement and ventilation without draughts.
4. Sound, dry flooring.
5. Good drainage.
6. Adequate lighting, both natural and artificial.
7. Good feeding and watering arrangements.

THE ENVIRONMENTAL REQUIREMENTS OF THE HORSE

Horses have suffered from a worrying increase in problems due to environmental stress within very recent times. There is certainly an increasing incidence of respiratory disease and a daunting number of causes have been suggested, many of them still questionable. It is generally accepted that the increase in the environmental and disease problems may be associated with the trend towards placing larger numbers of horses in close proximity within totally enclosed buildings. It is probably aggravated by the rapid movement of horses in substantial numbers from continent to continent by land and air. In order to provide the horse with the correct climatic environment in its housing and reduce the likelihood of disease, it is essential to have a basic understanding of the horse's physiological needs. The principal elements of the climatic environment which affect the horse are ambient temperature, relative humidity, ventilation rate and air movement.

Ambient temperature

A horse can readily tolerate a wide range, e.g. 40°F to 80°F (5°–27°C) of ambient temperatures without harm, depending on the degree to which the individual has been allowed to acclimatise and provided the atmosphere is free of damp and draught.

Temperature *per se* can be a misleading guide to the suitability of the environment and should never be used as the sole criterion. In practice, there is a risk that, in attempting to keep horses "warm", air flow and ventilation are restricted to a serious extent.

Relative humidity

The burden on the horse, especially on the respiratory system, is intensified if the air is damp (i.e. the relative humidity is near saturation) and there is condensation on the internal surfaces of the building. The worst conditions are those of low temperature and high humidity, the optimal conditions for the viability and inhalation of pathogens. Few stables are of sufficiently insulated construction and any restriction of ventilation causes severe condensation on the walls, roof and windows.

Ventilation rate and air movement

Ventilation rates are based primarily on: (1) the maximum rates to keep the building cool in warm weather; (2) the minimum rates to eliminate from the environment, even in the coldest weather, the humidity arising from the animals' exhalation and evaporation from their excreta.

Provided the minimum requirements are observed there will be no problem with gaseous exchange. Quite a wide range of ventilation rates will be required to satisfy the criteria — approximately a ratio of 10 to 1 from the hottest to coldest weather — and this must be achieved without extremes of draught or stagnation. Just what constitutes these extremes it is difficult to define accurately but a suggested range for air movement is from 0.15 to 0.5 m/sec (30 to 100 ft/min), at a minimal ventilation rate of $0.2 m^3/h/kg$ body weight (bwt) to $2.0 m^3/h/kg$ bwt ($0.05 ft^3/min/lb$ bwt to $0.5 ft^3/min/lb$ bwt).

The ventilation rates quoted here can only be achieved accurately if mechanical systems are installed and it is the area of air outlets and inlets that are important in traditional systems of natural air flow. Safe margins are, for the outlet, $0.1 m^2$ ($1 ft^2$) and, for the inlet, a controllable area of at least $0.3 m^2$ ($3 ft^2$)/horse. These areas are quite easily obtained but the location and design of the ventilators are critical.

Climatic environment and health

Recommendations, summarised in Table 3.1, provide a general guide to the required environment but, at present, they cannot be more precise. It is essential to produce an environment that helps the horse to maintain physiological well-being but there have been few investigations on which to base an objective approach. Reliance has to be placed on experience and opinion. The problem is relatively simple if horses are kept as individuals or in small numbers together in subdivided buildings. Difficulties arise when substantial numbers of animals are brought together, thus creating a disproportionately high cost of ventilation.

In the absence of objective information, the assumption is often made that ventilation is not an important feature; this has resulted in some of the most sophisticated buildings being built with little, if any, provision for ventilation. Horses kept in close quarters with one another require substantially more ventilation than those kept as individuals, or separated in small groups, because of the risks of cross-infection and a "build-up" of pathogens. There is also the greater likelihood of harmful effects by irritants, such as dust,

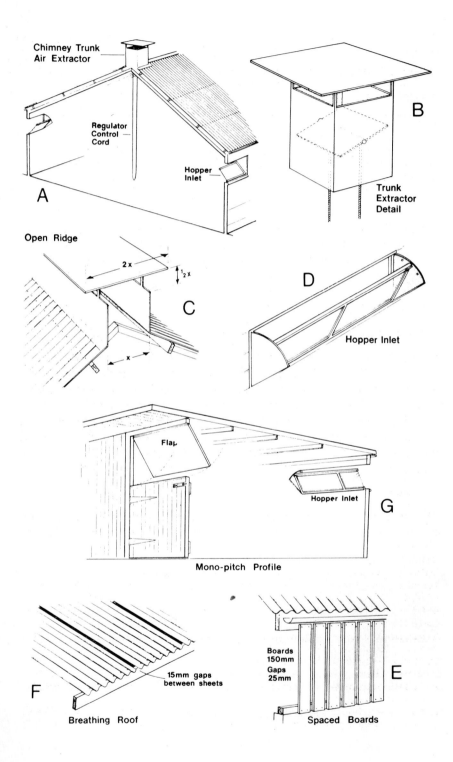

A

Chimney Trunk Air Extractor

Regulator Control Cord

Hopper Inlet

B

Trunk Extractor Detail

Open Ridge

2 x

½ x

x

C

D

Hopper Inlet

Flap

Hopper Inlet

G

Mono-pitch Profile

F

15mm gaps between sheets

Breathing Roof

Boards 150mm

Gaps 25mm

E

Spaced Boards

Table 3.1. Summary of the range of climatic environmental requirements for the stabled horse

	Metric	Imperial
Ambient temperature	0.30°C	32–85°F
Relative humidity	30–70%	30–70%
Air movement	0.15–0.5 m/sec	30–100 ft/min
Ventilation rate	0.2–2.0 m³/h/kg bwt	0.05–0.5 ft³/min/lb bwt
Outlet ventilation area	0.1 m²/horse	1 ft²/horse
Inlet ventilation area	0.3 m²/horse	3 ft²/horse

pollens, fungal hyphae and spores, causing mechanical and allergic damage. It seems reasonable to build a high degree of flexibility into the system so that a greater range of ventilation can be assured than that suggested as satisfying the physiological requirements.

MECHANICS OF VENTILATION

Building construction

It is essential that the structure of a building is thermally well insulated to ensure adequate ventilation. This helps to eliminate dampness created by condensation and assists in maintaining equable and uniform conditions, reducing peaks or troughs of ambient temperatures and humidities. Further, continuing air movement is promoted around the edge of the structure because the warm and stale air does not become chilled and therefore fall back towards the floor. It is not unduly expensive to install insulation into existing buildings and there are a number of economical methods, depending on the basic construction.

Air flow

The simplest ventilation technique, used since animals were housed, relies on the three natural forces of stack effect (i.e. rising warm and stale air),

Figure 3.1. Arrangements for natural ventilation of stables.
A: Use of extractor chimney trunks and hopper inlets for fresh air is suitable for all stables.
B: Extractor chimney trunk which may have a regulator or electric fan placed in the base.
C: Simple open ridge for extraction ventilation of covered yards; in yards up to 13 m wide *x* = 600 mm; in yards 13–25 m wide *x* = 600 mm.
D: Hopper window as fresh air inlet.
E: Spaced boards, giving draught-free ventilation.
F: "Breathing roof"—corrugated roof sheets fixed with 15 mm gaps for extractor ventilation.
G: Mono-pitch house showing hopper flap at back and ventilating flap at the front; overhang on roof protects horses from rain, sun and wind.

aspiration (i.e. wind force across the roof of the building sucking air out) and perflation (i.e. air blown from side to side and end to end of the building). A system based on these forces functions in the following ways.

Standard size buildings   A roof extracting trunk or chimney, of dimensions similar to or greater than, those shown in Table 3.1, provides the stack effect. Fresh air inlets replace the stale air and should be situated around the wall to give a uniform ingress of air (Figs. 3.1A and B).

With a suitable design of chimney, that allows the wind to blow freely across the top, the aspirating effect is encouraged and the perflating effect is ensured, providing there are inlets on both sides of the building to give a through draught when required. However, the inlets must be designed to prevent down-draught. The best control is, usually, an inward opening, bottom-hinged control flap constructed of timber or an alternative thermally insulated material. This construction deflects incoming air upwards in the cold weather if the flap is partly closed or, in warmer weather, the air can come directly into the building or blow across it. Ideally all ventilators should be fully controllable and their regulators easily reached. The inlets should be placed as low as practical without incurring the risk of interference by the horses. The best arrangement is one chimney trunk to each common air space but air inlets should be numerous to ensure a uniform intake of air around the building. Each stable should have its quota of half doors and windows, preferably of double glass, which can be opened for extra air flow in warmer weather. This system can be used to give a surplus of air flow with safety, if the air supply is draught-free, because the additional and faster air movement is well above the horse.

Buildings with special features   The arrangements described above may fail to function satisfactorily if the buildings are excessively wide or long or are boxed in by other buildings which impede natural flow. In these circumstances it may be necessary to use mechanical ventilation by installing extractor fans. This can be achieved with simplicity and economy by placing them at the base of the chimney. In this situation the fans are acting with, not against, natural forces and running costs are therefore minimal. Mechanical aids, thus installed, are preferable to complex pressurised ducting systems which are expensive and increase air turbulence and, consequently, the incidence of suspended dust particles. Such systems sometimes incorporate recirculation devices which may intensify the risk of cross-infection.

A ventilation system should remove instantly the horses' exhalations without stagnation, recirculation or cross-infection. This may best be achieved by a continuous upward movement of air within the building. Abrupt changes, such as are sometimes associated with on-off regulators of mechanical systems, should be avoided. Periods of "nil" air flow should be prevented and, so far as possible, all changes in ventilation rate should be gradual. It is also important to have a system that, even if mechanically aided, can still function naturally if the mechanical part fails. The principal system outlined above achieves these aims because the fans act as boosters to the natural flow.

The covered yard   In any large building containing a number of horses within a totally covered structure there is a considerable danger of environmental pollution. With the growing trend of housing horses in this way the ventilation system is absolutely critical to the horse's well-being. Whilst it is possible to give general guidance each building must be considered individually, so far as the ventilation and environmental control are concerned, because the air flow may be greatly influenced by the site chosen.

The aim of ventilating a large covered building is to ensure free air flow without draught at all times. The one satisfactory way of achieving good natural extraction ventilation is to install a wide open ridge, from 0.3–0.6 m (1–2 ft) wide, protected with a flap to keep most of the weather out but allowing the wind to blow freely across the ridge and thereby enhance the aspiratory effect (Fig. 3.1D). It is of advantage if there is some form of control to regulate the open area though this may be dispensed with where the buildings are in sheltered locations.

The control arrangement may be similar to that shown in Fig. 3.1B, or a hinged flap may be easier to fit and operate. An alternative and even better system is to use a series of the chimney trunks referred to and raise each one 1 m (3 ft) above the ridge. The movement of air through a chimney trunk is invariably superior to a continuous opening. There is also a further practical advantage. For example, in certain cases where the building is in a very sheltered position or is excessively wide, natural ventilation cannot cope satisfactorily at all times and an arrangement with chimney trunks is easily and economically mechanised by placing extractor fans in the throat of the trunk as described in the preceding paragraph.

Mention should be made of certain simpler techniques of roof ventilation. Corrugated roof sheets may be fitted upside down and with a gap of a few millimetres between the sheets. This method, described as the "breathing roof", is justifiably popular (Fig. 3.1F) but there is no possibility of regulation and there may be a slight ingress of rain or snow, so the method should be used only on sheltered sites. Another technique is to raise the overlapping on the roof sheets by fitting a batten between. Spacer battens of treated timber measuring 50 mm × 25 mm allow the air to pass between the sheets but the overlap will still prevent the entry of rain.

For the inlet of air through the walls or at the gable ends of the building the "Yorkshire" or spaced board (Fig. 3.1E) is a simple method. Traditionally this is a walling of 150 mm (6 in) boards with 17 mm (0.75 in) gaps between and is an excellent adjunct to roof ventilation when used around the sides, at the top half of the wall and at the gable ends above the doorways. "Hit and Miss" sliding boards are a form of controllable space boarding in which one set of boards slides over the others, and is recommended for buildings on an exposed site. Measures of restricting and individuality can be incorporated into the layout within the building as appropriate and these can be ventilated by measures similar to those required for a single box.

Mono-pitch house   An alternative type of house is the mono-pitch or lean-to building (Fig. 3.1G) an open-fronted shed, facing towards the warmest winds and the maximum sunlight and sloping from a low back to a high front. The pitch on the roof ensures a free flow of air towards the front and the generous air space promotes a good environment. The ventilation is completed with simple hopper flaps (Fig. 3.1D) in the back wall and doors at the front. An extension to the roof, on the open side, gives protection to the animals and their attendants. The division of the building by cross partitions prevents through draughts and provides a measure of isolation, helping the air to blow upwards. The cross partitions need not add to the cost because they can be load bearers for the roof. The building may be as basic or as elaborate as is wished but it is, fundamentally, a simple approach and avoids any real difficulties in environmental control.

Assessment of the environment

An objective assessment of the environment in stables may be achieved by taking a few measurements with quite simple instrumentation. The ordinary maximum and minimum mercury-in-glass thermometer is a considerable help as it gives the range of temperatures both within and outside the building. This has very much more meaning than an on-the-spot record with an ordinary wall thermometer. Even better is a recording thermohygrograph which gives a continuous chart record of both the air temperature and relative humidity. Air movement can be measured with the Kata thermometer (Fig. 3.2) and currents and gas levels in the air with the "Draeger" devices (Fig. 3.3). It is quite possible from the records of these reasonably simple and economic instruments to measure the total environment and calculate the ventilation rates.

THE SITING OF THE HOUSING

The best place for the siting of stables is on well drained land where there is a natural yet gentle fall away from the buildings. The housing should be

Figure 3.2. The Kata thermometer.

as protected as possible from cold and biting northerly and easterly winds with the main doors and at least a substantial proportion of the ventilating (opening) windows facing south or west. When this is not possible, a second best is to protect the stables by trees or other buildings. A good south-facing slope is the best possible location. Avoid low lying, damp and over protected locations where conditions can become extremely unhealthy in the damp weather of the autumn or the still, hot days of high summer. Low lying hollows can also be dangerous frost pockets in winter.

It may also be emphasised that care should be taken not to site the stables too close to domestic dwellings so as to cause a nuisance. Local planning approval and building regulations must be complied with to make sure the site, structure and drainage all comply with the law. Ignorance of these facts is serious as the general public are increasingly unsympathetic to animal smells and activities and speedily litigious.

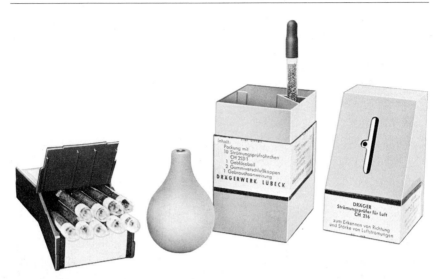

Figure 3.3. The Dräger devices for measuring air currents and gas levels.

THE HOUSING REQUIREMENTS

A stable block will consist of a series of boxes or stalls for the horses together with ancillary needs. For example, in a unit of a number of boxes there may be an isolation box for sick horses or foaling, a feed room, hay store, straw store, tack room, washing room with drying facilities and accommodation for the attendants, including a lavatory. There must also be some provision for clean manure storage situated conveniently but not too close to the boxes, and also office accommodation and garaging for vehicles and trailers. Obviously, the extent to which all these facilities will be required or can be provided will, to a large extent, depend on the nature and size of the establishment but when planning or purchasing stabling it is wise to have a check list such as this in mind. It is also important to consider carefully the duties and movement of the staff and animals between the various parts of the unit. Thus the fodder, straw and hay should be as centrally placed as possible to minimise movement of these heavy loads, and the manure store must be situated likewise with good access but strategically placed to avoid nuisance from smells, flies and insect breeding. A brief work-study exercise in which the movement of horses, attendants and materials are mapped out on a plan is well worth the effort and may save a great deal of energy being wasted later.

In general nowadays horses are housed in loose-boxes, though there are many stalls still in existence. Sometimes stalls are actually preferred, though it is more usual practice now to convert stalls into boxes. The essential dimensions of a loose-box for horses of most sizes is 4 m × 4 m (12 ft × 12 ft). Ponies can be placed in a smaller sized box of approximately 3.3 m × 3.3 m (10 ft × 10 ft). Stalls are smaller, usually 2 m (6 ft) wide and 3.3 m (10 ft) long to accommodate most horses. Foaling boxes are normally larger, up to 4.6 m × 4 m (14 ft × 12 ft).

Layouts

The simplest layout is a single row of boxes opening directly into the open air. It is a perfectly satisfactory design and can be embellished by extending a covered "overhang" on the boxes to give protection to the horses and the attendants, keeping the boxes drier and relieving them from excessive heating and glare from the sun (Figs 3.4 and 3.5). If it is constructed without

Figure 3.4. Prefabricated loose-box stables constructed with concrete panels. Note the clean finish and overhanging roof protection. (Courtesy of Marley Buildings Ltd., Guildford, Surrey.)

Figure 3.5. Single row of timber stabling with attractive appearance. Note wide doorways, excellent window areas and good roof overhang to protect the buildings and the horses. (Courtesy of Harlow Bros. Ltd., Long Whatton, Loughborough, Leics.)

supporting posts a hazard to horses and people is removed. It is advantageous if the overhang is extended through to the adjoining service buildings. It is a great help if most of the operations round the stables can be carried out under cover, and the design just mentioned is probably the healthiest arrangement for housing the horse, especially as it gives isolation between each box (Fig. 3.6).

When there are larger numbers of horses to be housed the boxes can then be arranged either on two, three or four sides of a square, finally giving a fully enclosed courtyard which serves as an invaluable exercising and service area (Fig. 3.7). These arrangements have the advantages of economising on movement and fencing and assisting in safety and security. The disadvantage is that some stables have a cold northerly or easterly facing aspect and will receive little or no sun on to the front of the boxes.

The totally enclosed unit

A more elaborate development is to place loose-boxes or stalls within a building with the adjacent passage inside it rather than as an outside covered passage. It is a design common in those magnificent stables erected in earlier ages than this and with all the ancillary needs for the attendant and the horse within the same unit, it forms an attractive and practical arrangement. It has some clear advantages. It provides better environmental control for the horses and the staff and all servicing can be carried out under much more comfortable conditions. On the other hand, there is a distinct hazard that infection may spread more rapidly and the precise method of achieving good environmental control may be more difficult. There is also the more modern and popular system of housing large numbers of horses in a *Barn*. In this case the total approach is rather different. The Barn is a large "covered yard", a lofty structure about 3 m (10 ft) to the eaves and rising to a ridge of about 8 m (25 ft). Within this building are placed rows of boxes, two or four rows depending on the width of the building. The boxes are of the usual size and there are adequately wide passageways between. So far as possible all ancillary services are also within or very close to this building. It is an arrangement much favoured in North America where the harsh winters often with deep snow make it very much more desirable and even essential compared with the more equable climate in the United Kingdom.

There is, however, a general feeling that the housing of considerable numbers of animals under one roof has contributed to the much greater incidence of respiratory disease in horses at the present time. Infection can spread rapidly and environmental conditions can be unsatisfactory. Thus the greatest care must be given to the planning of the ventilation system and to the choice of materials for the construction of the barns. Otherwise there is a tendency for the building to get excessively hot during the high summer

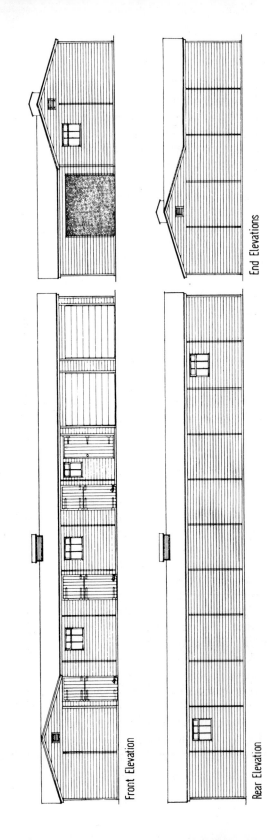

Front Elevation

End Elevations

Rear Elevation

Figure 3.6(a). Typical layout of a unit with two standard and one large loose-box together with tack room, hay store and garaging. Front, rear and end elevations.

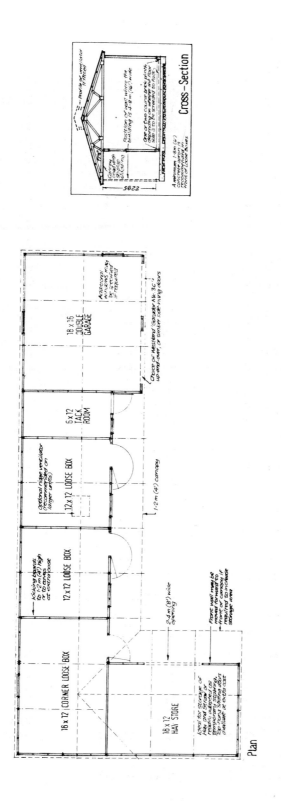

Plan

Cross-Section

Figure 3.6(b). Floor plan and cross-section. (Courtesy of Passmores Portable Buildings Ltd., Rochester, Kent.)

Figure 3.7. Double row of loose-boxes in prefabricated concrete facing into a well protected but open courtyard. Note the excellent drainage and the pleasing tiled roof. (Courtesy of Marley Buildings Ltd., Guildford, Surrey.)

conditions with long periods of sunshine, and likewise in the winter there may be excessive cold with condensation on the surfaces of the building producing an atmosphere which is highly charged with moisture and which places a greater strain on the respiratory system than does the "traditional" arrangement.

Isolation box   It is wise to have at least one isolation box for sick animals in any sizeable unit. Such a box is an essential if a horse is suffering from a contagious and infectious disease, but an isolation box is useful for all horses requiring special treatment. The box or boxes must be quite separate from the remainder of the establishment. The box should be readily accessible and the horse should not only be able to see people—and be seen—but should also be in sight of other horses if at all possible. An isolation box should be considerably larger than the ordinary box — about 5 m × 5 m (15 ft × 15 ft) is ideal, and with such special fittings as may be required for the handling of sick or injured horses. For example, it is desirable to have a strong beam or girder at a height of 3 m (10 ft) to support a horse on slings.

The feed room The feed room is used for storing fodder which is for immediate use, as opposed to the feed store, where the longer term supplies are kept. Its dimensions and that of the store will depend on the establishment's size. Daily feeds are prepared in this room and it will contain separate bins for the individual feeds, e.g. oats, bran, barley, etc. There will also be the need for a sink with hot and cold water and a tap to fill the buckets. In large establishments there may be chaff cutters and other machinery for the preparation of the feed.

Feed store The feed store usually adjoins the feed room and is also close to the storage areas for hay and straw. The store will contain feed in bulk bins, sacks or bags. A very convenient arrangement is to build the feed store above the feed room so that the larger quantities of feed can be hoisted up and then moved down as needed. The movement of food up to the store will only be done occasionally, but installation of a mechanical system will make this task far easier.

Hay and straw stores It is prudent to purchase good hay and straw, both essentials in good horse management, at a time when prices and availability are right and then store them in good conditions on one's own premises. A hay and straw barn is therefore needed in close proximity to the feed room and feed store and also sited in a handy position for off-loading the bales from the lorry. Storage may also be required for other forms of litter, such as peat, wood shavings, sawdust and shredded paper. These are either in sacks, bales or bulk but the storage requirements are similar to those of straw and hay, that is, a totally dry building with hard, dry floor and free ventilation throughout. There are grave dangers to horses from wet, damp, dusty or musty bedding and these faults must be avoided at all costs or there may be serious effects on respiratory disease and allergies.

Tack room The tack room (Fig. 3.8) is the most important service room in a group of stables. Within this room must be fitted the brackets, shelves and cupboards to take all the saddles, bridles, bits and so on associated with horses. It is also preferable to have closed cupboards and/or chests for the blankets and other clothing for the horses. The room should be well insulated and ventilated and have some form of heating.

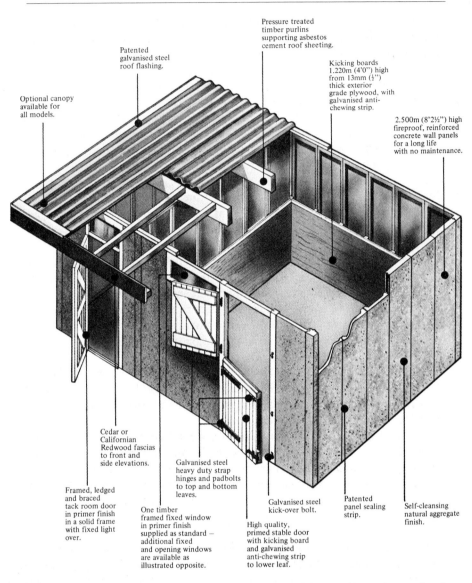

Pressure treated
timber purlins
supporting asbestos
cement roof sheeting.

Patented
galvanised steel
roof flashing.

Kicking boards
1.220m (4'0") high
from 13mm (½")
thick exterior
grade plywood, with
galvanised anti-
chewing strip.

Optional canopy
available for
all models.

2.500m (8'2½") high
fireproof, reinforced
concrete wall panels
for a long life
with no maintenance.

Cedar or
Californian
Redwood fascias
to front and
side elevations.

Galvanised steel
heavy duty strap
hinges and padbolts
to top and bottom
leaves.

Framed, ledged
and braced
tack room door
in primer finish
in a solid frame
with fixed light
over.

One timber
framed fixed window
in primer finish
supplied as standard —
additional fixed
and opening windows
are available as
illustrated opposite.

Galvanised steel
kick-over bolt.

High quality,
primed stable door
with kicking board
and galvanised
anti-chewing strip
to lower leaf.

Patented
panel sealing
strip.

Self-cleansing
natural aggregate
finish.

Figure 3.8. Three-dimensional drawing of a typical prefabricated loose-box and tack room. (Courtesy of Marley Buildings Ltd., Guildford, Surrey.)

Facilities for washing and cleaning   It may be possible to incorporate the washing and cleaning facilities within the tack room but it is much better to have a separate facility since the water and steam are not good for the clean stored clothing and tack. The needs in this room are for a large sink with hot and cold water, drying and airing facilities and saddle horses and bridle cleaning holders, together with a good assortment of cupboards to take all the cleaning materials.

Cover for drying bedding   After the morning's cleaning of the stables, there is a need to dry and air the clean litter and to do this throughout the year it is an invaluable help to have a high roofed cover without sides so the bedding can be laid out with unrestricted air flow through it. There must be a good overhang on the roof, and a raised floor on slats or a grille to provide air flow underneath.

Manure storage   The siting of a good manure store is all important. It must be easy of access to the boxes but also handy for loading off with good road access. If it is too close to the boxes it may create a smell and also increase the fly menace. The manure store usually consists of a concrete base with a fairly steep fall to drainage at the back so water and effluent does not run back on to the concrete access area. The rear and sides are usually sturdily constructed of blocks, bricks or reinforced concrete. Size will depend entirely on the period of storage envisaged. The more frequent the collection of the manure the better it will be as the nuisance is less and the store can be reduced in size.

Fences and gates   The whole establishment will require good fencing for the protection and safety of the horses and the staff. Post and rail fencing is ideal. Gates, which will often be used by mounted riders, should be fitted with hunting latches to facilitate opening.

A further requirement is for a mounting block, which is usually placed at the side of the yard with free access on all sides. Sometimes on the largest establishments weighing facilities are also incorporated, and an office is a necessity within which are all the facilities needed for records. This is also the best place for a cupboard that can be locked to keep medicines and poisons.

Stalls   Stalls, though rarely used nowadays, should be a minimum width of 2 m (6 ft) and a length of approximately 3.3 m (10 ft). There is also a passageway at the rear of not less than 2 m (6 ft), so that the minimum width of a building to take a single row of stalls would be 5 m (16 ft). All details relating to fittings and construction are similar to those with loose-boxes.

Roofs of stables   The roof construction may be of any suitable modern form but it is desirable to incorporate some form of thermal insulation. There are two ways of achieving this. The insulation can form an underlining to the exterior roof cladding, or there may be a flat ceiling formed. There is no great preference for either construction but ventilation does tend to be better if there is a pitch on the ceiling rather than a flat ceiling which holds the stale air.

There are a number of alternative constructional materials. A good inner lining is one of flat asbestos cement sheeting, giving a hard impervious material that can be easily cleaned and disinfected. Above this will be a generous thickness of glass fibre or other loose fill insulation and this will be sealed by a vapour seal of polythene to prevent damp penetrating it and condensing on the inside of the outer sheet. Alternatives to asbestos are oil-tempered hardboard or exterior grade plywood. A more modern approach is to use an inner lining of a plastic-faced rigid polyurethane board. This is very good as an insulator but is not as hardwearing as the asbestos or hardboard linings and separate insulation infill described earlier. So far as thicknesses are concerned, the loose fill insulation should not be less than 50 mm (2 in) but it is better to be 100 mm (4 in). The rigid lining board, being a better insulator, can be 37–50 mm (1.5–2 in).

The outer cladding of the roof is usually corrugated asbestos or metal — it makes little difference which is used.

The older traditional stable roofs are usually constructed of tiles or slates as the exterior cladding and have an inner lining of timber. This is perfectly satisfactory but is generally too expensive for modern use, using costly materials and requiring more substantial supports.

Walls   Traditionally constructed stables are usually built of brick, double thickness and uninsulated and this does not meet with the better need to have a reasonable degree of warmth and dryness. The modern stable is preferably of insulated construction and may be made of a number of different materials. If brick is used then it should be of cavity construction and it is of great advantage if the cavity is filled with insulation such as a polyurethane foam. Alternatively the wall may be constructed of concrete blocks, which may be cement rendered, using load bearing insulated construction, or of concrete panels or reinforced concrete poured *in situ*.

However, constructional procedures are resorting to various forms of prefabricated structures. Traditional methods are still perhaps the best of all but they are very expensive and cannot often be afforded or justified. On the other hand, there are several useful forms of timber and concrete prefabricated construction. In both cases it is best to have double linings, the interior lining being of stronger construction generally than the outer lining as the inner linings will receive the full impact of the horse.

Similar needs in relation to the strength of the materials apply to the choice of material used for all the internal divisions. These also may be constructed of brick or blocks, or by framing with the main framing built into the walls and the floors. The framing may be reinforced concrete or steel, which are certainly the most durable, or in timber.

Internal partitioning can be taken up to full height or only part of the way. Opinions on this item are fairly evenly divided. There is little doubt that from the point of view of environmental control full height partitioning is best since it keeps draughts down and enables each horse to have its own environment. On the other hand, horses prefer to be in close touch and see each other so the partitions can be solid up to about 1.3 to 1.6 m (4 or 5 ft) and the upper section can be formed of a metal grille. The overall height of the partition, including the grille, need not be more than 2.1 m (7 ft). This enables all horses to see each other but they cannot bite or interfere. It also allows for free air movement in the building which is an advantage if the building is rather poorly ventilated. The grille is usually made of vertical steel bars of 12 mm (0.5 in) diameter and not more than 75 mm (3 in) apart. Horizontal bars should not be used or the horse may develop a vice of biting them.

Internal finishes to walls and partitions must be able to stand up to the kicking of horses and be hygienic and easily cleaned. If cost is no problem, wall tiling is good. Another good finish is white or glazed bricks. But a granolithic rendering is cheaper and is most satisfactory, or even a simple cement and sand rendering. It is not entirely unsatisfactory to leave brick or concrete blocks uncovered. In the latter case, if the blocks have a smooth finish they may form almost as good a surface as a rendered one. Prefabricated concrete sections can also be kept very clean.

With prefabricated timber structures an internal lining that will withstand the horses' aggression may be achieved using either an 18 mm (0.75 in) board of exterior grade plywood or oil-tempered hardboard. These materials are also suitable for internal divisions when securely framed and fixed. Any reputation they may have for being insufficient to withstand the horses' attention is usually due to poor fixing or workmanship in their installation.

Windows   Windows are most important in the modern stable. They need to be fitted at a high level in order to prevent damage and also because the light penetration is best in this position. It is helpful if windows can serve also as part of the ventilation system and this is most satisfactorily achieved if the windows are inward opening, bottom hinged, with gussets to prevent draught blowing down on the horses' backs. The windows must be openable in stages so that they do not have to be either open fully or totally closed.

It is also a great advantage if the windows can be double glazed, preventing condensation and excessive heat loss. Windows on both sides of the box or stable enables some cross ventilation to be used when required. If windows must be lower down they must be protected from damage by the horses by protecting them by steel grilles. There is, however, no doubt that it is better still to insert some purpose-made inlet ventilators between the windows well baffled from draughts. If made of double timber construction they will keep dry and free of condensation and are easily controllable. Roof lights are essential in large barns to give good, natural light but they will also require to be of double construction if condensation is to be avoided. The most satisfactory material is translucent plastic but it is essential to make the inner sheets removable to facilitate cleaning from time to time.

Floors   It is impossible to exaggerate the importance of a really good floor surface. The general requirements are that the floor must be dry, reasonably smooth, non-slippery, non-absorbent and hard wearing. It must also have sufficient fall to take away the urine and any other fluids but not with so much fall that it leads to an increased danger of slipping and tends to encourage the bedding to be scuffed to one side. A fall of not more than 1 in 80 is recommended. In stalls, the fall is to the back but in boxes it is best towards the door so that any fluid drains out to a gullet in the passageway. In very few buildings are the falls on the floor correct; it is absolutely essential the floor is so sloped that, after washing, all the liquid flows by natural fall to the drains.

So far as the material of the floor is concerned, Adamantine clinkers or blue Staffordshire checkered pavings are the ideal but in most cases they are considered so expensive to use that these very hard wearing and traditional materials have to be replaced by concrete. Nevertheless it can be perfectly satisfactory if the greatest care is taken in its use. It should always be laid with a damp-proof membrane incorporated but on particularly damp and cold sites there is much to be said for the addition of an insulation layer. Thus the total floor construction from the ground upwards will consist of the following:

1. Ground, with all vegetation removed and well consolidated.
2. Hardcore, a thickness of 180 mm (6 in) of firmly compacted material.
3. Above, a thickness of 100–150 mm (4–6 in) of concrete forming the final surface.

The surface layer will be much improved in wearing qualities if a good granolithic topping is laid.

The additional features to note are that if it is felt desirable to incorporate an insulation layer, this should be placed just above the damp proof membrane. This, in turn, would be placed just above the hardcore. Most often the damp-proof membrane is of thick gauge polythene sheeting, whilst the insulation layer is usually of a proprietary insulation concrete used at about a 100 mm (4 in) thickness.

Drainage It is unnecessary and undesirable to have complicated drainage systems in stables. The first essential is to have a good fall on the floor of the box to take as much fluid as possible to the outside. Considerable care and skill are required to achieve this but it is worth all the trouble given to it. Outside the boxes will be the main drainage system with gullys to trap straw and other solid matter. Internal drain-inlets should be avoided since they are more difficult to clean and can contribute to unpleasant smells and pollution of the atmosphere. The design of the drainage system should ensure that careful consideration is given to easy clearing and rodding when, as inevitably will be the case, obstructions occur. Shallow and safe open channelling is vastly better than underground drains.

Doors Doors of loose-boxes and stables are best placed to one side of the box and where two boxes adjoin they should never be placed next to each other. By placing the door to one side of the box it is possible for the horse to keep cosier and away from the draught in the opposite corner.

Doors should not be less than a width of 1.3 m (4 ft) but 1.6 m (5 ft) is ideal so the horse can easily be led in and out without any injury. The height should not be less than 1.85 m (7 ft 6 in). The door openings should be made with all external edges rounded, as sharp edges can readily cause injury. Sometimes rollers are fitted on each side of the door opening to protect the horses when they move in or out. The doors themselves are side hung to open outwards—an essential since a horse while lying down or when ill or cast may prevent the door from being opened. External doors to boxes which open directly to the open air are invariably formed in two leaves, the lower section having a height of about 1.4 m (4 ft 6 in). Both leaves should be hung to close back against the wall when in an open position, so that no obstruction is formed. Doors and frames must be of heavy construction to withstand the rough handling that is inevitable. Frames should be at least 100 mm × 75 mm (4 in × 3 in). The most satisfactory door is of the framed, ledged, braced and boarded type. Framing should be at least 150 mm × 50 mm (6 in × 2 in) with a minimum of 25 mm (1 in) tongued and grooved boarding. The junction between the upper and lower sections is usually sloped to help keep the box drier and draught free. Sheet metal should be fixed to the top edge of the lower door, both to protect it and to discourage biting by the horse. This must be of good quality and firmly fitted, without protruding nails or screw heads, to remove any danger of the horse injuring itself. Latches on doors should be of a type that does not project to remove any danger of injuries to horses or attendants.

Fittings and fixtures

In every box or stall the following are required: mangers, water points, hay racks or net and tying up rings.

Mangers   Mangers for the horses' feed are usually placed at a height above the floor of approximately 900 mm (3 ft). The various alternative materials for their construction are: vitreous enamel, galvanised steel, stainless steel, timber and salt glazed channels. They can be fitted either along the face of a wall or in the corner. There is no doubt that in spite of its expense stainless steel is the nearest material to perfection as its life is without limit and it is easy to clean. Mangers may be obtained that are combined with a hay rack or a water trough but the latter combinations are not recommended as the food tends to foul the water trough.

Fixed mangers can be difficult to clean adequately. To overcome this problem the manger can be fitted with a waste pipe or the manger may be made so it is removable from its framework to facilitate cleaning.

The manger is usually fitted on the wall of the box opposite the door. Though this means the attendant has the longest distance for feeding and servicing, it is a position best for the horse and the care and safety of the operator. The manger should be fitted towards the opposite corner to the door, as the horse will then be out of a draught when tied up.

There are those who advocate no fitted equipment but the use of all movable feeding and watering equipment. Movable metal or timber feeding troughs are used, with water from a bucket and hay being fed either loose on the floor or from a hay-net tied to one of the rings. The advantages of this system are that the cleaning of the box is easier and the horses feed off the floor, which is the natural position for normal feeding.

It is often advocated that the space beneath fitted mangers should be filled in to prevent a horse injuring itself by raising its leg sharply or if it should be cast with its head underneath. Filling in completely is good but just pannelling it in leaves an area which is potentially an area for dirt or vermin and is not recommended.

Hay rack   Hay racks can either be combined with the manger or be entirely separate. The hay rack must be securely fixed to the wall. There has always been some dispute as to its best siting. If it is at a low level, the horse eats in a normal position but if it is placed higher at, say, 1.5 m (5 ft) from the floor dust and seeds can enter the horse's eyes. As an alternative to the rack, hay nets can be used, in which case they may be secured to a ring placed at the side of the manger.

Water   Every box or stall requires fittings for clean, fresh and wholesome water. The water can be provided in a bucket, or an automatic water trough may be fitted. If buckets are used then there should be a proper bucket holder, this being secured to the wall. Troughs containing 2–3 gallons are made of galvanised or stainless steel and incorporate a ball valve, drain tap and overflow.

All pipes and fittings associated with the water supply systems must be well insulated against freezing.

Rings Rings are required for both boxes and stalls for tying up horses. Two or three rings per box or stall are usually adequate and there may also be an additional ring required for hay nets. One ring can be placed at about 1 m (3 ft) in height for tying the horse to the manger whilst another is put at a height of 1.5–1.55 m (5 ft to 5 ft 6 in) for general tying up purposes.

A horse may be secured at the manger by a headstall and headrope, the latter being attached to the "D" at the back of the nose-band. The headrope may be tied to the manger ring but it is safer if it passes through and is then secured to a "log" so that it just touches the floor when the horse is standing up to the manger. This log ensures there is no slack on the rope so there is little danger of the horse getting its leg over the rope and becoming cast. One should be positioned close to the hay rack or hay net folder or ring so that the horse can eat its hay while being cleaned or treated. Rings must be very securely fitted and the best fitting is to have the bolts pass right through the wall and then be secured on the other side.

One other fitting that may be used is a salt lick holder. It is of benefit to the horse to have one of these special blocks available and such a holder is fitted either over or to one side of the manger.

Electric points and lighting At least one electric point is required in each box — many of the major duties associated with the grooming and cleaning of the horses are carried out at night time, but it is much better to have two lights on each side of the box. Suspended lights should be placed at a minimum level of 3 m (10 ft). The best fittings are watertight, bulkhead covers, fitted securely to the wall and fitted with heavy prism glass or strong grilles. They should be placed 1.6–1.9 m high (7–8 ft). These arrangements are the easiest to keep clean, are safe against splashing and disinfectants and provide good light in any part of the box without excessive shadows.

Hygiene and disinfection Great care should be taken in the design and construction of stables to ensure that the surfaces of the building can be readily and routinely cleaned and disinfected. To minimise the risk of pathogenic micro-organisms becoming established in the fabric of stables there should be a regular programme of depopulating the accommodation and subjecting it to a thorough sterilisation. It is only possible to do this effectively if all internal surfaces have a smooth and uncluttered finish and the materials used will withstand pressure cleaners, strong detergents and a full range of disinfectants such as formaldehyde, organic acids and phenolics.

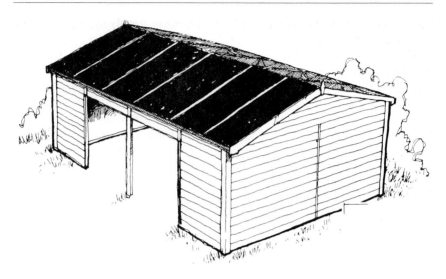

Figure 3.9. A Field Shelter measuring 9.2 × 4.9 m (30 ft × 16 ft) for the protection of horses in the field together with some storage of fodder and bedding if required.

Miscellaneous

Field shelters   A very useful aid to the protection and care of horses in the field is a wooden shelter (Fig. 3.9). These are usually simple timber structures, most if not all of the front being open and of about 5 m (16 ft) depth. They are normally sold in widths of 6.6 m (20 ft) upwards and may be used in addition as stores for some of the essentials of management as well as protecting the stock.

Fire hazards   The risk of fire in the stables is one that must be considered. Material can be chosen for the construction that is as fire resistant as possible but much of the material around the accommodation, such as hay, straw and timber, is combustible. Fire prevention is a professional matter and all those planning or using stables are strongly advised to consult the local fire prevention officer in the setting up of fire alarms, extinguishers and fire check doors.

Improving existing buildings

Those who have studied this chapter may feel that the present accommodation for their horses falls short in one or more respects of the standards suggested here. The author's experience suggests that the most serious deficiencies are with ventilation, insulation and drainage. The methods described in this chapter have, however, been presented in the knowledge that it is more likely the average reader will be concerned with improvements rather than making a clean start from scratch. It is therefore a comparatively straightforward matter to add some ventilators, such as are described here, or line the inside of a poor wall with a tough internal lining with some insulation behind it. The perfect stable is not often found but every step taken towards it is likely to produce a dividend that eases management and improves the condition, health and well-being of the horses.

4 Stable Environment in Relation to the Control of Respiratory Diseases

A. F. CLARKE

INTRODUCTION

Historically the association between the stable environment and the horse's respiratory well-being is firmly established. Miles (1880) likened the horse's lungs to "a second stomach, and respiration but another form of digestion. . . . If the stomach receives impure and unwholesome food disease is induced; if the lungs receive effluvia, instead of pure atmospheric air, then also the system becomes diseased". Percivall (1853) noted that in moving the horse " from his native air (i.e. at grass) to that of the stable . . ., soon or late, he is pretty certain to pay the penalty for it" in terms of respiratory disease.

The physical environment of the stable can affect the incidence and severity of respiratory disease by:

(i) altering the horse's systemic or local resistance to infection;
(ii) increasing the size of challenges from infectious micro-organisms, parasites or allergenic particles by enhancing their proliferation and/or surval and/or spread;
(iii) changing the rate at which pathogens are deposited within and cleared from the respiratory tract.

The ideal stable environment, in the context of preventing respiratory disease, must be based on consideration of all of the above factors. To date the few approaches to the air hygiene of stables have been based on non-empirical recommendations which have largely come from extrapolations from the housing of the agricultural species. However, this approach ignores

Horse Management 2nd edition
ISBN: 0-12-347218-0 case

several fundamental and important differences between the housing and performance requirements of horses compared with other farm animals.

Horses are kept at a lower stocking density than farm animals; they usually have individual living areas, and a much greater labour input per animal is present in stables than with farm animal housing. With farm animals, stock may be housed with "all-in all-out" systems, only animals of similar ages and previous "experiences" being mixed. A further advantage of farm animal management over that of the horse is that specific pathogen-free herds and flocks may be kept in strict quarantine with no animals being returned to the farm once they have been transported away. Spread of infectious respiratory disease has been associated with the national and international travel of horse to competitive events, studs and sales (Powell, 1985). Likewise increased incidences of respiratory disease coincide with the seasonal influx of yearlings into racing yards (Burrell et al., 1985). The "dealing spirit" is usually high with most horse people who keep an eye open for that promising youngster or bargain horse which is snatched up and quickly spirited home.

However, the most critical difference between the horse and farm animals are the demands that are placed on the respiratory system. Levels of respiratory disease which are unacceptable in the short or long term for horses need not alter the production (i.e. body weight gains) of farm animal species (Wilson et al., 1986). Indeed, with intensively housed livestock some degree of respiratory disease is almost unavoidable. The horse's short-term production, i.e. athletic ability, and long-term welfare are, however, very dependent on its respiratory well-being.

This chapter considers environmental factors which may predispose, induce or prolong respiratory disease in the horse. It considers factors which affect: (i) airborne pollutants such as dust and noxious gases; (ii) the survival and spread of both pathogenic and non-pathogenic micro-organisms; and (iii) the likelihood of given challenges inducing disease in the horse and the fate of these pathogens after they enter the body. Emphasis is placed on the interaction between the horse and its environment rather than considering the two as independent entities. This approach requires some consideration of basic equine physiology and immunology. It is also important in establishing environmental criteria for the horse's respiratory well-being that other systems are considered and the horse is appreciated as being more than a set of lungs.

The chapter is divided into two main sections. The first section treats the determinants of stable environment and respiratory disease in the horse; the second considers the practical aspects of controlling respiratory disease.

Before beginning the first section the demands of respiratory function in the horse will be considered. These demands highlight the importance of respiratory disease and its prevention for the horse.

THE DEMANDS OF RESPIRATORY FUNCTION

At rest the average 550 kg horse has a tidal volume of approximately 5 litres, respiratory rate of 12 per min, a minute volume of 60 litres, heart rate of 32 per min and an oxygen uptake of 1.8 litres per min. At the gallop the tidal volume increases 2.6-fold, respiratory rate 10-fold, minute volume 26-fold, heart rate 5.6-fold and oxygen uptake 33-fold (Hornicke et al., 1983). Respiratory and locomotor cycles are locked into a 1:1 phase in the cantering and galloping horse, i.e. with one step phase there is one respiratory cycle (Attenburrow, 1983). The galloping horse has less than half a second to complete one respiratory cycle. Although tidal volumes rarely exceed one third of the horse's vital capacity, maximum flow rates of above 60 litres per sec occur with the galloping horse.

Trotting horses by comparison have a 1:1 synchronisation of respiration and stride frequency at submaximal exercise but at maximal exercise change over to a 1:3, 1:2 or 1:1.5, 1:1.25 relationship which allows deeper tidal volumes of 20 to 25 litres. Karlsen and Nadaljak (1965, cited by Hornicke et al., 1983) reported oxygen uptake of 64.2 litres per min in trotters at a velocity of 710 m per min with pulmonary ventilation of 1211 litres per min and a respiratory frequency of 68 per min. The galloping horse thus uses quicker, shallower breaths than the trotter. However, both need to move large volumes of air efficiently in short periods of time to maintain maximum performance potentials.

Minor levels of subclinical airway disease may be associated with poor performance in thoroughbred racehorses. One such condition is associated with lower respiratory tract inflammation and may be diagnosed on endoscopy and microscopic examination of tracheal washes (Burrell, 1985). Using the tracheal stethoscope developed by Attenburrow (1976) it is possible to demonstrate that the normal inspiratory/expiratory ratio of the galloping horse is lost with lower airway disease. With less severe levels of inflammation there is a lengthening of the expiratory time with the one-to-one respiratory cycle/locomotor cycle being maintained. However, with more severe changes as the horses start to show clinical disease, the respiratory cycle may impinge into the next locomotor cycle so that there is a 1:2 ratio or with the most severe cases a 1:3 ratio.

Minor degrees of respiratory disease are unlikely to be of immediate significance to the pony used for hacking. With increased severity of disease light exercise becomes distressing and with the worst cases even a comfortable retirement is not possible. The diagnosis and prevention of mild forms of respiratory disease is therefore of importance not only to the equine athlete. These forms of disease are most likely precursors of more severe degrees of disease which develop later in life (Viel, 1985; Clarke, 1987).

Respiratory infections which may be life-threatening in the extreme will also affect horses' performances if they compete with less severe forms of the infection. Exertion while suffering a respiratory infection may also be associated with the development of myocardial lesions (Steel, 1963; Stewart *et al.*, 1983). Horses suffering short-term bacterial and viral infections are also more prone to develop long-term hypersensitivity airway disease such as chronic pulmonary disease (CPD) (Gerber, 1973). Extra attention to the stable environment is therefore warranted in the face of outbreaks of infectious respiratory disease.

The demands of respiratory function are not without their "costs" to the horse. Huston *et al.* (1986) showed that training and strenuous exercise have a deleterious effect on alveolar macrophage function. This was associated with increased cortisol levels within the airways during strenuous exercise. Liggitt *et al.* (1985) showed that the stress of transport affects pulmonary macrophage function. These changes may be important in explaining increased incidences of respiratory disease in horses exposed to these conditions. Horses involved in athletic competition suffer exercise induced pulmonary haemorrhage (EIPH) (Clarke, 1985). This haemorrhage originates from rupture of capillaries in the dorsocaudal area of the diaphragmatic lung lobes. This makes an ideal site for secondary infections to develop.

DETERMINANTS OF STABLE ENVIRONMENT AND EQUINE RESPIRATORY DISEASE

Physics of airborne particles

The dynamics of an airborne particle will determine:

(i) its ability to remain airborne and persist in the environment;
(ii) its site of deposition within the respiratory tract;
(iii) the most appropriate sampling technique for its collection.

A falling particle will reach a constant velocity when the resistance to particle motion is equivalent to gravitational forces on the particle. This is the terminal or settling velocity of the particle and is described by Stoke's Law:

$$Vt = \frac{(fp - ft)\ g\ D^2 P}{18\ uf} \tag{1}$$

Vt = terminal velocity, fp = density of particle, ft = density of air, g = gravitational acceleration, Dp = diameter of particle, uf = viscosity of air.

It is conventional to describe a particle in terms of its aerodynamic diameter rather than its terminal velocity. The aerodynamic diameter is the diameter of a unit-density sphere having the same terminal velocity as the particle in

Table 4.1. Terminal velocities of unit density spheres

Particle diameter (μm)	Terminal velocity (μm/sec)	Example	Time to fall 2 metres
50	70,000	Pollen grain or large fungal spore	28 sec
5	740	Aspergillus/Penicillium type spore	45 min
1	33	Actinomycete spore	16 h

Examples of unit density spheres and their terminal velocities. Small particles are capable of reaching the alveolar membrane in the lung and also remain in the air for horses to inhale for many hours unless cleared by ventilation.

question. The difference between the actual and the aerodynamic characteristics of a particle depend on its density, its size and its shape. A large flat particle will behave aerodynamically as if it is considerably smaller than its actual size. This is because of its increased buoyancy. Terminal velocity increases with the square of the particle's actual diameter (equation (1)). Table 4.1 shows the relationship between the aerodynamic diameter and ability of particles to remain airborne. 1 μm particles such as single actinomycete spores remain airborne for long periods of time.

Stoke's Law may also be used to describe the behaviour of a particle moving in an airstream such as within the respiratory tract or through an air sampler. When an obstacle such as a turbinate bone, a tracheal or bronchiole bifurcation or a microscope slide is placed in an airstream, smaller particles are able to follow the gas flow lines. The larger particles owing to their inertia are unable to change direction quickly enough to remain airborne and are thus impacted (Fig. 4.1). Long, thin fibres such as asbestos fibres line up with airstream lines. They are therefore able to deposit lower down the respiratory tract than would be expected by their actual size. Inhaled particles which are aerodynamically small enough to evade impaction in the upper respiratory tract are carried further towards the alveolus with decreasing velocity as the total cross-sectional area for air movement increases with the branching of the airways. At the alveolar level, deposition of inhaled particles occurs by sedimentation and Brownian motion.

Dynamics of inhaled aerosols

The likelihood of an aerosol challenge of a pathogen-inducing disease is dependent on:

(i) the amount of active pathogen retained in the tract at any given time;
(ii) the pathogenicity of the particle;
(iii) the susceptibility of the host.

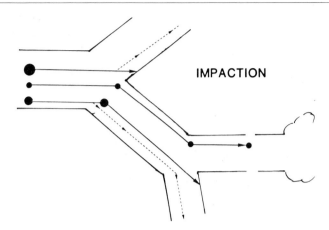

Figure 4.1. Impaction of particles in moving air streams. Smaller particles are able to follow airflow lines (broken lines) around bends. Larger particles are unable to change direction quickly enough to follow airflow lines and are then impacted.

The amount of inhaled aerosol retained in the respiratory tract depends on an equilibrium developing between the amount of aerosol deposited and the efficacy of clearance mechanisms.

Deposition of inhaled particles   Deposition describes the initial adhesion of inhaled particles which are not exhaled. The site of deposition of a respiratory pathogen need not be the site of its action. For infectious diseases deposition in the upper respiratory tract may be all that is necessary to allow colonisation at this site and entry to the host prior to systemic spread, e.g. Equine Herpes Virus Type 1 (EHV1) or herpes.

Conventionally inhaled particles are described as being either respirable (i.e. capable of reaching the alveolar membrane) or non-respirable (i.e. those deposited higher in the respiratory tract). This division is not absolute (Fig. 4.2). The deposition of particles is governed by the laws of probability. An aerosol challenge of small particles will lead to deposition throughout the upper and lower respiratory tracts. Equally, a large challenge of particles which have a small probability of reaching the lower respiratory tract will lead to some of these particles reaching the alveolus.

The deposition and clearance parameters of inhaled particles have not been described in the horse. Studies of the deposition and clearance of mono-dispersed aerosols in horses have been carried out by the author. These used radiolabelled mono-dispersed aerosols. Preliminary results from these studies would suggest that the pulmonary deposition of inhaled aerosols in the horse is similar to that of other species (Albert *et al.*, 1968). This confirms the

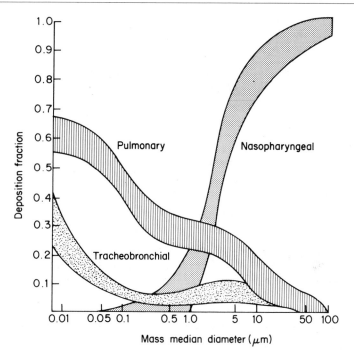

Figure 4.2. Regional depositions of inhaled aerosols as a function of the mass median aerodynamic diameter of the aerosol after Task Group on Lung Dynamics (1966).

theoretical considerations of deposition in the lungs of different species described by Stauffer (1975).

Diseases such as chronic pulmonary disease (CPD) which are associated with bronchoconstriction and narrowing of airways because of inflammation and increased presence of mucus, lead to increased deposition of particles in the central airways away from the more peripheral airways. Depending on disease status, clearance rates (due primarily to mucociliary clearance) of particles deposited in the airways are consistent with individuals but vary considerably between individual horses. This phenomenon has been observed in other species and may be related to individual variations in susceptibilities to disease. The horse does, however, appear to have a potential to increase mucociliary clearance in the face of a dust challenge. The author has observed clearance rate constants up to 2.5 per h in response to an acute exposure to mouldy hay. This level is higher than those reported in other species but may be in part related to the large size of the lungs in horses compared with those in other species.

Factors other than the aerodynamic diameters of particles regulate their deposition within the respiratory tract including air flow rates, tidal volumes, respiratory pauses and regional variations of lung function (Brain and Valberg, 1979). A horse returns from work with tidal volumes and respiratory flow rates above those seen at rest. The arytenoids are abducted until the respiratory rate decreases following exercise. The horse is subject to many variations of respiratory function which ensure thorough distribution of inhaled aerosols. In this context it is interesting to note that sand has been observed in the trachea of trotting horses after racing on a sand-based track (Speirs *et al.*, 1982).

Clearance mechanisms   The respiratory tract possesses a variety of clearance mechanisms whereby deposited particles may be physically removed or at least have their pathogenic potential neutralised. Particles deposited in the airways may be removed via mucociliary clearance, via mucus, or if they are highly soluble they may pass to the bronchial circulation and lymphatics. Clearance of particles at the alveolar level may be via the alveolar macrophages or, depending on the endothelial and epithelial permeability of the contaminant, by absorption. The equine respiratory tract also possesses a local mucosal immune system (Mair, 1985). This system is made up of extensive mucosal lymphoid tissue, a mucosal mast-cell population throughout the tract, immunoglobulin-producing mucosal plasma cells, locally produced free immunoglobulins within the tract and a population of free cells within the tract including lymphocytes and macrophages.

The importance of local as distinct from systemic immunity has been highlighted in studies on Equine Influenza Virus (Rouse & Ditchfield, 1970) and *Streptococcus equi* (Strangles) infections in horses (Galan and Timoney, 1985; Timoney and Eggers, 1985). Following infection with *Streptococcus equi* infection there is (i) a local production of IgA and IgG in nasopharyngeal mucus, and (ii) no association between levels of bactericidal antibody in serum and resistance to reinfection. These observations are of critical importance to the future approaches to vaccination against respiratory disease and antigen presentation.

Retention   Retention describes the amount of "aerosol" present in the respiratory tract at any one time. The likelihood of a pathogen-inducing disease increases as the amount of "active" pathogen present increases or the resistance of the host decreases. Absolute or relative increases in the amounts of active pathogen may occur as a result of:

(i) increased concentrations of inhaled pollutants;
(ii) decreased efficiency of respiratory defence mechanisms;
(iii) increased susceptibility of the host.

The effect that an alteration of clearance can have on retained pathogen is presented in Fig. 4.3. Assuming that a normally efficient pulmonary

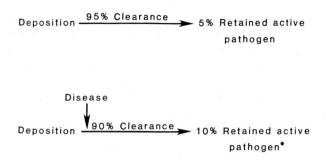

Figure 4.3. Alterations to pulmonary clearance mechanisms markedly increase the levels of retained active pathogen within the lung.* A 5% decrease in clearance leads to a doubling of retained pathogen in this example (Clarke, 1986).

"clearance" removes 95% of an inhaled pathogen, 5% of active pathogen remains in the lung. If "clearance" is now decreased by only 5%, this leads to an effective doubling of retained pathogen.

The development of chronic pulmonary disease (CPD) as a sequel to bacterial and viral infections may result from:

(i) decreased mucociliary clearance and increased retention of mould spores;
(ii) damaged epithelial surfaces allowing greater antigen penetration, or
(iii) altered immune responses.

These are conditions which have all been associated with infectious respiratory disease in the horse (Bryans, 1981a, b; Allen and Bryans, 1985). The stress of transporting horses affects pulmonary macrophage function (Huston *et al.*, 1986) and may partly explain the aetiology of pleuropneumonia in horses which is often associated with transport as well as bacterial and viral infections and strenuous exercise (Sweeney *et al.*, 1985).

Respiratory pathogens

Stabled horses are exposed to a wide range of respiratory pathogens. These include fungal and actinogmycete spore clouds, bacterial and viral aerosols, noxious gases such as NH_3, plant material and dust mites (Clarke, 1987). These contaminants may be pathogenic in a number of ways. They may

(i) cause infection,
(ii) be allergenic,
(iii) behave as primary irritants,
(iv) destroy tissue (e.g. parasitic migration),
(v) simply "clog" or impede the respiratory defence mechanisms so increasing the system's susceptibility to other pathogens,
(vi) act as toxins (e.g. 3-methylindole).

These mechanisms of pathogenicity are not mutually exclusive for any given contaminant. Spores of the fungus *Aspergillus fumigatus* are capable of causing both lung infections and allergic respiratory disease (Austwick, 1966). Irritant enzymes are released from the spores on contact with moist mucosal surfaces. The small size of the spores ensures that on inhalation a high proportion will reach the alveolar membrane and require removal by pulmonary macrophages potentially diverting the activity of the latter away from other respiratory pathogens, such as bacteria and viruses.

Viability of an inhaled pathogen is a prerequisite for an infection to develop. However, the viability of a contaminant is not necessary for allergenic, irritant or "clogging" reactions to occur.

There is considerable research at present into the conditions which affect the viability of aerosols of bacteria and viruses (Donaldson, 1978). This research has potential in alleviating the spread of infectious disease by aerosol dispersal. However, these studies involve (i) nebulising the micro-organism thus subjecting it to unnatural "stress", (ii) further "stressing" the micro-organisms with impingers which collect aerosol at high air flow rates, (iii) assessing viability on artificial growth media or cell culture after collection. While micro-organisms may not be "viable" using such culture techniques they may remain viable and infective if allowed to deposit in their ideal culturing surface, i.e. the respiratory tract. "Non-viability" in the laboratory does not necessarily equate to loss of infectivity in the animal. The trends in the laboratory can almost certainly be equated to survival trends of airborne microbes in the field. However, further research is necessary into the viability and virulence of microbes aerosolised in respiratory tract secretions before the practical implications of the laboratory findings to date become clear. The significance of the airborne survival of bacterial and viral aerosols is discussed further in the sources and sinks sections and practicalities of control sections.

The virulence of the virus or the bacterium is also of importance. The minimal infective dose may be quite small so that if, for example, 1000 particles are aerosolised and 99% die because of environmental conditions, the 10 potential pathogens remaining viable may still be sufficient to induce disease. Infectious respiratory pathogens may be spread by routes other than aerosol transmission.

Noxious gases may either behave as irritants, e.g. ammonia (NH_3) and hydrogen sulphide (H_2S), or asphyxiants, e.g. methane and carbon dioxide (CO_2). While the levels of these gases are in general not as high in stables as in the more intensive livestock housing, significant levels of NH_3 may occur in poorly ventilated stables with deep litter bedding and poor drainage.

The pathogenicity of one contaminant may be exacerbated by another contaminant. These inter-relationships need not involve the same pathogenic potentials of the contaminants involved. Perhaps the best known of these synergistic actions involves the close association between respiratory tract infections and the subsequent development of hypersensitivity airway disease. This association is highlighted at several points in this chapter.

Lungworm (*Dictyocaulus arnfieldi*) is frequently implicated with chronic respiratory disease in horses (Round, 1976). There are also several gastrointestinal parasites which migrate through the lungs during their normal life cycles producing respiratory disease, especially in young animals (Clayton, 1981). These include *Parascaris equorum*, *Strongyloides westeri* and *Strongylus vulgaris* (Duncan, 1985; Turk and Klei, 1984). *P. equorum* has also been implicated in hypersensitivity airway disease in the horse (Mirbahar and Eyre, 1986).

A possible role of ingested toxins in the aetiology of equine respiratory disease has been suggested by Breeze *et al.* (1984). These workers induced obstructive pulmonary disease in horses by oral administration of 3-methyl-indole. This would appear to be a condition limited to horses stabled in research institutions. Side effects of pharmacological agents such as anti-inflammatory drugs could potentially have a role in the aetiology of equine respiratory disease. While they may not be directly involved, altered immune responses with these drugs could potentially decrease a horse's resistance to respiratory disease (Lees and Higgins, 1985).

Kinetics of airborne contaminants

The airborne concentration (C) and therefore the respiratory challenge of a pathogen is proportional to the rates of its release (R) into, and rates of its clearance (q) from, the atmosphere (Fig. 4.4).

Under conditions of constant release of pathogens and constant clearance rate an equilibrium concentration of the pollutant will be established and may be described by the formula

$$R = Cq \tag{2}$$

R = release rate (particles/cc/min)
C = concentration
q = clearance

There are several components which make up the clearance of airborne pollutants from stables (Fig. 4.4). These include:

(i) ventilation (qv): natural ventilation is often overlooked with horse housing;

(ii) sedimentation and impaction (qs): leads to deposition of particles at a rate proportional to their aerodynamic diameter. For many of the respiratory pathogens, e.g. actinomycetes and small fungi, this is not a significant clearance mechanism;

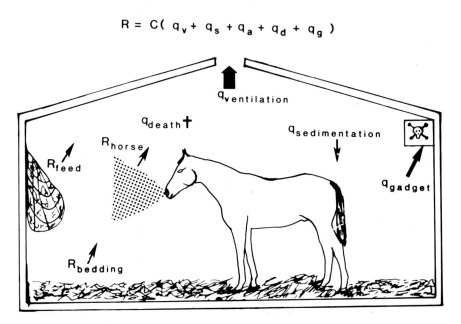

Balance of airborne contaminants

$$R = C(q_v + q_s + q_a + q_d + q_g)$$

Figure 4.4. The airborne concentration (C) of a pathogen is proportional to the rate of its release (R) into and rate of its clearance (q) from the air (Clarke, 1986).

(iii) the breathing animal (qa): this is unlikely to be a significant clearance mechanism in well ventilated boxes at the low stocking densities associated with horse housing;

(iv) death of the pathogen while airborne (qd): this will decrease the respiratory challenge of potential infectious pathogens;

(v) gadgetry including air filters, ionisers (qg).

All of the clearance mechanisms can be expressed in terms of air changes per hour. The importance of the different clearance mechanisms may vary with each pathogen (see next section).

The equilibrium state described by equation (2) is based on constant and continuous release rates and illustrates the relationship between the airborne concentration of pollutants and rates of release and clearance. However, such a situation does not exist in the real world of the stable where release rates of contaminants vary at different times of the day. Figure 4.5 shows the relationship between stable activity and dust levels over a 24-hour period.

During quieter times of the day, equilibrium concentrations do develop. An example of this is seen in Fig. 4.5; the effect of closing the barn door

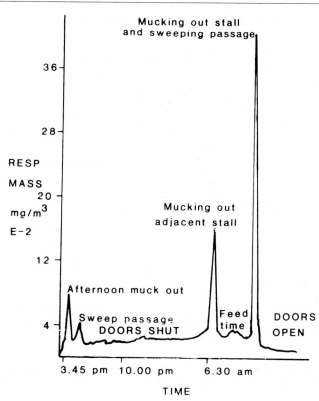

Figure 4.5. Respirable mass over a 24-hour period in one of 16 boxes in a barn. The main peaks are associated with mucking out. The effects of ventilation (doors open vs doors shut) on steady state levels of respirable dust in the barn are highlighted in this figure (Clarke, 1986).

is to increase the concentration of airborne dust over that when the barn door is left open. While equilibrium concentrations describe the respirable challenges in steady release it is important to evaluate environments in terms of peak release rates of contaminants. In the case of respirable fungal spores a well ventilated box with a heavily contaminated bedding may have an airborne concentration in the centre of the stable which is small. Thus potential respirable challenges may be much greater than airborne concentration at a given time suggests. True indications of levels of contamination of bedding can only be gained with readings taken at mucking out. The horse is also inquisitive—it will sniff and nuzzle at bedding.

In theory, removing the horse from the stable during periods of peak activity would help decrease exposures to dusts. Ventilation will also undoubtedly decrease the overall challenge of airborne particles presented to the horse by a contaminated bedding. However, whether this reduction is sufficient to prevent respiratory disease is another matter. The importance of assessing the air hygiene of buildings in terms of concentrations of contaminants that the occupant is likely to actually inhale rather than airborne concentrations in the centre of the box will be returned to later in this chapter.

The sources and sinks of airborne contaminants

The airborne contaminants of stables arise from many sources and their clearance from the air has several potential pathways. Airborne contaminants may therefore be considered in terms of their sources for release into the air and sinks (clearance pathways) out of the air.

Stable dusts—fungal and actinomycete spores   The main source of respirable dust in stables is the horse's feed and bedding. High respirable challenges of dust are usually associated with either or both of these sources being moulded (Clarke, 1987). Occasionally fungi may grow on the walls of stables. This occurs especially with the seepage of water through walls as a result of damaged or poorly designed drains.

Ventilation is the main clearance mechanism for fungal and actinomycete spores. The relationship between fungal contamination of source materials in stables and ventilation as they relate to numbers of airborne particles are presented in Fig. 4.6. The small aerodynamic diameter of mould spores ensures that sedimentation (qs) is of minor importance. The viability of a mould spore is not a prerequisite for its pathogenicity. Death of the spore while airborne will remove its infectious potential. However, its allergenic, irritant and "clogging" potentials will be retained.

Bacteria, viruses, mycoplasma, chlamydia, etc.   The main sources of these airborne respiratory pathogens are the horses themselves. The coughing and sneezing horse releases infectious aerosols into the air. Infectious organisms also originate from sources other than horses. Secondary invaders such as *Escherichia coli* and primary pathogens such as *Rhodococcus equi* may originate from bedding. This latter organism is associated with severe pneumonia in foals (Smith, 1982). It is basically a soil saprophyte which requires a warm climate in order to proliferate in soil and manure. It apparently invades the lung on small dust particles, though ingestion is a further possible portal of entry to the body. The disease is unknown in mild climates such as the United Kingdom and its incidence may be reduced in contaminated areas by damping down dust. This bacteria is very resistant to desiccation, ultra violet light and many chemical disinfectants (Hillidge, 1986). Once aerosolised the main sink for dust clouds of *R. equi* is ventilation.

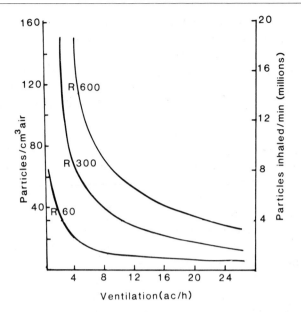

Figure 4.6. The effects of release rate (R, particles/cm³ air per hour) and ventilation rate (qv, ac/h) on the concentration of particles in stable air and inhaled by horses.

Aerosols of non-infectious bacteria are also present in animal houses. They usually arise from the animal's bedding and its coat (Wathes *et al.*, 1983). While they are not infectious, high challenges may compromise the respiratory tract defence mechanisms. Airborne concentrations of these bacteria in reasonably well ventilated stables with clean, well managed beddings will be lower than with other livestock housing because of the different stocking densities.

Pathogens of systems other than the respiratory tract may be spread via the airborne route. Examples of such pathogens include Salmonella and *E. coli* (Curtis, 1983).

The survival of aerosolised viruses and bacteria may only be a matter of seconds. Clearance rates of up to 1000 per h which are far in excess of those possible by ventilation have been recorded (Donaldson, 1978). The viability of aerosolised viruses is especially sensitive to relative humidity. Viruses with a lipoprotein envelope tend to survive best at low relative humidities, whereas non-enveloped viruses tend to be unstable in dry conditions but survive best at high relative humidity. The survival of Equine Rhinovirus 1 is improved with increased relative humidity, whereas Equine Herpes Virus 1 and Equine Arteritis Virus have decreased survival rates with increased relative humidities.

There is usually an increased death rate of airborne bacteria associated with increased temperatures. Haemolytic streptococci, for example, survive poorly in dry air but are relatively stable with high humidity.

Thus, with aerosols of infectious bacteria and viruses clearance by death of the pathogen while airborne may be potentially a lot greater than clearance by ventilation. However, the practical implications of these observations may be limited. In attempting to alter environmental factors to the detriment of one pathogen the airborne survival of another could be enhanced. Moreover, in most stables relative humidity and temperature are going to be mainly dependent on ambient conditions. The long-term control of humidity and temperature in stables would be technically difficult and highly expensive requiring completely closed mechanically operated housing systems. Also, in a practical sense high humidity has other inherent problems such as condensation and moulding of plant-based beddings in situ. Even if optimal atmospheric conditions could be set to decrease the airborne survival of these pathogens, it would not eliminate transmission by means other than aerosol dispersion.

Noxious gases   The most common noxious gases found in animal houses are ammonia, hydrogen sulphide, methane and carbon dioxide (Nordstrom and McQuitty, 1976). These are generally only seen in high levels with heavy stocking density and especially with slurry pits. The most common noxious gases in stables are ammonia and hydrogen sulphide and their main source is bedding, especially where deep-litter management is practised. The only clearance mechanism for noxious gases is ventilation (qv). Proper ventilation will help prevent the noxious gas production in the first place.

Environmental criteria for horse housing

There are many criteria against which the stable environment can be evaluated. These include the control of air temperature and air speed at horse height, the control of relative humidity and prevention of condensation and the maintenance of tolerable concentrations of gas, dust and airborne micro-organism.

These factors must be assessed for each housing situation and the primary limiting factors should be given highest priority. Ultimately this will be judged on the well-being and performance of the horses being housed.

Thermal environment   There are two ways in which environmental temperature may affect the respiratory well-being of the horse. First, the survival of airborne microbes and development of moulds in beddings can be temperature dependent. Second, cold or heat stress may affect the systemic or local resistance of the horse to disease.

Horses have a wide thermal tolerance, with an ideal temperature range being between 10–30°C. However, with acclimatisation temperatures as low as − 10°C may be tolerated (McBride *et al.*, 1983). In an unheated building with low air movement the only horses likely to experience cold stress are newborn foals or foals whose metabolic rate is low because of disease or malnutrition. These animals can undoubtedly be stressed by cold in unheated buildings, and an extra source of heat, such as radiant heat lamp, is necessary in such situations. Increasing air movement significantly increases the critical lower temperature for the newborn foal. Mature horses are unlikely to experience a systemic stress of cold in a dry stable unless air movement is exceptionally high. However, over the centuries, a short coat has proved to be aesthetically more pleasing than a long coat, and a short coat dries more easily than a long coat after exercise. Clipping the coat will aid in overcoming these problems. However, this necessitates minimising exposure to cold and provision of rugs. While rugging will help minimise the effects of cold another stimulus to the "breaking" of the summer coat, decreasing daylight hours, may be partially overcome by providing artificial light. This technique is used to stimulate the normal reproductive cycling of mares early in the stud season.

It is important to ensure that adequate ventilation is achieved without excess air movement (draughts) at horse (or foal) height. Basically, a draught is too much ventilation in the wrong place causing enhanced heat loss and chilling in cold weather. A draught may arbitrarily be defined as a wind speed of > 0.40 metres per second at animal height. The arbitrary nature of this definition is appreciated if we consider that the chilling effect of a wind blowing on a neonate is considerably less than that blowing onto an adult horse wearing a rug. Draughts may be prevented with the correct design and positioning of inlets (see section on ventilation).

Flooring and bedding materials affect heat loss by conduction from individual animals. The thermal properties of flooring for livestock have been described by Bruce (1979) using a heat loss simulator. The conductive losses were greatest through concrete floors. A deep bedding will help decrease these conductive heat losses by providing insulation as well as minimising air movement at ground height and thus decreasing convective heat losses as well.

There is no experimental evidence known to this author which associates thermal environment with changes to local respiratory tract resistance to disease in the horse. The horse certainly has a very efficient turbinate system which is capable of humidifying and heating air on inhalation (Clarke, 1985). Extrapolating results from other species would indicate that extremes of temperatures would be necessary for such changes to occur.

Relative humidity and condensation   Condensation occurs when a cool surface is surrounded with warm moist air (Wathes, 1981). Condensation may also occur within a structure depending on diffusion of water vapour through the material. When the latter occurs it is termed "interstitial condensation". Whether condensation occurs on the surface or within a structure depends on the layer at which the ambient temperature is equal to or less than dewpoint temperature. If the temperature at which condensation occurs is <0°C hoar frost will form. As the temperature rises this frost will thaw and water will appear on the surface of the structure. Condensation can stimulate fungal growth on walls and beddings. Interstitital condensation may cause wet rot in roof timbers, and where hoar frost occurs this will damage insulation materials and roof timbers and also cause increased heat loss from the building. A vapour check provides a useful barrier to prevent interstitial condensation within insulation. This may be made most simply from 0.1 mm of polyethylene sheet which is placed on the warm side of the insulation. The vapour check must have a good air seal.

Pattern staining on ceilings provides evidence of condensation (either surface or interstitial) but may also occur with rain entering through a damaged roof. There should be no delay in the investigation of any such staining which occurs around light fittings in stables. Four metal shoes ensure effective conductivity for electricity.

Condensation is more likely to occur in intensively housed agricultural livestock than in well ventilated stables. This is because the higher stocking densities are associated with increased vapour release. However, under conditions of clammy, foggy weather and sudden weather changes (e.g. autumn day) condensation in stables may be unavoidable. This is because walls warming at a slower rate than the surrounding air serve as a surface for condensation to occur. This will typically occur early in the morning and condensation as well as occurring in loose-boxes will occur in other out-buildings where there are no horses. This latter observation associated with the appropriate weather changes helps to differentiate condensation associated with climatic changes to those which occur because of overstocking and poor ventilation. Insulation, painting with condensation resistant paint and adequate but not excessive ventilation will help to limit condensation. The practical benefits of insulation are discussed further in the section dealing with ventilation.

The effect of changes in relative humidity on the respiratory well-being of the horse have not been studied. The effects of relative humidity on the survival of aerosols of infectious bacteria and viruses varies and has been discussed. The effects of relative humidity on a horse's thermoregulatory capability is likely to be small in all but the most humid and hot climates (e.g. 80% relative humidity, temperature >30°C).

Noxious gases, respirable dusts and airborne micro-organisms The safe levels or Threshold Limiting Value (TLVs) for these airborne contaminants based on direct or indirect pathogenicity to the respiratory tract of the horse are unknown.

Ammonia, hydrogen sulphide, methane and carbon dioxide are the most common noxious gases found in animal houses. Of these ammonia is the most common in stables in significant levels. The absence of slurry pits and low stocking-density and air space volumes of stables preclude problems with methane and carbon dioxide respectively. While TLVs for noxious gases in horses are unknown, guidelines may be taken from human TLVs based on 8-hour, 5-day-a-week exposures. Current safe limits for ammonia and hydrogen sulphide are set at 20 and 5 ppm respectively (American Conference of Government Industrial Hygienists, 1972). However, it must be emphasised that even minor degrees of respiratory tract disease which could be induced by lower levels of noxious gases are potentially significant to the horse. Draeger tubes (Draegar, Chesham) offer a simple technique for those working in the field to assess noxious gas levels in animal housing (Fig. 4.7). Where concentrations above those described earlier are recorded, remedial action is necessary. In stables this usually involves attention to drainage and the management of bedding.

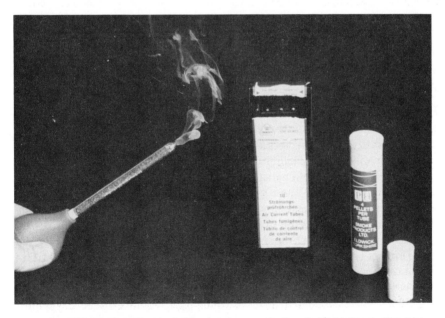

Figure 4.7. Smoke tubes (Draeger, Chesham) or smoke bombs (P. H. Smoke Products, Bingley) are useful for assessing ventilation and air flow patterns within buildings.

The current TLV value for human exposure to dust is $10\,mg/m^3$ even though levels of $5\,mg/m^3$ have been associated with serious effects on the respiratory function of grain elevator workers (Emarson *et al.*, 1985). In horses a 1-hour exposure to the dust shaken off mouldy hay and straw has been reported to induce symptoms of CPD in previously asymptomatic cases (McPherson *et al.*, 1979). 12 mg of *Aspergillus fumigatus* antigen or 12 mg of *Micropolyspora faeni* antigen administered via a nebuliser will also induce CPD. These represent challenges which actually induce disease; they are not threshold limiting values. TLVs for respirable dust for the horse, though yet to be established, are likely to be considerably less than those for other species, including those apparently "acceptable" for man. These TLVs are also likely to alter considerably within and between individuals. Infectious respiratory diseases, for example, lower the horse's TLVs to mould spores.

Levels of airborne bacteria and viruses which cause an infection to develop in horses which have not been challenged by other routes are unknown. The highly contagious nature of most respiratory tract infections and their ability to be transmitted by routes other than aerosol dispersion makes their environmental control very difficult.

Logically respiratory challenges should be minimised at all times. This is especially so since with most airborne pollutants a graded response to various levels of challenge is likely. There are basically two ways to reduce concentrations of airborne pollutants in stables:

(i) increase clearance rates;
(ii) modify sources materials and release rates

Ventilation

Ventilation plays an integral role in the control of environmental factors of animal housing. Environmental temperature and levels of water vapour, noxious gases, dust and microbes are all influenced by ventilation. The provision of fresh air will aid in the killing of airborne microbes by the "open air factor" (Druett, 1973). Incompatabilities may exist in the ventilation rates necessary to meet more than one of these criteria. Optimal conditions for the efficient conversion of food to meat may require maintenance of higher environmental temperatures at the expense of some degree of lung damage associated with decreased ventilation (Sainsbury and Sainsbury, 1979). Horses are not expected to be efficient meat producers and so the ventilation of stables may in the first instance concentrate on maintaining "fresh air" for the horse's respiratory well-being. The low stocking densities of horse housing dictates that in extreme conditions where debilitated horses or young foals require extra heat, this must be introduced into the stable from external sources (e.g. heat lamp) rather than relying on closing the building up and using the heat generated from other horses.

It is most important that ventilation is not used to try to overcome inherent management problems of stables. A horse may still inhale significant levels of contaminants such as mould spores from mouldy hays or beddings which are in well ventilated stables. Deep litter beddings may also promote the production of noxious gases such as ammonia. Some of the most significant respiratory pathogens of the horse are a result of the level of contamination of feeds and beddings which are introduced into the stable, or arise from deep littering of beddings, or improper drainage. These problems can be avoided with careful selection of feeds and beddings, avoidance of deep litter beddings and ensuring proper drainage. The consequence of the high stocking densities used with livestock other than horses is that airborne contaminants cannot be avoided at source, e.g. noxious gases in slurry pits. In this latter situation ventilation becomes the main controlling factor for challenges to the housed livestock. Such is not the case with horses where many of the most dangerous airborne contaminants are consequential to management practices rather than to large numbers of livestock being kept in close confinement.

The ventilation of horse housing is still of critical importance under the best management regimes. Ventilation will (i) aid in the control of condensation, (ii) lower the level of airborne contaminants which cannot be controlled by other methods (e.g. bacterial and viral aerosols generated by coughing horses) and (iii) aid in preventing the moulding of plant-based bedding materials.

Assessing the ventilation of buildings Recommendations for the ventilation of buildings are often given in terms of volumetric air movement, e.g. m^3 of air per hour. However, the clearance of contaminants from the air in a building is dependent on the number of times the air is changed in a set time, e.g. air changes per hour (ac/h). Air change rates are dependent on the size of the building and the volumetric movement of air from the building. Air change rates can be assessed from the clearance of a marker released into the building. Sulphur hexafluoride SF_6, an inert gas, or smoke particles, make ideal tracers and their clearance can be monitored using an autoanalyser or an airborne particle counter respectively (Clarke, 1986).

The theoretical requirements of inlet and outlet sizes to provide adequate ventilation of buildings is described in the next section. Computer programs are also available for such theoretical assessments (Jones *et al.*, 1987). These techniques do not, however, consider the distribution of openings in buildings; a factor which is of critical importance and discussed further later in this chapter.

When expensive analysers or particle counters are not available, smoke tubes or smoke bombs offer a simple practical technique to at least partially assess the ventilation of buildings (Fig. 4.7). They will highlight patterns of air movement within a building and can be used to assess the functional status of ducts and venting within buildings. A wind speed anemometer is another useful and relatively inexpensive device. This instrument may be used to assess air movement within a building and detect draughts and assess the efficacy of baffling and vents.

Natural forces of ventilation   There are three natural forces of ventilation:

  (i) The stack effect (rising warm air);
 (ii) aspiration (i.e. wind blowing across the roof of a building sucking air out);
(iii) perflation (i.e. air blown from side to side and end to end of the building) (Sainsbury, 1981).

The ventilation of a building is most tested in still air conditions, i.e. when the only driving force for ventilation is warm air rising off the horse. Thus, theoretical considerations of providing ventilation should be based on the assumption that windless conditions prevail. However, such conditions are rarely maintained for long periods of time. To ensure efficient ventilation of horse housing year round all natural forces of ventilation must be harnessed (Clarke, 1986).

Theoretical principles   The relationship between the levels of airborne contaminants and ventilation is curvilinear (Fig. 4.6); ventilation being directly related to the reciprocal of the concentration of airborne contaminants. Thus concentration to contaminants increases sharply at low ventilation rates. Figure 4.6 shows that with low release rates, i.e. when source materials are clean, the level of airborne contaminants do not increase sharply until ventilation falls below 4 ac/h. In considering the theoretical principles of ventilation in still air conditions a target of 4 ac/h with the top door of the loose-box or the main barn doors closed should ensure adequate ventilation all year round. This is especially the case if the doors are only closed in inclement weather.

The principles that govern the natural ventilation of livestock housing in still air conditions have been described by Bruce (1978):

$$\frac{1}{(A_{in})^2} + \frac{1}{(A_{out})^2} = \frac{K.D\Delta T}{T.Q^2} \tag{3}$$

A_{in} = inlet area (m^2)

A_{out} = outlet area (m^2)

T   = ambient temperature (°K)

ΔT   = temperature difference between inside and outside (°C)

D   = vertical distance between inlets and outlets (m)

Q   = ventilation rate m^3/sec

K   = constant = 7 m/sec^2.

Equation 3 may be simplified if it is assumed that inlet areas should have twice the area of outlets to . . .

$$(A_{in})^2 = \frac{5\,T\,Q^2}{7\,D\Delta T} \tag{4}$$

The driving force for ventilation by the stack effect is dependent on the temperature difference between the inside and outside of the building. In steady state conditions the temperature gradient (ΔT) between the inside and outside of the stable is given by

T   = Hn/(hb + hv)   (5)

Hn   = sensible heat loss from horses (watts)

hb   = building heat loss (W/°C) conducted from building
              through walls, roof and floor

hv   = heat convected from the building by ventilation (W/°C)

Heat loss from buildings is determined by the surface area of the building and the insulation of walls and roof. This is defined in terms of "U" values (W/m2.°C). A typical U value of an uninsulated building is 2.0. Once the *roof* and *walls* have been insulated this may be realistically reduced to 0.4. It is important to note that the walls and roof must be insulated to obtain maximum benefits from insulation.

Table 4.2 describes the situation with a typical individual box or barn which are either well insulated or uninsulated. The figures assume a heat loss of 800 watts per horse in cool conditions. Units are expressed on a "per horse" basis but external surface areas are calculated on the assumption that the barn or row of boxes contain 10 horses.

Table 4.2. Requirement for natural ventilation of a typical barn and horsebox

	Box	Barn
Dimensions (per horse)		
volume (m³)	50	85
surface area of building[a] (m²)	41	43
height from inlets to outlets (m)	1.0	2.0
ventilation rate at 4 ac/h (m³/sec)	0.055	0.094
ventilation heat loss at 4 ac/h (W/°C)	67	114

	Box		Barn	
	Insulated	Uninsulated	Insulated	Uninsulated
"U" value of walls and roof (W/m²°C)	0.4	0.2	0.4	2.0
Building heat loss (W/°C)	16	82	17	86
Temperature gradient (°C) at 4 ac/h (from Fig. 4.2)	8.4	5.2	6.0	4.1
Required inlet area/horse (m²)	0.27	0.34	0.38	0.46
outlet area/horse (m²)	0.14	0.17	0.19	0.23

[a] Assuming 10 horses in a row of boxes of a single barn.
Effects of dimensions, air space per animal and thermal insulation of a typical box and barn on building heat loss, ventilation heat loss and the size of air inlets and outlets required to achieve 4 air changes per hour in still air conditions (Webster et al., 1987).

Table 4.2 demonstrates the effect of insulating walls and roof on building heat loss, temperature gradient and inlet and outlet areas to maintain 4 ac/h (based on equation (4)). Figure 4.8 illustrates the relationship between ventilation and temperature rise ΔT in insulated and uninsulated buildings. It should be apparent that the main virtue of insulating horse housing is to make natural ventilation work properly rather than to keep horses warm. This is in contrast to the situation seen with intensively housed livestock where the higher stocking densities are associated with higher heat production.

The air space allowances in Table 4.2 are very generous and so for most boxes and barns these outlets will provide more than adequate ventilation.

Distribution of openings   The distribution of openings within horse housing is as critical as the size of the openings. This is especially the case with large barns. The theoretical guidelines above assume perfect mixing of air within the building and a distribution of openings which ensures that all air can move freely in and out. This is valid for well designed individual boxes and small barns housing up to 10–12 horses, but is almost impossible to achieve with large barns. Large barns used for horse housing have individual boxes, a factor which further complicates air movement within the building compared with the use of barns for other agricultural livestock, where no such partitioning exists.

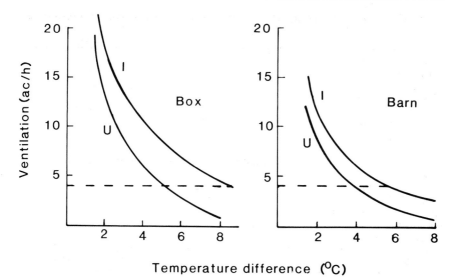

Figure 4.8. The relationship between ventilation rate (ac/h) and the temperature difference (°C) between inside and outside of insulated (I) and uninsulated (U) buildings having the dimensions and characteristics stated in Table 2 (Webster *et al.*, 1987).

Where the top doors of loose-boxes are left open the theoretical inlet areas required for natural ventilation are normally exceeded (see Table 2). However, when these doors are closed the natural ventilation of most boxes is severely limited. This is because most boxes have only one other opening besides the top door and this opening is usually inadequate in size. The lack of additional openings, such as substantial back wall vents, precludes proper mixing of air within stables, even with top doors open (Clarke, 1987; Daws, 1967). For a box with a monopitched roof there should be at least two openings besides the door; one in the front wall and one in the back. Boxes with peaked rooves should ideally have an extra opening consisting of a capped chimney in the roof. Extra openings may be placed in side walls depending on the prevailing weather conditions. Baffling of such openings will inhibit the entrance of rain or snow and decrease draughts.

The complex air movement patterns which are necessary for thorough mixing of air in large barns demands a more widespread distribution of openings than with loose-boxes. This is more easily achieved using large areas of Yorkshire boarding which ensures a thorough mixing of air. Capped ridges and "breathing rooves" will further add to the natural ventilation of big barns. A full use of all forces of natural ventilation, i.e. stack effect, perflation, and aspiration, should be made with all types of horse housing. This should ensure that the boxes and barns are well ventilated in all conditions, including situations where boxes are empty or barns are not full to capacity. Further discussion on design features and methods of altering buildings with emphasis on ventilation and draughts and on the use of natural landscape and prevailing weather conditions is presented by Clarke (1987).

Mechanical ventilation and gadgetry   Mechanical assistance of ventilation by fans may be beneficial in buildings which are excessively wide or long (Sainsbury and Sainsbury, 1979). Breakdowns or electrical failures are potential inherent weaknesses of such systems. Fans should normally only be installed where all avenues of natural ventilation which are practical have been taken.

Air filters may be used to provide an effective form of ventilation in closed buildings. The volumetric throughput of air and filtering efficiency are the limiting constraints on these machines. Air filters are most likely to be of benefit in intensive animal housing where natural ventilation is limited in an attempt to maintain temperatures and so provide efficient conversion of feed to meat or eggs (Curtis, 1983; Carpenter *et al.*, 1986). However, they are unlikely to provide benefit in stables which are well ventilated by natural means and which have clean bedding materials. If air filters are used they must be cleaned and serviced regularly. This is a chore which in many stables is often overlooked.

The present author has found no beneficial effect of ionisers on dust levels in stables. A wide range of beneficial effects of negative ions have been claimed in various species (Wehner, 1969). This ranges from stimulation of ciliary beat (Wehner, 1969) to elimination of bacterial foci on the surface of the trachea (Robinzon *et al.*, 1983).

Jensen and Curtis (1976) reported that ionisers reduced air pollutant levels in piggeries. However, pigs raised in ionised environments did not have performance differences to those held in non-ionised atmospheres. Negative and positive ions have been shown to enhance the killing of airborne bacteria (Phillips *et al.*, 1964) and inhibit the germination of fungal spores of Penicillium species (Pratt and Barnard, 1960). Studies on the effects of ionisers on the clearance of air of dust generated from mouldy hay showed no beneficial effects (Edwards *et al.*, 1985). The above findings would indicate that ionisers may have a beneficial effect in preventing the airborne spread of infectious disease primarily by decreasing the survival of airborne pathogens. However, as has already been stated, the viability of allergens is not essential for this aspect of their pathogenicity to be manifest. If ionisers have beneficial effects on the equine respiratory system, especially in relation to hypersensitivity airway disease, it is likely to lie in their intrinsic biological effects on the horse. Any such effects remain to be proven.

"Investment" in mechanical gadgetry should be approached with care and given lower priority compared with building alterations which will improve natural ventilation and management procedures which decrease levels of airborne contaminants.

PRACTICALITIES OF CONTROLLING RESPIRATORY DISEASE

The respiratory well-being of stables horses must be adjudged not only on the incidence of respiratory disease but also on the severity and duration of the episodes which occur. Respiratory diseases may be divided into the acute and the chronic (Fig. 4.9). Furthermore, most of these conditions occur in degrees ranging from those which are life-threatening, e.g. *Rhodococcus equi* infection in foals, to mild subclinical diseases such as lower respiratory tract inflammation. Mild degrees of respiratory disease may be made manifest only when the horse is put under exertion. Subclinical airway disease not only has short-term implications for athletic performance but may also be the precursor of more debilitating diseases in later life, such as chronic pulmonary disease (CPD). In approaching the control of respiratory disease, clinical and subclinical disease must be considered. Where episodes of disease occur the aim must be to minimise the duration and severity in affected individuals.

Rest is critical in minimising the effects of respiratory disease to the horse. The clinical manifestations of covert respiratory disease have been described

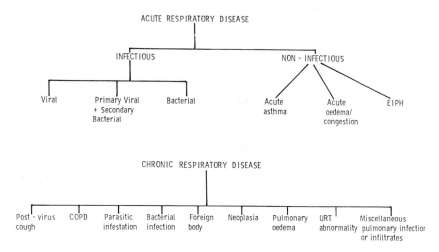

Figure 4.9. Some of the more common acute and chronic respiratory conditions of the horse.

by Greet (1985). However, many bouts of subclinical respiratory disease, especially infections, go unnoticed. Daily taking of temperature should ensure that pyrexic horses are not worked. Other signs of covert respiratory disease include poor performance, lower food intake, and gulping during exercise. This gulping is associated with the swallowing of mucus brought up from the trachea during exercise.

Infectious respiratory disease

Infectious respiratory disease may be transmitted via

  (i)   airborne diffusions of aerosols generated ruing normal breathing, sneezing or coughing;
  (ii) short-range direct transmission in droplet form following coughing and sneezing;
  (iii) direct contact between animals, i.e. nose to nose;
  (iv) indirect contact between animals, e.g. via handlers, harness, grooming equipment, horse boxes, picking at grass previously contaminated by an infected horse or carrier, etc.

When an aerosol of infectious bacteria or virus is released by a coughing or sneezing horse, the likelihood of a horse in a shared air space or down wind of the infected horse becoming infected will depend on

  (i)   the survival of the pathogen while airborne;
  (ii) the pathogenicity of the micro-organism; and
  (iii) the susceptibility of the horse being exposed.

Basically environmental control alone is unlikely to greatly influence the spread of highly contagious respiratory pathogens within a population of *susceptible* individuals. This is especially the case where the horses are themselves the main source of the pathogen which in turn cannot exist outside the host they infect for long periods of time, e.g. herpes, influenza and strangles. Environmental control is likely to be successful with diseases where the agents survive, and possibly even replicate, outside the host. *Rhodococcus equi* infection of foals is such a disease. *R. equi* is an actinomycete which proliferates in soil and faeces. The disease is more prevalent in dusty areas, presumably because of the requirement of particulate matter to transport the particles into the lungs. By constant removal of faeces and damping down dusty areas in paddocks or avoiding such areas with foals, the incidence of this disease may be markedly decreased in endemic areas.

Immunity Clearly the avenue providing the greatest potential for the successful control of equine infectious respiratory disease is the alteration of the immunological status of the horse. Immunity, either natural, as the result of a field infection, or induced via vaccination, provides the main defence of all animals against infectious disease. Unfortunately, with infectious respiratory diseases the immunity to natural infection is not necessarily long-lasting. The same is true of the immunity induced by many of the currently available vaccines which are not necessarily effective nor long-lasting. In diseases where a systemic spread of the agent is involved natural immunity is generally strong and long-lasting. However, where disease results from damage to tissues which the pathogen first encounters (i.e. respiratory tract epithelium) and especially where short incubation periods are involved, immunity is not usually long-lasting.

These features are exemplified by Equine Herpes Virus Type 1. Following respiratory disease associated with infection caused by subtype 1 and 2 the resistance of the horse to another bout of similar disease is short-lived. A horse may become reinfected and again shed virus into the environment within 3–4 months of recovery from an initial bout of respiratory disease. However, the length of resistance to abortion associated with subtype 1 infection is much longer compared with that to respiratory disease caused by the same organism (Allen and Bryans, 1986).

Present research into vaccines is being aimed at the long immunity associated with viruses which have systemic spread and long incubation periods. This work involves cloning genes which specify antigenic fragments of respiratory pathogens such as influenza virus into vaccinia virus in an attempt to produce effective and enduring immunity to the former (Panicali et al., 1983). The use of live though attenuated vaccinia virus for such uses is not without hazards (Anon., 1986). Another main area of development of vaccines for the equine viral diseases is into the use of immune stimulating complexes (Iscoms). These complexes are formed by a detergent treatment of virus and subsequent addition of an adjuvant (Quill A) and are particularly immunogenic compared with viral glycoproteins (Russell and Edington, 1985).

The ability of respiratory pathogens to remain latent in carrier animals for long periods of time further complicates the control of infectious respiratory disease. Chronic carrier states have been identified with many infectious agents, including EHV (Allen and Bryans, 1986) and strangles (Reif et al., 1981). The genome associated with the latent state of herpes virus is resistant to both chemotherapy and immunotherapy (Allen and Bryans, 1986). This ability to remain latent in the host for months or years will make total disease eradication highly unlikely and complicate programmes for the control of herpes infections by vaccination.

Vaccines already play a major role in the control of infectious respiratory disease, e.g. influenza. The development of new more effective vaccines will undoubtedly lower the overall incidence of infectious respiratory disease in horses. However, the problems of latency and the ability to mutate is likely to keep viruses one step ahead of researchers for many years to come.

The nature of the horse immune response to EHV-respiratory tract infections and the presence of carriers is the basis for one of the most frustrating features of viral infections to horse owners, trainers and veterinarians. This is the occurrence of insidious "rumbling" respiratory problems within groups of horses. The typical picture is that of a yard of horses with a low incidence of respiratory disease affecting individual horses randomly and intermittently. The condition is not life-threatening but disruptive to training regimes and planning participation in competitive events and requires close scrutiny to temperature-taking to ensure that pyrexic horses are not worked hard.

Control programmes for infectious disease   Control programmes should aim at:

(i)   decreasing the incidence of disease, and if disease occurs;
(ii) contain its spread;
(iii) limit its duration and prevent secondary complications.

The approaches which are taken will depend on the types of ages of horses involved and the diseases involved. Young foals exemplify this situation. They are more easily cold stressed than older horses so provision for additional heating is appropriate in considering the stable environment for foals. Colostrum intake within the first hours of life is also of critical importance to the foals in protecting them against all types of infectious disease. Once the foals reach weaning age weaning should be carried out with minimal stress. Where groups of mares and foals are involved this may be done by gradually removing mares over several days.

Isolation   It is possible to keep groups of breeding mares and young horses reasonably well isolated from the "problems" of the outside world (Bryans, 1981a,b). This includes not only isolation from horses but also potential carriers, including lorries, cars, personnel from other studs, etc. In fact, on farms or studs where EHV1 abortion is likely to occur, young horses and mares are ideally kept well isolated from each other. However, horses which are used for competition cannot be isolated. Isolation and quarantine are unlikely to be successful in controlling diseases which are pandemic or endemic within a population of horses. EHV1 is an example of such a disease. However, for diseases which are not endemic within a population and occur as isolated epidemics, isolation and quarantine should be instigated quickly, e.g. Equine Viral Arteritis (Mumford, 1985).

Quarantine If suitable stabling and staffing is available a mandatory quarantine period of 4 weeks for new horses entering a stable is good practice. This time period should be adequate to overcome recrudescence of latent viruses in horses which have gone through the "stress" of the sale ring or transport (Allen and Bryans, 1986). Where large numbers of horses are moved into a stable, e.g. yearlings moving into a racing stable from the sales, a type of reverse quarantine may be practised. This involves keeping a small number of horses whose continued performance is considered essential away from the main body of horses. This should lessen the chance of important competitors having to be laid up at critical times. A mandatory quarantine period should be practised on studs especially where EHV1 abortion could be a problem.

"Stable dust" and infectious respiratory disease The importance of minimising the challenges of respirable dusts will be more fully discussed in the section on hypersensitivity respiratory diseases. Suffice it to say at present that the presence of high dust burdens can increase the duration of respiratory disease associated with infection. Such challenges are also likely to lower the horse's resistance to infections in the first place. In general terms, stables should be well ventilated, individuality of air spaces should be maintained and food and bedding or management systems which are likely to be associated with increased challenges of dusts to horses should be avoided.

Disinfection Disinfection involves the destruction of viable micro-organisms and parasites that may infect animals when they are not associated with their hosts (Curtis, 1983). There are three modes of disinfection; these are mechanical, physical and chemical.

Successful disinfection of a housing area will require mechanical cleaning at the very least. This is because the presence of organic material, such as bedding, faeces, pus and other discharges, will interfere with chemical and physical disinfection. After bedding material, etc. has been removed, hosing or scrubbing of surfaces especially with the addition of a detergent or caustic soda will further decrease the levels of contaminants.

Physical disinfection may be carried out by heat, desiccation and radiation. The sun provides a ready, natural source of heat and ultraviolet radiation. Steam cleaners provide one of the most effective methods of disinfection with wet heat being more effective in killing micro-organisms than dry heat. The efficacy of steam cleaning is greatly enhanced with the addition of a detergent and/or a chemical disinfectant to the steaming solution.

Of the chemical disinfectants the halogen compounds, i.e. those containing iodine and chlorine, and formaldehyde, are effective against the widest spectrum of microbes. However, problems with disinfectants may arise because of their corrosive effects on building materials, the release of odours or irritating fumes and long residual activities. The selection and use of chemical disinfectants is described more fully by Curtis (1983).

With regard to routine stable management, boxes which are not in use should be thoroughly cleaned out. Such times arise at the end of a hunting or racing season, for example, or when horses are spelling at pasture. This provides an ideal opportunity to disinfect boxes as well. Boxes which are not in use should be left open to ensure that they dry out thoroughly. Designing boxes to ensure that all natural forces of ventilation are used ensures that "stagnation" will not occur when boxes are empty.

On a more regular basis, mechanical cleaning of dust from exposed surfaces such as window ledges and cross beams, should be carried out. This dust is described as "secondary dust" because activity within the stables or gusts of wind may resuspend the particles so that they become respirable again.

Chronic pulmonary disease

The clinical signs of a horse suffering chronic pulmonary disease (CPD) include chronic coughing, dyspnoea, double expiratory effort, nasal discharge and exercise intolerance. CPD is probably best considered as being a disease complex with the overt manifestation just described being the most severe form or end stages of this complex (Viel, 1985). CPD is a disease of the lower airways, which is associated with bronchospasm, mucosal inflammation and mucus plugging. Minor levels of lower respiratory tract (LRT) disease which cause a horse to perform badly may not be detected on clinical examination of the horse at rest. The diagnosis of these degrees of disease require endoscopic examination and ideally tracheal washings (Clarke, 1987).

Overt CPD is a well known sequelae to bacterial and viral infections; it is associated with exposure to mouldy hay and straws and is more common in horses stabled in "badly ventilated" accommodation (McPherson *et al.*, 1979). Covert LRT disease is typically seen in practice in two situations. The first is where horses are exposed to mouldy feeds or bedding; the second is as a consequence to infectious respiratory disease. In the latter cases, levels of dust which are of no apparent consequence prior to infection are capable of extending the horse's recovery time and initiating hypersensitivity airway disease.

In the short term, pharmacological agents can give at least partial relief to the symptoms of CPD. Sodium cromoglycate may be of use where susceptible (though asymptomatic) horses are likely to be unavoidably exposed to mould spores. A 5-day course of this drug typically provides 30 days of protection (Thomson and McPherson, 1981). The long-term answer lies in improving the horse's environment. This basically involves minimising the horse's exposure to mould spores. There are two major sources of these spores in stables: (i) feed (usually hay), and (ii) the bedding. In this section the factors which affect the moulding of these source materials are considered initially along with a simple objective technique to assess such moulding. The various source materials used as beddings and feeds are then discussed along with techniques that minimise the risk of horses being exposed to mould spores from these materials.

Moulding and its assessment

The moulding of hay and straw is primarily dependent on the moisture content at baling (Gregory *et al.*, 1963). Self-heating of bales occurs with baling at high water contents (35–50%). It is under these conditions that thermophlic and thermotolerant actinomyetes and fungi develop. These species are very prolific and their spores are small. This makes them ideal respiratory pathogens. The magnitude of the response to an aerosol challenge is proportional to the dose. A horse eating from a hay net containing mouldy hay may inhale 10^7 spores per breath (Fig. 4.10). The small aerodynamic size of the spores added to the fact that many become airborne as chain and individually ensure that they are deposited in large numbers at all levels of

Figure 4.10. Back-lighting highlights the cloud of spores released from mouldy hay as the horse pulls a mouthful of hay from its hay net. Counts of 4000 particles per cc of air are typical in such spore clouds. Assuming a 2.5 litre tidal volume, 10^7 particles would be inhaled per breath.

the respiratory tract down to the alveolus. Once airborne these spores remain so for long periods of time. This latter feature ensures that horses under "dust free" management may be exposed to significant levels of contaminants generated well away from their "clean" environment. Such situations may arise from several sources including (i) mouldy hay and straw in adjacent boxes, (ii) feed rooms or hay storage areas, (iii) dung heaps, (iv) less common sources such as grain dryers.

The need for an objective assessment of hay and straw was highlighted by Clarke and Madelin (1987) who found that over 50% of the hay and straw they examined which had been selected using "the horse's eye and nose" was significantly contaminated with mould spores. Objective techniques involving culture procedures employing a range of media and incubation temperatures, multistage impaction instruments and particle counters are available. These are time consuming and the necessary equipment and back-up staff beyond the reach of most practices and small diagnostic laboratories. Clarke and Madelin (1987) describe these techniques along with a simple rapid method using a hand-held sampler (Equigiene, Wrington, Avon) for the assessment of moulding of source materials in practice (Fig. 4.11).

This simple procedure involves the collection of dust from the source material being tested on to a microscope slide. The sample is graded according to the presence of plant debris, pollen grains, fungal spores (size and morphology), dust mites and their excreta.

Figure 4.11. Typical sampling position for hand-held slit sampler (Equigiene, Wrington) for sampling hay.

Microscope slides are pre-coated with a drop of gelatin-based mountant. A biscuit of hay or straw is then agitated and the sampler held in the dust cloud for 3 seconds (Fig. 4.11). Source materials such as wood shavings and grains may be shaken in a large paper bag prior to sampling the air directly from the bag. In a stable bedding material may be agitated and tested *in situ*.

The primary dust constituents of hay and straw which has been baled with low moisture content (15–20%) is plant material. This includes pollen grain, plant hairs and leaf segments (Fig. 4.12). If fungal spores are present they will primarily be the so-called "fine weather airspora" (i.e. *Cladosporium, rust uredospores, Helminth osporium* and *Alternaria*). They have large easily recognisable spores (Fig. 4.13) and are not very prolific. These species grow as saprophytes or parasites on plant material as it stands in the field. They are more typical of "clean" straw than hay since the straw tends to stand dead in the field for longer periods prior to baling allowing greater colonisation.

Baling hay and straw with 20–30% water content leads to temperature of 35–45°C developing and significant levels of mould contaminants developing. The worst contamination occurs with moisture contents of 35–50% at baling which is associated with spontaneous heating up to 60°C (Fig. 4.14). Samples collected show large numbers of small respirable spores (Fig. 4.15). The moulding process is dynamic. There is a continuous change in the

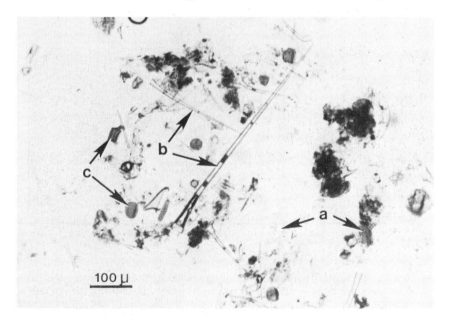

Figure 4.12. A clean hay sample. a = plant material, b = plant hair, c = pollen grains (Clarke, 1986).

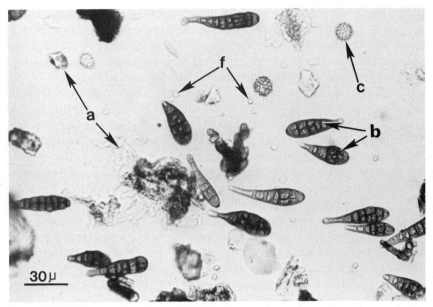

Figure 4.13. "Fair weather air spore", typical of non-heated straw. a = plant material, b = alternaria spore, c = rust spore, f = aspergillus/penicillium type spore (Clarke, 1986).

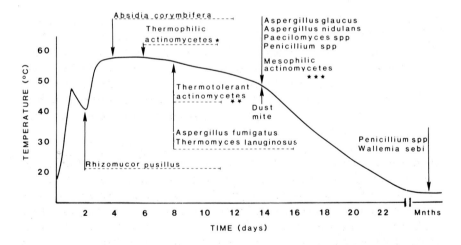

Figure 4.14. The heating and moulding of hay baled with 46% water content. The initial temperature peak is thought to be due to plant metabolism; thereafter the temperature increase is due to microbial activity. As the hay dries and the temperature drops (assuming that combustion has not occurred) the mould species which prefer more temperate conditions proliferate. The broken lines indicate the times when spores of the various species are seen in their peaks (After Gregory *et al.*, 1963).

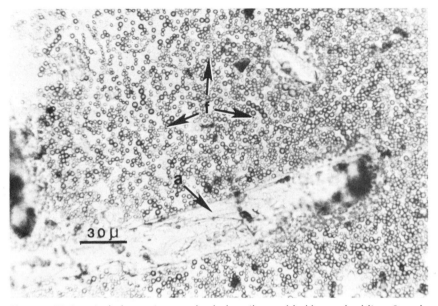

Figure 4.15. A typical photomicrograph of a heavily moulded hay or bedding. Sample a = plant material, f = small respirable fungal spores of aspergillus/penicillium type (Clarke, 1986).

predominant mould species during moulding. Some of the spores from species which develop early are broken down as the moulding continues so that the dust from these source materials contains fungal spores as well as smaller irritant and allergenic particles of metabolites and pieces of broken down spores. Strong and dependent synergistic relationships exist between the various species. The initial development of the mucoraceous species utilises soluble sugars increasing the pH, a requirement for the thermophilic actinomycetes which develop a few days later.

Dust mites which forage on fungal spores add a further complicating factor (Fig. 4.16). The foraging of the mites may decrease the overall number of spores, however, the spores are replaced by dust mite faeces which are basically ''processed'' fungal antigens. The dust mites themselves may be found in hay which has been stored for several months or they may disappear with the only evidence of their previous presence being their faecal pellets. A hay sample which has large numbers of dust mite or their faecal pellets must be considered as being a potential respiratory pathogen.

Beddings

The association between contaminated beddings and chronic pulmonary disease (CPD) in the horse is well established (Thomson and McPherson, 1984).

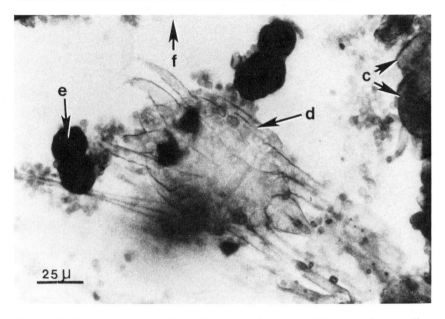

Figure 4.16. Photomicrograph of sample from a contaminated hay sample. c = pollen grains, d = dust mite, e = mite excreta, f = aspergillus/penicillium type spore.

More recently an association has been made between bedding and covert subclinical airway disease. Burrell *et al.* (1985) reported in a study of respiratory disease in two groups of thoroughbreds, one group bedded on straw and one on newspaper, that while the overall incidence of respiratory disease was similar in both groups, the average duration of disease was longer with horses bedded on straw than with those bedded on shredded newspaper. Clarke *et al.* (1987) found that the incidence and degree of LRT inflammation as adjudged by amounts of mucopus in the trachea was higher in a group of 2-year-old thoroughbreds housed in stables with contaminated wood shavings compared with a similar group of 2-year-olds on "clean" shavings. While there was an association between LRT inflammation and environment in this study, there was no association between environment and PLH grading or between mucopus grading and pharyngeal lymphoid hyperplasia (PLH) grading.

Table 4.3 presents analyses of different bedding materials prior to their use in stables. The alternative beddings to straw are generally much cleaner than straw when they are first put down, one exception being contaminated wood shavings. Table 4.4 shows the effects of the air hygiene of various beddings under different management practices. The particle counts are presented in terms of increases in airborne particles seen at mucking out of

Table 4.3. Aerobiology of fresh bedding materials

	Equibed[c]	Fenpol[d]	Diced[e] newspaper	Shavings (range)	Straw (range)	Tissue[f]
Particle per mg[a]	19	4	78	148 873	1490 28,100	53
Proportion of particles(%)[b]:						
Fungal and actinomycete particles	ng	ng	ng	5 96	90 100	ng
Plant material	ng	ng	ng	95 4	8 trace	ng
Other	100	100	100	ng ng	2 trace	100

[a] Assessed using aerodynamic particle sizer.
[b] Assessed using May Impactor.
[c] Equibed—Melcourt Industries, Tetbury, Glos. (absorbent synthetic bedding).
[d] Fenpol—Raymond Barnes Bloodstock, Newmarket (non-absorbent synthetic bedding).
[e] Diced newspaper—Shredabed, Exeter.
[f] Tissue bedding—F. H. Lee, Bolton, Greater Manchester.

Table 4.4. Air hygiene of stables using various beddings and management procedures

	Equibed[a]	Shavings[a]	Shavings[b] dirty	Shavings deep litter	Straw[a]	Paper[a]	Paper deep litter
Particles/cc air[c]	1.5	6–104	603	400–1223	167–724	25–100	62–1296
Description of mould contamination	Negligible	Negligible	1° mould spores	1° mould spores very heavy	1° mould spores heavy/very heavy	Moderate	Very heavy

[a] Daily mucking out.
[b] Daily mucking out in poorly ventilated heavily insulated (hot and humid) barn.
[c] Increases at mucking out, i.e. time of maximum dust release from bedding using RION KCOIA 5-channel particle counter. This is time of maximum release of particles from bedding into the stable.

boxes. This is the time of maximum release of contaminants from bedding and while airborne levels of spores may be lower at other times of the day, the horse may still inhale large numbers of spores as it sniffs at a contaminated bedding.

Tables 4.3 and 4.4 show that in deep-litter management systems or in hot humid environments that plant-based beddings may mould heavily. This problem has been more fully described elsewhere (Clarke, 1987). Noxious gases build up in deep litter. Deep or semi-deep litter management of bedding may also be associated with the build up of non-respiratory pathogens. The eggs and larvae of endoparasites and bacteria such as *E. coli, Salmonella and Pseudomonas* may survive for long periods of time in beddings (Curtis, 1983). However, in well ventilated boxes with daily removal of manure and turning over of bedding, these problems are usually avoidable.

The use of deep litter bedding systems in stables is therefore best avoided. Potential hazards, including the respiratory pathogens, which may arise from such systems are undesirable and easily prevented.

The use of chemical additives such as weak solutions of thiabendazole to inhibit moulding of plant-based bedding materials *in situ* is one promising area of research. However, the use of any chemicals must follow tests to ensure that they themselves are not potential respiratory pathogens. The synthetic bedding materials do not provide the substrate for mould contamination that the plant-based materials do. It is also of critical importance that beddings are not left down in boxes which are not in use. Empty boxes should be thoroughly cleaned out (see section on Disinfection).

Straw Straw continues to be the most widely used bedding material today. The level of contamination of straw varies considerably. A clean straw is usually characterised by the presence of large-spored fungal species (Fig. 4.13) with small numbers of the small-spored species which are capable of reaching the alveolar membrane. The dust from "self-heated" straw is characterised by large numbers of respirable actinomycete and fungal spores. In general terms using straw bedding is likely to be associated with greater risks of respiratory disease than with other beddings. The selection of straw using either the simple technique or a culture analysis (cited earlier) will decrease the risk of large challenges. However, the cleanest of straws have levels of mould spores which are greater than those of the alternatives. Even though these levels of spores are not high, they are ideally avoided since the TLV for these contaminants are not known and the response to respiratory challenge is likely to be graded.

Plant-based beddings These beddings include wood shavings, peat, shredded or diced newspaper and treated tissue paper.

Wood shavings are the most commonly used alternative to straw and they provide an excellent bedding when managed properly. Apart from the treated tissue paper, wood shavings seem to provide the greatest resistance to microbial degradation and mould contamination of all the beddings. This is most probably associated with the higher levels of lignin in the wood shavings compared with other beddings. In properly ventilated boxes and with daily mucking out and turning of beddings, these plant-based beddings offer excellent alternatives to straw-based beddings (Clarke, 1987). Damping down of wood shavings beddings with a watering can decreases the amounts of dust given off by such a bedding. This dampening does not seem to encourage moulding of the shavings in well ventilated boxes where bedding is turned daily. Furthermore, this dampening may help prevent feet from becoming too dry. However, wood shavings that are kept in storage prior to use should be kept dry because of the risk of heating and moulding.

Synthetic beddings   The latest developments in livestock beddings involve the use of synthetic beddings. These are very clean when put down and do not promote fungal growth *in situ*. They are also thermally efficient for a horse lying on them.

Synthetic beddings may be either absorbent or non-absorbent. The absorbent synthetic bedding (Equibed) is shredded rayon-type material with a resilient spring texture. The non-absorbent equivalent is Fenpol (Fenpol Ltd., c/o Raymond Barnes Bloodstock, Newmarket). This consists of lengths of an expanded polystyrene-type material made to simulate straw. Urine will drain away quickly from this bedding leaving a thermally efficient dry bedding.

The problem of disposal of these materials should be considered. The simplest and safest disposal involves the use of skips which are regularly removed to a landfill site.

Dung heaps   Heaps of rotting soiled bedding materials should be avoided near stables. These heaps are potential sources of large numbers of mould spores for horses housed nearby (Fig. 4.17). Spoiled bedding is best placed in a skip or trailer and removed from close proximity of the stables every 3–4 days for burying or burning.

Hay and feeds

Mouldy feeds provide a constant source of respirable challenge for the horse as it stands eating them. The horse's nostrils are positioned for maximum pathogen uptake as it eats from a hay net or grain manger (Fig. 4.10). Ventilation will minimise dust levels within the stable that arise from the hay net. However, significant levels of spores may still be inhaled from the dense

Figure 4.17. Mound of rotting bedding material at the back of a stables. Note the close proximity to air vents.

spore clouds that arise as mouthfuls of hay are pulled from the hay net. This is evidenced by the fact that asymptomatic CPD horses at pasture (conditions of maximum ventilation) will become symptomatic when fed mouldy hay.

Table 4.5 presents analysis results of various types of hay which were in use at stables within the United Kingdom and Ireland. There is clearly a wide range of variation in the degree of moulding present in these hays. In one survey of hays used for feeding horses over 50% were found to be heavily moulded (Clarke and Madelin, 1987). The importance of a technique for the objective assessments of hay and other feeds has been discussed earlier in this chapter.

Heavy moulding may occur in all types of hays. In recent years the practice of importing hay into the United Kingdom (especially lucerne and clover hays) has become widespread. The high price premiums paid for such hays does not ensure that they are "clean". The succulent nature of lucerne and clover hays make their curing and baling at suitable low moisture contents to prevent moulding more difficult than meadow grasses such as ryegrass. This is further compounded in mild climates such as in the United Kingdom. However, it has been the author's experience that lucerne hays imported into the United Kingdom from much warmer climates such as America, may still be heavily

Table 4.5. Aerobiology of hays and silages

	Rye (clean)	Rye (dirty)	Lucerne (clean)	Lucerne (dirty)	Horsehage[c]	Silage[c]
Particle per mg[a]	980 →	65,190	840 →	39,270	44	19
Proportion of particles(%)[b]						
Fungal and actinomycete particles	30	99	55	99	5	5
Plant material	70	ng	45	ng	95	95
Other	ng	1	ng	1	ng	ng

[a] Assessed using aerodynamic particle sizer.
[b] Assessed using a May Impactor.
[c] Sealed bags.

contaminated. Horse trainers may be paying a premium for having LRT inflammation induced in their horses if their hays are not objectively tested.

The soaking of hay and dampening of hard feeds has been the traditional approach to decreasing the respirable challenges of dust from the horse's feed. This approach is well described in the veterinary literature. White (1822) writing on broken wind stated that the hay should be "best quality . . . always wetted with water''. Table 4.6 shows the effect of soaking mouldy hay for 5 minutes and 24 hours on the levels of dust released on agitation. Five-minute soaking can be associated with poor penetration of water throughout the hay. This results in small pockets of dry hay which still release spore clouds. The duration of soaking is not as important as the thorough wetting of hay. From stable management and daily routine points of view, hay is best put in to soak in the morning and fed in the evening when the next morning's hay is put in to soak. It is essential that the strings of the bales are cut and that the bale is fully immersed to ensure that no areas remain dry.

When hay is soaked most of the dust particles present (including spores) adhere to the surface of the hay due to changes in surface tensions. There is an old wives' tale often described of fungal spores increasing in size due to absorbing water so that their aerodynamic diameters are increased. This is supposed to lead to these spores being deposited in the upper airways rather than lower down in the respiratory tract. However, Table 4.6 shows that this is not the case. If there was a shift in size of the spores similar numbers of

Table 4.6. The effect of soaking moulded ryegrass hay on particle release

	Fresh	5-min soak	24-hour soak
Respirable particles per mg source material	45,000	1650	525

spores would be released with each of the soaking times, with their aerodynamic diameters increasing with increased soaking time.

The soaking of hay is not always successful in preventing hypersensitive airway disease (McPherson *et al.*, 1979). Pieces of hay fall to the floor as the horse eats its hay. This hay will dry out and viable spores may be released to be inhaled by the horse or to "seed" plant-based bedding materials such as wood shavings or shredded newspaper. Mouldy hay and grain has also been associated with mycotoxin-related diseases and systemic and local fungal infections (Asquith, 1983; Austwick, 1966).

An alternative to soaking of hay is to use a mechanical cleaner (Dust Cure, Bridgwater, Somerset). This machine agitates hay and straw and vacuums away airborne spores. The efficiency of the dust removal is high and a final steam process ensures that hay is dampened without being sodden. Nutritional additives can also be added at the steaming phase. This technique provides a highly cost effective method of preparation of "dust-free" feeds when comparing the cost of the basic raw material hay to haylage substitutes.

The use of preservatives, such as propionic acid, on hay prior to baling will decrease the total spore burden of hay baled with high moisture contents. However, with high moisture contents concentrations of spores of the thermophilic and thermotolerant moulds may still develop which will induce respiratory disease (Kotimaa *et al.*, 1983). This is due in part to the poor distribution of chemical through the bales and the ability of some moulds to tolerate and metabolise proprionic acid (Kotimaa *et al.*, 1983; Lord and Lacey, 1978).

Barn drying of hay is another method of trying to overcome moulding under otherwise unfavourable conditions. The hay is baled within a few days of cutting and is then stacked into a barn with air vents through the stock. Air movement through these vents may be increased with fans. The air vents allow the hay to dry out and lowers the temperatures which may be reached in the bales. While the barn drying decreases the total amount of spores, significant moulding may still occur. The degree of moulding will, as with all hay, be primarily dependent on moisture content at baling.

With the inherent problems of feeding hay, the search for and use of alternatives to hay has become widespread. Complete cubed diets offer one alternative and their use along with clean bedding has been successfully used in the environmental control of chronic pulmonary disease (Thomson and McPherson, 1984).

Silage offers one of the most promising alternatives to hay. A drawback with traditionally made silage is that it may cause diarrhoea. However, the ensiling of more mature grass, termed "haylage", has largely overcome this problem. Haylage is available commercially (Horsehage, Exeter; Harris Hay, York) and some farmers now produce suitable "big bale" silage. It is important to ensure that the packing around the silage or haylage is air-tight and when bales are opened they should be used within 3 days because of the risk of rapid fungal colonisation. Botulism has been associated with the feeding of contaminated "big-bale" silage. The risk of botulism from big-bale silage may be minimised by using only air-tight bags and avoiding bales which (i) have a pH of >0.5, (ii) give off an ammoniacal smell when opened, or (iii) which contain dirt (Cuddeford, 1986).

Storage of hay and straw

Condensation on hay and straw may encourage growth of fungi *in situ*. If this occurs a carpet of moulds develops over the surface of the hay or straw. Muggy, humid conditions and closed, poorly ventilated buildings are associated with this problem. Covering block sections of stacked bales with a "breathing cover" such as canvas and making sure the storage area is well ventilated will help to overcome these problems. If this moulding occurs outside contaminated bales should be discarded. Workers doing this work should wear effective face masks.

Parasites

Lung worm (*Dictyocaulus arnfieldi*) is frequently implicated with chronic respiratory disease of the horses. There are also several gastro-intestinal parasites which migrate through the lungs during their normal life cycles producing respiratory disease, especially in young animals. These include *Parascaris equorum* and *Strongyloides westeri*. *P. equorum* has also been implicated in hypersensitivity airway disease of horses. The control of parasites involves pasture management and regular use of appropriate anthelmintics (Duncan, 1985). In relation to stable management droppings should be regularly and thoroughly removed and deep litter beddings avoided. This is because the eggs and larval stages of most parasites are long-lived and resistant to desiccation, the only possible exception being the larval stages of *S. westeri* which are susceptible to desiccation and therefore levels in a stable may be minimised in dry beddings.

ACKNOWLEDGEMENTS

The author is indebted to the Home of Rest for Horses for their generosity in funding a 3-year study into the air hygiene of stables and equine respiratory disease. This chapter could not have been completed without the support provided by Mr Robert Sangster through Swettenham Stud.

REFERENCES

Albert, R. E., Spiegelman, J., Lippmann, M. and Bennett, R. (1968). The characteristics of bronchial clearance in the miniature donkey. *Arch. Environ. Health* **17**, 50–58.

Allen, G. P. and Bryans, J. T. (1986). Molecular epizootiology, pathogenesis and prophylaxis of Equine Herpes Virus-1 infection. *Prog. Vet. Microbiol. Immun.* **2**, 78–144.

American Conference of Government Industrial Hygienists (1972). Threshold limiting values for substance in work room air with intended changes for 1972. American Conference of Government Industrial Hygienists, PO Box 1937, Cincinnati, Ohio.

Anonymous (1986). Recombinant vaccinia. *Vet. Rec.* **119**, 413.

Asquith, R. L. (1983). Biological effects of aflatoxins: horses. In: *Aflatoxin and Aspergillus flavus in corn.* (U. L. Diener, R. L. Asquith and J. W. Dickens, eds), pp. 62–66. Auburn University, Alabama.

Attenburrow, D. P. (1976). Measurements of respiratory sounds produced in the respiratory tract of the horse at exercise, together with some applications. FRCVS thesis.

Attenburrow, D. P. (1983). Respiration and locomotion. Proc. 1st Int. Conf. Equine Exercise Physiol., September 1982, pp. 17–22.

Austwick, P. K. C. (1966). The role of spores in the allergies and mycoses of man and animals. In: *The Fungus Spore* (M. F. Madelin, ed.) pp. 321–337 Butterworths, London.

Brain, J. D. and Valberg, P. A. (1979). Deposition of aerosol in the respiratory tract. *Am. Rev. Resp. Dis.* **120**, 1325–1373.

Breeze, R. G., Brown, C. M. and Turk, M. A. (1984). 3-methylindole as a model of equine obstructive lung disease. *Equine Vet. J.* **16**, 108–112.

Bruce, J. M. (1978). Natural ventilation through openings and its application to cattle building ventilation. *J. Agric. Engineering Res.* **23**, 151–167.

Bruce, J. M. (1979). Heat loss from animals to floors. *Frm Bldgs Prog.* **55**, 1–4.

Bryans, J. T. (1981a). Application of management procedures and prophylactic immunisation to control of equine rhinopneumonitis. Proc. 26th Ann. Con. Am. Ass. Equine Pract., Anaheim, pp. 259–272.

Bryans, J. T. (1981b). Control of equine influenza. Proc. 26th Ann. Con. Am. Ass. Equine Pract., Anaheim, pp. 279–287.

Burrell, M. H. (1985). Endoscopic and virological observations on respiratory disease in a group of young thoroughbred horses in training. *Equine Vet. J.* **17**, 99–103.

Burrell, M. H., Mackintosh, Mary E., Mumford, Jennifer A. and Rossdale, P. D. (1985). A two-year study of respiratory disease in a Newmarket stable: Some preliminary observations. Proc. Society for Vet. Epidem. and Preventive Med., pp. 74–83.

Carpenter, G. A., Smith, W. K., MacLaren, A. P. C. and Spackman, D. (1986). Effects of internal air filtration on the performance of broilers and the aerial concentrations of dust and bacteria. *Br. Poult. Sci.* **27**, 471–480.

Clarke, A. F. (1985). Review of exercise induced pulmonary haemorrhage and its possible relationship with mechanical stress. *Equine Vet. J.* **17**, 166–172.

Clarke, A. F. (1986). Air hygiene of stables and Chronic Pulmonary Disease of the horse. Ph.D. thesis, University of Bristol.

Clarke, A. F. (1987). Chronic pulmonary disease—a multifaceted disease complex in the horse. *Irish Vet. J.* **41**, 258–261.

Clarke, A. F. and Madelin, T. M. (1987). A simple technique for assessing respiratory health hazards from hay and other source materials. *Equine Vet. J.* (in press).

Clarke, A. F., Madelin, T. M. and Allpress, R. G. (1987). The relationship of air hygiene in stables to lower airway disease and pharyngeal lymphoid hyperplasia in two groups of thoroughbred horses. *Equine Vet. J.* **19** (in press).

Clayton, Hilary M. (1981). Clinical aspects of ascarid and lung worm infections in horses. Proc. 26th Ann. Con. Am. Assoc. Equine Pract., Anaheim, pp. 29–32.

Cuddeford, D. (1986). Alternative feedstuffs for horses. *In Practice* **8**, 68–70.

Curtis, S. E. (1983). Environmental management in animal agriculture. Iowa State University Press, Iowa, pp. 323–357.

Daws, L. F. (1967). Movement of air streams indoors. In: *Airborne Microbes* (P. H. Gregory and J. L. Monteith, eds), pp. 30–59. 17th Sym. Soc. Gen. Microbiol., Cambridge University Press.

Donaldson, A. I. (1978). Factors influencing the dispersal, survival and deposition of airborne pathogens of farm animals. *The Vet. Bull.* **48**, 83–94.

Druett, H. A. (1973). The open air factor. In: *Airborne Transmission and Airborne Infection* (J. F. Hers, Ph. and K. C. Winkler, eds), pp. 141–149. Oosthoek Pub. Co., Utrecht.

Duncan, J. L. (1985). Parasitic diseases. In: *Equine Surgery and Medicine*, Vol. 1 (J. Hickman, ed.) Academic Press, London.

Edwards, J. H., Trotman, D. M. and Mason, O. F. (1985). Methods of reducing particle concentrations of Aspergillus fumigatus conidia and mouldy hay dust. *Sabauroudia* **23**, 237–243.

Emarson, D. A., Vidal, S. and Chan-Yeung, M. (1985). Rapid decline in FEV in grain handlers: Relation to level of dust exposure. *Am. Rev. Respir. Dis.* **132**, 814–817.

Galan, J. E. and Timoney, J. F. (1985). Mucosal nasopharyngeal responses of horses to protein antigens of *Streptococcus equi. Infect. Immun.* **47**, 623–628.

Gerber, H. (1973). Chronic pulmonary disease in the horse. *Equine Vet. J.* **5**, 26–33.

Greet, T. R. C. (1985). The respiratory tract, In: *Equine Surgery and Medicine*, Vol. 1 (J. Hickman, ed.), pp. 247–296. Academic Press, London.

Gregory, P. H., Lacey, Maureen, Fenstenstein, G. W. and Skinner, F. A. (1963). Microbial and biochemical changes during the moulding of hay. *J. Gen. Microbiol.* **33**, 147–174.

Hillidge, C. J. (1986). Review of *Corynebacterium (Rhodococcus) equi* lung abscesses in foals: pathogenesis, diagnosis and treatment. *Vet. Rec.* **119**, 261–264.

Hornicke, H., Meixner, R. and Pollman, U. (1983). Respiration in exercising horses. Proc. 1st Int. Conf. Equine Exercise Physiol., September 1982, pp. 7–16.

Huston, L. J., Bayly, W. M. and Liggitt, H. D. (1986). The effects of strenuous exercise and training on alveolar macrophage function. 2nd Int. Conf. Equine Exercise Physiology Abstract.

Jensen, A. H. and Curtis, S. E. (1976). Effects of group size and of negative air ionization on performance of growing-finishing swine. *J. Anim. Sci.* **42**, 8–11.

Jones, R. D., McGreevy, P. D., Robertson, A., Clarke, A. F. and Wathes, C. M. (1987). A survey of the designs of racehorse stables in the South-West of England. *Equine Vet. J.* (in press).

Kotimaa, M., Mustonen, M. and Husman, J. (1983). The effect of ADD-H preservative (ammonium propionate) on the moulding of baled hay. *Maataloustieteellinen Aikakauskirja* **55**, 371–383.

Lacey, J., Lord, K. A., King, H. G. C. and Manlove, Rosemary (1978). Preservation of baled hay with propionic and formic acids and a proprietary additive. *Ann. appl. Biol.* **88**, 65–73.

Lees, P. and Higgins, A. J. (1985). Clinical pharmacology and therapeutic uses of non-steroidal anti-inflammatory drugs in the horse. *Equine Vet. J.* **17**, 83–96.

Liggitt, D., Bayly, W. and Bassaraba, R. (1985). Challenges to the equine lung mucosal defense system. Proc. 4th Ann. Sports Medicine for the Equine Practitioner, Washington State University, Pullman, pp. 70–86.

Lord, K. A. and Lacey, J. (1978). Chemicals to prevent the moulding of hay and other crops. *J. Sci. Fd. Agric.* **29**, 574–575.

Mair, T. S. (1985). Respiratory immunology and hypersensitivity in the horse. Ph.D. thesis, University of Bristol.

McBride, G. E., Christopherson, R. J. and Sauer, W. C. (1983). Metabolic responses of horses to temperature stress. *J. Anim. Sci.* **57**, Supply. p. 175.

McCormack, J. A. D., Clark, J. J. and Knowles, L. C. (1984). Survey of naturally ventilated buildings for beef cattle. *Farm Buildings Progress* **78**, 31–35.

McPherson, E. A., Lawson, G. H. K., Murphy, J. R. Nicholson, J. M., Breeze, R. G. and Pirie, H. M. (1979). Chronic obstructive pulmonary disease (COPD): Factors influencing the occurrence. *Equine Vet. J.* **11**, 167–171.

Miles, W. J. (1880). *Modern Practical Farriery.* Mackenzie, London.

Mirbahar, K. B. and Eyre, P. (1986). Chronic obstructive pulmonary disease (COPD) in horses. *The Vet. Annual*, 26th Ed., pp. 146–155. Scientechnica, Bristol.

Mumford, J. A. (1985). Preparing for equine arteritis. *Equine Vet. J.* **17**, 6–11.

Nordstrum, G. A. and McQuitty, J. B. (1976). Manure gases in the animal environment. University of Alberta Research Bulletin **76**, 1.

Panicali, D., Davis, S. W., Weinberg, R. L. and Paoletti, E. (1983). Construction of live vaccines by using genetically engineered pox viruses: biological activity of recombinant vaccinia virus expressing influenza virus hemagglutin. *Proc. Natl Acad. Sci.* **88**, 5364–5368.

Percivall, W. (1853). *The Disease of the Chest and Air passages of the Horse*, p. 1. Longman, Brown, Green & Longman, London.

Phillips, G. B., Harris, G. J. and Jones, M. W. (1964). The effect of air ions on bacterial aerosols. *Int. J. of Biometerology* **8**, 27–32.

Powell, D. G. (1985). International movement of horses and its influence on the spread of infectious disease. Proc. Soc. for Vet. Epidem. and Preventive Medicine, pp. 90–95.

Pratt, R. and Barnard, R. W. (1960). Some effects of ionized air on *Penicillium notafum. J. Am. Pharm. Assoc.* **49**, 643–646.

Reif, J. S., George, J. L. and Shideler, R. K. (1981). Recent developments in strangles research. Observations on the carrier state and evaluation of a new vaccine. 27th Proc. Am. Assoc. Equine Pract., pp. 33–40.

Robinzon, B., Liffshitz, E., Pyrzak, R. and Snapir, N. (1983). Effect of negative and positive air ions on the chicken tracheal surface morphology: Study with scanning electron microscopy. *Avian Diseases* **27**, 531–538.

Round, M. C. (1976). Lungworm infection (*Dictyocaulus arnfieldi*) of the horse and donkey. *Vet. Rec.* **99**, 393–395.

Rouse, B. T. and Ditchfield, W. J. B. (1970). The response of ponies to Myxovirus Influenza A equi-2. III. The protection effect of serum and nasal antibody against experimental challenge. *Can. J. Comp. Med.* **34**, 7–12.

Russell, P. H. and Edington, N. (1985). *Veterinary Virus.* Burlington Press, Cambridge.

Sainsbury, D. W. B. (1981). Ventilation and environment in relation to equine respiratory disease. *Equine Vet. J.* **13**, 167–170.

Sainsbury, D. W. B. and Sainsbury, P. (1979). *Livestock Health and Housing*, 2nd Ed. Bailliere Tindall, London.

Smith, B. P. (1982) Problems of *Corynebacterium equi* pneumonia in foals. *J. Reprod. Fert.* Suppl. **32**, 465–468.

Speirs, V. C., van Veenendaal, J. C., Harrison, I. W., Smyth, G. B., Anderson, G. A., Wilson, D. V. and Gilbo, B. (1982). Pulmonary haemorrhage in standardbred horses after racing. *Aust. vet. J.* **59**, 38–40.

Stauffer, D. (1975). Scaling theory for aerosol deposition in the lungs of different mammals. *J. Aerosol. Sci.* **6**, 223–225.

Steel, J. D. (1963). *Studies on the Electrocardiogram of the Racehorse.* Australian Med. Publishing Co. Ltd., Sydney.

Stewart, J. H., Rose, R. J., Davis, P. E. and Koffman, K. (1983). A comparison of electrocardiographic findings in racehorses presented either for routine examination or poor racing performance. Proc. 1st Int. Conf. Equine Exercise Physiol, September 1982, pp. 135–143.

Sweeney, C. R. Divers, T. J. and Benson, C. E. (1985). Anaerobic bacteria in 21 horses with pleuropneumonia. *JAVMA* **7**, 721–724.

Task Group on Lung Dynamics (1966). Deposition and retention models for internal dosimetry of the human respiratory tract. *Health Physics* **12**, 173.

Thomson, J. R. and McPherson, E. A. (1981). Prophylactic effects of sodium cromoglycate on COPD in the horse. *Equine Vet. J.* **13**, 243–246.

Thomson, J. R. and McPherson, E. A. (1984). Effects of environmental control on pulmonary function of horses affected with chronic obstructive pulmonary disease. *Equine Vet. J.* **16**, 35–38.

Timoney, J. F. and Eggers, D. (1985). Serum bactericidal responses to *Streptococcus equi* of horses following infection or vaccination. *Equine Vet. J.* **17**, 306–310.

Turk, M. A. M. and Klei, T. R. (1984). Effect of ivermectin treatment on eosinophilic pneumonia and other extravascular lesions of late *Strongylus vulgaris* larval migration in foals. *Vet. Pathol.* **21**, 87–92.

Viel, L. (1985). Diagnostic procedures, prognosis and therapeutic approaches of chronic respiratory diseases in horses. *Can. Vet. J.* **26**, 33–35.

Wathes, C. M. (1981). Insulation of animal houses. In: *Environmental Aspects of Housing for Animal Production* (Clark, J. A., ed.), p. 379. Butterworths, London.

Wathes, C. M., Jones, C. D. R. and Webster, A. F. J. (1983). Ventilation, air hygiene and animal health. *Vet. Rec.* **113**, 554–559.

Webster, A. J., Clarke, A. F., Madelin, T. M. and Wathes, C. M. (1987). Effects of stable design, ventilation and management on the concentration of respirable dust. *Equine Vet. J.* (in press).

Wehner, A. P. (1969). Electro-aerosols, air ions and physical. medicine. *Am. J. Phy. Med.* **48**, 119–148.

White, James (1822). *The Veterinary Art*, 13th Ed. Longman, London.

Wilson, M. R., Takov, R., Friendship, R. M., Martin, S. W., McMillan, I., Hacker, R. R. and Swaminathan, S. (1986). Prevalence of respiratory diseases and their association with growth rate and space in randomly selected swine herds. *Can. J. Vet. Res.* **50**, 209–216.

5   Nutrition of Horses

D. R. WISE

INTRODUCTION

The nutritional requirements of horses are less precisely understood than are those of farm livestock. However, there is an increasing body of information, much of it emanating from research conducted in the United States. The American National Research Council (NRC) summarised the nutrient requirements of horses (1978). Advances in knowledge since then have been reviewed by Ott (1982) and by Hintz (1985). I have drawn very heavily on these publications in preparing this chapter as well as upon references to other work cited in them. The chapter has been written as a largely unreferenced review, the only references given in the text being those unobtainable from any of the publications mentioned above.

THE EQUINE DIGESTIVE SYSTEM

Horses are specialised herbivores adapted to life on open grasslands. The stems and leaves of grasses, legumes and herbs have cell walls composed of cellulose, hemicellulose and lignin. These three components represent the crude fibre fraction of feed. Cellulose and hemicellulose are polysaccharides which are sugars linked together to form chains. In this respect, they resemble starches. However, the sugar molecules in starches are joined together with linkages which can be broken down by vertebrate digestive enzymes, while the union of sugars in cellulose and hemicellulose is composed of linkages which cannot be split by such enzymes. A herbivore is thus confronted with a food which consists of between 20 and 50% dry weight of a material which its own digestive enzymes cannot handle. Furthermore, this refractory material, sited in cell walls, surrounds the more readily digestible nutrients to be found within the cells. Different herbivore

Horse Management 2nd edition
ISBN: 0-12-347218-0 case

species have developed different strategies for coping with the problem. Geese, for example, manage to extract enough nutrients from the vegetative parts of plants merely by eating very large quantities of plant material. Their gizzards squeeze the cell contents out of the cell walls. The former material is digested while the latter is passed rapidly out in the faeces. It only takes a meal from six to eight hours to pass through a goose. The efficiency of digestion of the food as a whole is very low but, by eating large quantities, the goose is able to derive sufficient nutrients to sustain it. Ruminant animals have evolved an almost opposite strategy. They have developed a complex foregut of reticulum, rumen, omasum and abomasum, capable of containing a very large volume of semifluid content in which bacteria and protozoa attack the food. Bacterial and protozoal enzymes are able to cleave the linkages of cellulose and hemicellulose. The bacteria and protozoa have first choice of the feed and it is only when the digesta pass further down the gut that the ruminant's own digestive enzymes are brought to bear. This ability to harness micro-organisms in the service of one's digestion has obvious advantages but also disadvantages which are somewhat less obvious and will therefore be highlighted below.

The ruminant "bacterial vat" is a bulky structure and the weight of the ruminant gut and content represents approximately 15% of the animal's total weight as opposed to 5% in an animal with a typical simple stomach. This is an obvious disadvantage in terms of evolution in that the extra "dead weight" in the gut compromises predator avoidance. A second disadvantage of allowing micro-organisms first choice of the feed is that a price has to be paid. Bacteria convert all carbohydrates, which comprise not only the fibrous components of feeds but also the starches and sugars, to volatile fatty acids, primarily propionic and acetic acids. The animal can absorb these acids and use them as alternatives to glucose as energy sources. However, in the process, the bacteria also produce methane gas which cannot be absorbed and thus represents wasted energy. Obviously, wasting some energy from cellulose and hemicellulose is much better than deriving none at all. However, the same methane energy loss is applicable to the carbohydrate cell contents of plants which, in a non-ruminant animal, would have been digested and absorbed as glucose with no methane production.

There is one further and, perhaps, even more serious drawback to the ruminant digestive system. It has been stated that ruminants can digest cellulose and hemicellulose. However, they and their digestive micro-organisms cannot use lignin, the third component of crude fibre. Lignin is not a defined chemical entity but a complex of totally indigestible organic compounds and is to be found in plant walls in close association with cellulose and hemicellulose. The proportion of lignin in grass increases with plant maturity and increasing lignin decreases the ruminant's ability to digest cellulose and hemicellulose. Furthermore, the appetite of a ruminant is depressed by increasing the lignin level of its diet because the rumen is unable to pass poorly digested material to the abomasum. Thus, as the quality of the feed declines so does appetite until the animal starves to death even though feed quantity may not be limiting. It is for this reason that antelopes and wildebeest can die in large numbers in drought conditions in Africa while zebra are, at the same time, apparently thriving. The statement, therefore, that the ruminant digestive system is the one best suited to cope with fibrous feed is only partly true. What is true is that the ruminant can derive a higher proportion of energy from a given weight of fibrous feed than can any other animal.

It is now appropriate to discuss the equine strategy for coping with a herbivorous diet. The previous discussion of geese and ruminants was not irrelevant in that the horse has an intermediate approach to digestion and one that has generally been less well researched than those of poultry and ruminants. The horse does contain a dilatation of its gastrointestinal tract, the large caecum, which acts as a vat for microbial fermentation. However, this is smaller than that of the ruminant and is situated at the hind end of the gut. In equines, therefore, the animals' own digestive enzymes take precedence over the microbial enzymes. Equines have small stomachs and have evolved to feed for prolonged periods rather than to take large, spaced meals. The ruminant can eat and swallow a large amount of feed without chewing and store it in the foregut before returning it to the mouth for mastication at its leisure. The horse has to chew each mouthful of feed as it takes it and cannot return it to its mouth for reprocessing once it has been swallowed. On fibrous feeds of low digestible energy, therefore, it has to spend a very substantial part of its day in eating enough to obtain sufficient nutrients. The physiological stimulus for gastric emptying would appear to be the chewing of more food in the mouth. Because the microbial vat of the horse is at the distal end of its gut, the cell contents of plants which are squeezed from the cell walls by chewing can be digested chemically and not microbiologically. Readily digestible carbohydrates are thus absorbed as glucose with no associated methane energy loss. Digestible protein is converted to amino acids, the amino acids absorbed having the same pattern as those in the diet. When protein is fed to ruminants, a substantial part is converted to ammonia and then synthesised by bacteria and protozoa and incorporated as microbial protein. The latter is, in turn, digested by the ruminant's own enzymes, converted to amino acids and absorbed. In the latter case, the pattern of amino acids absorbed is not the pattern in the original feed but more nearly reflects that of the microbial protein. Because the "bacterial vat" of the horse is relatively smaller than that of the ruminant the weight of the horse's gut plus content represents approximately 10% of total weight, less than that of a ruminant but double that, say, of man. Fibre digestion in the horse is less efficient than in the ruminant in that the proportion of a given quantity of cellulose and hemicellulose digested is lower. Feed passage time in equines, however, is quicker than in ruminants, 36–48 h compared with 72–96 h, and equine appetite does not fall to the same extent as that of ruminants when forage quality declines as crude fibre levels increase.

RULES OF EQUINE FEEDING

Before discussing levels of required nutrients and ration formulation for equines, it is considered desirable to highlight the principles of feeding that follow from an understanding of the equine gut.

1. The natural herbivorous diet of the horse contains between 20–50% crude fibre, depending upon plant maturity. At the higher crude fibre levels, insufficient nutrients will be available to provide energy surplus to that required for pure maintenance. Energy in excess of maintenance is needed for growth, reproduction and physical work. It follows that feeds other than natural must be supplied to enable the horse to achieve the aims of today's owners. These feeds are usually so-called concentrate feeds containing higher levels of digestible energy and, sometimes, of digestible protein, than typical pasture or forages. However, it should be remembered that, in the wild, horses are perfectly capable of breeding and surviving without concentrates. They have access to wide areas of range and are not confined to fenced paddocks, they breed at a natural time of the year in order to exploit grass when it is at its most nutritious, they grow more slowly and they are not required to carry passengers and indulge in feats of athletic prowess.

2. The upper limit to daily feed intake in adult mature horses is probably somewhere between 2.8 and 3.0% of body weight when expressed in terms of dry weight of feed. For an animal of 500 kg eating a diet of 90% dry matter this represents 15.6–16.7 kg of feed daily on an as-fed basis. In growing horses, the upper limit in foals is approximately 4%, falling with age and, in ponies, perhaps as much as 6%. However, there is no reason whatsoever why one should wish to satisfy the upper limits of appetite of equines if one can provide them with sufficient nutrients in a lesser bulk. This is particularly the case for high performance horses. It has previously been stated that the weight of an average horse's gut and content represents 10% of its bodyweight. This figure is subject to dietary manipulation and the lower it is, within physiological limits, the better would be anticipated athletic performance. Most of the weight is water. Horses with a high dry matter intake also have a high water

intake. As an example, the consumption of 1 kg of carbohydrate concentrate requires a water intake of approximately 2 kg while that of 1 kg of hay requires 3.5–4.0 kg of water.

While it may be desirable in high performance horses to limit voluntary water intake by dietary manipulation, this in no way implies that horses should not have free access to water at all times, even during spells of prolonged exercise and immediately after it.

3. The lower limit to feed intake will be determined by two factors; nutrient and, in particular, energy requirements and also by the minimum level of roughage consistent with proper digestive function. The subject of energy requirements will be discussed in a later section. There is no very reliable evidence to suggest that crude fibre is necessary to proper equine digestion but, because forage feeds are considered natural to horses, most authorities recommend that, where possible and consistent with adequate energy intake, forages should comprise at least 50% of the diet. This is certainly sensible and safe advice and there is no very good reason to query it unless it becomes incompatible with other aims such as that of achieving good racing performance. All authorities mention some daily forage requirement, the lowest being 0.4 kg per 100 kg body weight when fed as hay. Clearly, minimum levels will to an extent depend upon the crude fibre level of the forage and the digestibility of the crude fibre.

It is perhaps more pertinent to consider the risk to equine welfare of failing to provide sufficient forage than to concern oneself with minimum forage requirements. The greater the forage component of the ration, the greater the proportion of digestion undertaken in the large gut by microbial means. B vitamins are synthesised by large gut bacteria and can be absorbed and used by the host animal. Ruminants are independent of dietary sources of B vitamins and vitamin K since their microbes synthesise them. The same is probably true for horses but it is possible that, where caecal function is reduced by lack of influent material, B vitamin synthesis may become inadequate. This is only a theoretical possibility but should be borne in mind, particularly in high performance horses with increased B vitamin requirements needed for their extra energy metabolism.

One further factor that suggests that low levels of forage intake may be contra-indicated is the albeit disputed and conflicting evidence that stabled horses deprived of long hay are more likely to eat their beds and chew wood. The form of presentation of dry forage does not appear to influence its digestibility to any great extent. It may be fed long, chopped or cubed. Voluntary dry matter intake will tend to be greatest when it is pelleted.

4. One of the more commonly quoted rules of feeding is to feed little and often. In view of the comments made previously about the horse's small stomach, the stimulus required for its emptying and the fact that equines evolved as almost continous grazers, this would appear to be eminently sensible advice. There are occasional reports in the literature that, when complete pelleted feeds of mixed forages and concentrates are used, one or two meals per day are as well digested as are more frequent smaller feeds. Notwithstanding such information, it would seem prudent to limit meals of high dietary density, those of concentrated carbohydrate sources, to no more than 0.5 kg of dry matter per 100 kg body weight per meal and, if necessary for energy intake, to feed up to four meals per day.

The timing of feeding in relation to work should be considered. Horses take approximately 40 min to eat 1 kg of hay and 10 min to eat 1 kg of cereals or concentrate pellets. The very time taken to eat the hay would indicate that most of it should be fed after work with the last feed. It is unwise to work horses within 1.5 h of their completion of a concentrate feed. In the case of racehorses, no hay and only very small concentrate feeds should be offered on race days before the race. It may be necessary to muzzle some horses or to take other action to prevent them from eating their beds on such occasions. It should be realised, however, that while hay deprivation on the day of a race will probably mean that hay will no longer be in the stomach and thus not impeding the diaphragm, it will be residing in the hind gut along with the large quantity of water required for its digestion. Most of it will not have left the body until 36–48 h have elapsed.

5. Two rules of feeding which do not necessarily follow from the preceding section but are nonetheless important are the following:

(a) feed according to condition of horse and to the work required;
(b) make no sudden dietary changes.

Notwithstanding any criteria laid down in the later sections on the subject of feed or energy requirements, the experienced horse feeder will alter the diet to suit the condition of the individual animal. Different animals of different size will vary in their response to a given level of feed intake. The rule of feeding according to condition is more easily written than practised and considerable experience is necessary. The facility to weigh horses on a regular basis is highly desirable but not usually available. However, it is very much easier to ensure that feed level mirrors the work required. If a horse becomes lame or cannot be worked for reasons of bad weather, its intake of digestible energy should be reduced along the lines recommended in a later section. It is normal to achieve this by reducing concentrate intake and increasing forage intake while leaving total dry matter intake constant — in other words by reducing the energy density of the total diet. In the United Kingdom, many horsemen reduce energy density by decreasing oats and increasing bran. While this does reduce the energy density of the concentrate part of the ration, it is usually a very expensive way of achieving a reduction.

While changes of proportions of standard dietary ingredients according to changed energy requirements are reasonable, sudden introductions of new dietary ingredients are unwise. Digestive enzymes can adapt to new feeds but only over periods of days and not hours. A sudden switch, for example, from oats to molassed sugar beet pulp would be contraindicated. Equally, a large introduction of skim milk powder as a protein supplement can have adverse consequences. While foals are obviously well adapted to digest milk sugar, many adult horses have completely or partially lost their lactase enzymes. Lactose in such circumstances may cause uncontrolled microbial fermentation in the gut.

APPETITE AND PALATABILITY

The main determinant of appetite is energy requirement. A horse will eat to satisfy its energy demands. Thus, when rationing equines, one should meet such demands while ensuring that the digestible energy level of the feed is balanced with appropriate levels of protein, minerals and vitamins. On palatable diets such as spring pasture over-consumption of energy may occur, leading to fat deposition. This is not necessarily desirable in modern horses but the ability to store fat to be used in times of subsequent nutritional stress was obviously an advantage in evolutionary terms. However, under modern conditions, the advantage may become a disadvantage. Ponies, in particular, which probably evolved in harsh nutritional environments, are very prone to excessive fatness and laminitis when given high planes of nutrition, possibly because of their relative insensitivity to insulin.

If the energy density of the diet is low, appetite may be limited by the sheer bulk of the food before energy requirements are satisfied. It has been previously stated that upper levels of daily dry matter intake in mature horses lie in the range 2.8–3.0% of body weight. However, the bulk limitation does not have to be in the form of dry matter. Carrots have a high level of digestible energy on a dry matter basis but only contain about 10% of their weight as dry matter. The appetite constraint here would be imposed by the 9 kg of water ingested with every 1 kg of carrot dry matter. It should be recognised that brood mares in late stages of pregnancy can consume less bulk than equivalent non-pregnant animals for the obvious reason that the unborn foal and placenta occupy space normally available to the digestive tract.

While appetite is determined largely by energy demand on normal diets of average palatability in healthy equines, it may be depressed by a variety of factors. Ill health and nutritional deficiencies are possible causes of low levels of feed intake. The most likely equine deficiencies to have this effect in the United Kingdom are those of salt and vitamin A although, theoretically, several others are possible.

The palatibility of the food, too, may limit voluntary intake. Individual animals tend to have their own likes and dislikes. If one is attempting to feed high energy intakes to hard working horses, it is necessary to take note of these fads. Often it is difficult to persuade such animals to eat sufficient energy to perform the work required of them. In general, the horse's preferred feed is to be obtained from good quality grazing. Of the concentrate feeds, oats are often but not invariably the preferred cereal of equines and can be fed whole, bruised or crushed. Horses normally use their teeth to sufficient effect that whole oats are nearly as well digested as those fed bruised or rolled. This does not apply to barley which is a harder grain and should be rolled or flaked before feeding. In Britain, maize is normally fed only in flaked form. Oats contain approximately 5% oil which is liable to rancidity once the grain is crushed. Crushed oats thus become progressively less palatable with storage. It has previously been stated that probably the most palatable feed for horses is young grass. The cell contents of young grass are high in sugars and most horses appear to have a sweet tooth. Molasses or molassed sugar beet pulp are thus often appreciated by horses and can be used to enhance the palatability of concentrate feeds. Bran, a food beloved of British horse owners, is more palatable when fed as a mash than when fed dry. Bran mashes can be useful for tired horses after heavy work when they may have to be tempted to eat.

Some horses are greedy feeders and bolt their concentrate feeds without proper mastication. Such animals are subject to colic and digestive disturbance. Even if such disorders are avoided, much of the food may pass through them undigested. Various strategies may be used to overcome this habit or its adverse consequences. The energy density of the concentrate feed may be diluted with bran which also opens its texture. Alternatively, the concentrate may be mixed with chopped hay, making it much more difficult to swallow without chewing. Another approach is to place large round stones of a size impossible to swallow in the food bowl, thus slowing down the horse's rate of ingestion.

ESSENTIAL NUTRIENTS

Food is a mixed package of water and dry matter which contains nutrients. The moisture content of feed seldom contains sufficient water so an extra source should be supplied *ad libitum*. The dry matter component of feed may be divided into organic and inorganic or mineral fractions. It is in the organic fraction that are to be found the carbohydrates, fats, proteins and vitamins. Only a proportion of the organic matter is digestible, the residue being of no nutritional use to the animal. When considering nutrients, it is logical to start with a consideration of energy since, as had been previously stated, it is the requirement for energy that largely determines appetite.

While animals have a requirement for energy, it can be obtained from a variety of nutrients and a given quantity will have more or less the same effect on the animal from whatever source it comes. Energy from carbohydrate is absorbed as glucose or as volatile fatty acids while that from fat as free fatty acids. Within the body, however, the different sources can be used for equivalent activities or converted from one to the other, surplus being stored as fat. There is no basis in fact for the old horseman's belief that barley makes a horse fat and oats fit when intakes of digestible energy from each are similar.

Carbohydrates are the prime source of energy in typical equine diets. The vegetative parts of plants contain sugars within their cells and crude fibre in their cell walls. Both are carbohydrates. Concentrated carbohydrate sources, starches, are to be found in the seeds or storage organs of some plants as in the case of cereal grains. When proteins are oxidised, they provide an amount of energy to the animal equivalent to an equal weight of digestible carbohydrate. Thus, while protein is essential for many synthetic functions in the body, it should not be forgotten that it must also be regarded as a source of energy. The most potent source of energy is fat (or oil), a given quantity of digestible fat providing some 2.25 times the energy of an equal quantity of digestible carbohydrate or protein. In Britain, energy is usually measured in Megajoules (MJ) and, for the horse, expressed as digestible energy (DE). In cattle, energy is expressed as metabolisable energy (ME) which is approximately 80% of DE. The equivalent conversion figure for horses is not precisely known but would tend to be higher. In the United States, energy is expressed in Megacalories (Mcal) which can be converted to MJ by multiplying by 4.184. The recommended levels of energy and other nutrients for maintenance, work, growth, pregnancy and lactation will be provided in a later section.

Protein is digested in the gut and absorbed as amino acids. Animals have need of approximately 20 amino acids of which 10 are so called essential amino acids which can only arrive through the gut wall. The non-essential amino acids can be made from other amino acids within the body provided that there is a large enough pool of total amino acids to work with. Dietary protein, therefore, must be adequate to provide this pool. The dietary requirements of equines for essential amino acids have not been fully elucidated. A certain amount will be provided by the absorption of amino acids of digested microbial protein in the hind gut. Much of the essential amino acid requirement of cattle is acquired from the digestion in the small intestine of amino acids derived from bacterial and protozoal protein produced in the rumen. Protein requirements can even be spared in ruminants by feeding non-protein nitrogen sources such as urea which the rumen microflora convert to protein. These processes in horses are certainly less efficient. It is therefore almost certain that dietary sources of essential amino acids must be provided. Vegetable proteins are often very unbalanced in that their pattern of amino acids differs from that of mammalian tissue and hence requirement. The most limiting essential amino acid of typical vegetable protein is lysine, followed by methionine. Of all vegetable protein sources, soya bean is the best for these two acids. Animal proteins such as fish meal, meat meal and skim milk are generally good sources of lysine and methionine and can thus be used as supplements in critical diets such as those formulated for weaned foals.

Animals have no general dietary requirements for fat other than for small quantities of essential fatty acids which are needed for synthetic activity and which cannot themselves be synthesised in the body. Linoleic acid is the simplest of the essential fatty acids and horses can make the more unsaturated fatty acids, linolenic and arachidonic, from it. The oils in oats, maize, soya bean and linseed are all rich in linoleic acid as is fresh herbage. Normal horse diets are thus unlikely to be deficient in essential fatty acid. Despite the absence of gall bladders, equines are able to digest quite high levels of fat. Up to 15% soya bean oil and 10% animal fat have been successfully fed.

There are some one dozen vitamins, organic compounds incapable of synthesis within the animal but needed for essential functions, which must be absorbed through the gut in adequate but very small amounts. Vitamin C is not required by equines which synthesise their own. The other vitamins can be divided into fat and water soluble groups. The latter group, comprising B vitamins, can be produced by the microflora of the large gut as can the fat soluble vitamin K. The combined dietary sources coupled with what is produced and then absorbed from the large gut of these vitamins means that deficiencies of them in equines are unlikely. Notwithstanding, they are relatively inexpensive and non-toxic and, particularly in horses achieving high levels of physical performance, there may be a case for judicious supplementation. They should also be included in the diets of young foals which will not have established the full range of microbial gut synthesis. Thiamin and folic acid are the two B vitamins which are possibly most likely to become limiting—the latter only when no green feed is available. Vitamin B12, cyanocobalamin, is often given by injection to horses. Some veterinarians believe that it acts as a stimulant and, while this may be so, there is no reason to suppose that horses are ever deficient in this vitamin which they synthesise for themselves, given a dietary source of cobalt. Horses are known to be very substantially less subject to cobalt deficiency than are ruminants on the same range conditions. There is currently a practice of feeding high levels of biotin, far higher than normal physiological requirements, to animals with hoof problems. The scientific basis for this is disputed but, given the value of many horses, it is scarcely surprising that biotin use for specific problem animals continues pending resolution of the scientific debate.

The fat soluble vitamins, A, D and E, may become limiting or deficient in British horses in late winter. Fresh green feed is a potent source of β carotene, a vitamin A precursor, and grazing animals will never be deficient. Furthermore, they will be able to store body reserves sufficient for two to three months. However, β carotene is destroyed during hay making. Carrots and grass meal are good sources of β carotene and synthetic vitamin A is cheap and readily available. There are no good natural sources of vitamin D available to horses other than fish liver oil which was often used in the past. However, it is liable to rancidity and synthetic vitamin D is now preferred when needed. In the summer, the vitamin is synthesised in the skin of animals under the influence of ultraviolet light. There is not much ultraviolet light or even sunlight in the United Kingdom during winter. Furthermore, ultraviolet does not penetrate window glass. Supplementary vitamin D should thus be available to housed horses in winter. Vitamin D is one of the most toxic vitamins and the supply of even five to ten times the recommended level is liable to produce pathological change in the form of metastatic calcification of soft tissues. Vitamin E is quite widely distributed in feedstuffs of vegetable origin but is largely destroyed in grass during haymaking and may decline in cereals during storage. Winter supplementation with synthetic vitamin E along with A and D is thus sensible, particularly in the case of growing stock and those used for breeding or high performance and which are largely stabled.

The remaining essential nutrients to be mentioned are the minerals or inorganic components of the ration. These can be divided into major and trace minerals depending upon the quantities required in the diet. There are seven major minerals which will be discussed first. Sodium and chlorine are best dealt with together. Common salt is the most satisfactory supplement, containing approximately 40% sodium and 60% chlorine. Vegetable diets tend to be sodium deficient and salt requirements vary greatly, depending upon the degree of sweating. Generally, therefore, salt should always be provided as a lick since horses seem to have nutritional wisdom with respect to sodium and can adjust intake to requirement. It should not be assumed from this that they have specific appetites for other such nutrients. It has been shown, for example, that they cannot balance intakes of calcium and phosphorus given free choice of these minerals. Calcium and phosphorus are two other major minerals which can conveniently be dealt with together in that the relationship of one to the other in the diet should be fixed within a narrow band. Recommended Ca to P ratios for equines lie in the range 1.1–1.6. Growing and breeding animals have higher requirements for these minerals than do mature animals due to the fact that a high proportion of calcium and phosphorus is in the form of bone. Calcium requirements as a percentage of dietary dry matter range from 0.8% in young foals to 0.3% in non-breeding mature horses. The equivalent figures for phosphorus are 0.5 to 0.25%. When these levels are contrasted to those commonly found in typical horse rations, it will be immediately apparent that calcium in particular may often be marginal or inadequate even in mature animals. Typical east of England hay samples contain 0.3% calcium and 0.23% phosphorus. Legume hays usually contain four times more calcium and slightly more phosphorus. The calcium and phosphorus levels of oats are 0.1% and 0.38% while those of bran are 0.15% and 1.3%. It will thus be seen that a 50:50 oat grass hay ration will only provide 0.2% dietary calcium. The phosphorus in vegetable concentrates, such as cereals and soya bean meal, is phytate bound and is only about two thirds as available to horses as is phosphorus in grass or hay. Nevertheless, the calcium to phosphorus ratio is still very adverse in cereal grains and even more so in bran. The cheapest and most readily available calcium supplement is limestone flour, containing approximately 35% calcium. Steamed bone flour and dicalcium phosphate contain both calcium and phosphate, 24% Ca, 12% P and 20% Ca, 18.5% P respectively. It should be noted that horse owners often feed trace mineral vitamin supplements in a limestone base and think that this is also providing a calcium supplement without appreciating that, in this case, the limestone is merely a vehicle for the trace nutrients and does not provide significant levels of calcium.

The dietary magnesium levels for horses should be approximately 0.15%. The only circumstances in which problems may occur are in horses grazing fertilised spring grass in which magnesium is very poorly available. There have been reports of hypomagnesaemic tetany (grass staggers) in such circumstances. However, cattle would appear to be more vulnerable in this respect than equines. It has been suggested that the addition of 5% magnesium oxide (calcined magnesite) to the salt lick at this period will protect against this eventuality. There is one further possible use for magnesium supplementation. Ross (1986) claimed that horses with sweet itch, a hypersensitivity to midge bites, would seek out and readily consume magnesium salts and this would cure their problem. Horses in the same pasture but without sweet itch would not consume the magnesium. This surprising observation does not appear to have been independently substantiated.

Horses need a dietary potassium level of approximately 0.6%. Potassium levels of forages are substantially in excess of this as are those of molasses and oil seed meals. Cereals are slightly below. However, potassium is most unlikely to be a limiting nutrient in equines fed other than very atypical rations. Sulphur, the last of the major minerals to be mentioned, is only required in the form of sulphur-containing amino acids and thus requirements are met if the dietary protein levels are adequate.

Several of the trace minerals have, on occasions, been shown to be deficient in horse rations. Horses need iodine, cobalt, copper, iron, manganese, zinc and selenium. The dietary iodine requirement is estimated at 0.1 ppm. Adult horses have never shown signs of iodine deficiency but brood mares in areas where goitre is known to occur have given birth to weak or stillborn foals with enlarged thyroid glands as have mares fed excess iodine, often in the form of seaweed meal. Cobalt is needed for the synthesis of vitamin B12 but horses need far less than ruminants and are extremely unlikely ever to become deficient. The copper requirement of horses is in the region of 9 ppm. Molybdenum and zinc can interfere with copper absorption. Some United Kingdom pastures are marginal in copper and others are associated with copper deficiency in cattle due to their high molybdenum levels. It is reasonable to suspect such pastures as being potentially problematical for horses. In the USA, copper deficiency has been reported to cause exostoses above and below the joints in foals and may also cause osteochondrosis. Neither zinc nor iron is likely to be deficient in normal rations for horses although extra iron may be needed following severe blood loss. Selenium is an essential nutrient for horses. The dietary requirement does not exceed 0.2 ppm and may be as low as 0.1 ppm. This mineral is deficient in many areas of the world including parts of the United Kingdom. Selenium deficiency problems, often associated with muscle disorders, are being increasingly diagnosed in ruminants in Britain and have been reported in horses in other parts of the world.

ENERGY AND PROTEIN NEEDS ACCORDING TO THE PHYSIOLOGICAL STATE

Maintenance

This can be defined as a condition of no weight change and minimal activity in a thermoneutral environment. The energy required for maintenance is generally related more to surface area than to weight and maintenance energy is thus usually expressed in relation to metabolic body weight, $W^{.75}$, where W equals the weight of the horse in kilograms. The digestible energy (DE) needed for maintenance $= 0.65 \, MJ/W^{.75}/day$. Table 5.1 shows the relationships between weight, metabolic weight and maintenance energy per

Table 5.1. Maintenance energy requirements

Weight (kg)	Metabolic weight (kg)	DE (MJ) for maintenance
200	53	34
400	89	58
600	121	79
800	150	98
1000	178	116

day for horses of varying weight. The maintenance protein requirement is 4.65 g of digestible crude protein (DCP) per MJ of energy. In other words, a 600 kg horse would require $79 \times 4.65 = 367$ g of digestible crude protein per day for maintenance. It is interesting to contrast this with what it would receive in the way of DCP when given various maintenance diets. A typical east of England hay sample made from mature grass would have an energy level on a dry matter (DM) basis of 7.5 MJ of DE/kg and a DCP level of 28 g/kg. To satisfy maintenance energy needs, the horse would need to consume $79/7.5 = 10.5$ kg of such hay. However, this would would only provide $10.5 \times 28 = 294$ g of DCP, leaving a 20% shortfall in protein. The effect of a protein deficit is a decreased appetite. The horse, therefore, would lose condition on such hay. However, if one offered 2 kg DM of oats (12.9 MJ DE and 84 g DCP/kg DM), the horse would also need to consume $\frac{79 - (2 \times 12.9)}{7.5} = 7.1$ kg of hay to satisfy its maintenance energy needs. Its DCP intake under such circumstances would be $(7.1 \times 28) + (2 \times 84) = 367$ g, precisely matching requirements. Good quality grass hay, as opposed to that above, would have a typical analysis of 9.5 MJ DE and 65 g DCP/kg. Maintenance energy of the 600 kg horse would be provided by an intake of $79/9.5 = 8.3$ kg of this good hay on a dry matter basis, providing $8.3 \times 65 = 540$ g DCP. This is approximately 47% more than required but within reason and with the possible exception of racehorses, protein surplus is not harmful. In summary, therefore, the only maintenance diets likely to be limiting in protein are those in which attempts are made to meet total requirements with poor quality forage alone.

Work

When a horse is required to take exercise, its energy requirement obviously increases to an extent which depends upon the length and severity of the exercise. The extra energy over that required for maintenance is illustrated in Table 5.2. While energy requirement increases sharply with exercise, protein requirement rises only slightly or not at all unless sweating occurs. The horse is peculiar in that protein is initially excreted in large amounts in sweat, causing the lather seen with sweating. However, when sweating is prolonged,

Table 5.2. Feed energy required for various types of exercise

Activity	Requirement (MJ/100 kg/hour)
Walk	0.2
Slow trot/canter	2.1
Fast trot/canter	5.2
Canter + some galloping/jumping	10.0
Gallop	16.3

the initial protein concentration of the sweat reduces by half within half an hour and reduces by a further half in two hours. To accommodate this protein loss, it is conservatively suggested that one should provide 4.65 g DCP/MJ energy fed as is the case for maintenance. Thus, if one opts to feed more of the standard maintenance ration for work, there is no need to consider altering its protein concentration. If, however, one opts to feed the same quantity of feed as at maintenance, then one should increase its protein concentration by the same proportion as one increases the energy concentration. Many laymen, even experienced horsemen, consider that dietary protein concentrations should be markedly increased to meet the requirements of hard work. Unfortunately, some horse feed compounders manufacture feeds to accommodate these foibles. There is recent evidence that racehorse performance declines as dietary protein levels become excessive.

It is worth considering how work is likely to influence the feed requirements of a hunter of 600 kg, hunting two days a week. On five days, suppose that it spends 22 hours per day in the stable, one hour at the walk and one hour at a slow trot or canter. It would require 79 MJ DE/day for maintenance, $0.2 \times 6 = 1.2$ MJ DE for walking and $2.1 \times 6 = 12$ MJ DE for the faster work. This totals 92.8 MJ DE/day, an increase of 17% over maintenance. On a hunting day, suppose that it spends 16 hours in the stable, three hours travelling and at the meet, two hours walking, two hours at a fast trot or canter and one hour cantering with some galloping and jumping. The horse would continue to require 79 MJ DE/day for maintenance. Assume that the three hours of travelling and at the meet are as energetic as walking. The extra energy would thus be $5 \times 0.2 \times 6 = 6$ MJ $+ 2 \times 5.2 \times 6 = 62.4$ MJ $+ 1 \times 10.0 \times 6 = 60$ MJ. This makes a grand total of 207.4 MJ, 2.6 times maintenance needs. The horse could only achieve this intake of energy from oats if it ate 2.6% of its body weight, 16 kg DM of oats. This would clearly be ridiculous. However, if one averages the daily energy requirements needed over one week, the result is $\frac{(5 \times 92.8) + (2 \times 207.4)}{7} = 125.5$ MJ DE. One would probably have to overfeed slightly on non-hunting days because the horse may have neither the time nor inclination to eat 125.5 MJ of feed energy after hunting and should not receive more than a small amount of concentrates before. However, for the sake of this example, suppose that one keeps to a daily allowance of 125.5 MJ and attempts to achieve it by feeding oats and hay to provide 2% of body weight or 12 kg DM between them. If one feeds half oats (12.9 MJ DE/kg) and half hay (8.5 MJ/DE/kg), one would provide 128.4 MJ DE/day, near enough to that required. There would be plenty of protein in such a ration.

Racehorses in work probably warrant more thought than many receive in respect of their feeding. Quite apart from the fact that muscle and skeletal growth are not complete in the younger ones, making mineral, vitamin and possibly even limiting amino acid supplementation important, one should consider the effect of diet on fast work for all racehorses. Diets of low energy density (high fibre) will greatly increase gut fill and hence dead weight while those of high density will act in the opposite direction. It would thus be helpful to performance, at least in theory, to pack as much energy into as small a bulk as consistent with proper digestive function. Trainers might thus give more consideration to minimising hay dry matter intake, to feeding only hay of maximum digestible energy and to adding up to 10% of fat to rations. They should also avoid surplus protein.

Pregnancy

Pregnancy does not materially increase energy or protein requirements above those of maintenance until the final three months. Thereafter, energy intake should be increased by 15% and digestible protein by 30% over minimal maintenance needs. It should, however, be remembered that many maintenance rations contain surplus digestible protein. In quantitative terms, digestible protein fed per MJ of energy should rise from 4.65 g at maintenance to a minimum of 5.25 g in late pregnancy. Mares should not be allowed to become even slightly thin during late lactation otherwise subsequent rebreeding and lactation will be adversely affected. Extreme fatness should obviously be avoided.

Lactation

Mares probably achieve maximum milk yields when holding a constant weight or gaining weight slightly. Peak yields occur at approximately eight weeks post foaling and thereafter decline. Yields vary from mare to mare, probably averaging 3% of body weight/day at peak. Energy needs are increased by 80 to 90% over those of maintenance at peak lactation. It should perhaps be noted that these could be met by very high quality grazing alone were lactation to be synchronised with the grass growing season. This also tends to illustrate how much surplus energy is available to non-breeding animals turned out on spring grass and explains why many become excessively fat in consequence.

The protein level of mare's milk is approximately 3.1% after 1 week of lactation falling to 2.2% at peak yield. Dietary protein requirements are roughly 2.5 times greater at peak yield than at maintenance. On this basis, digestible protein fed per MJ of digestible energy should initially be 7.3 g in early lactation falling to 6.4 g at peak and after. These protein levels cannot readily be achieved on oat grass hay diets, given the usually disappointing DCP levels of British hays. A protein supplement should thus be included in the concentrate part of the ration. It is sensible that such a supplement should have a well balanced amino acid profile as lysine tends to be low in cereal protein. Feed compounders making pelleted diets may opt for an animal protein supplement such as fish meal. Those wishing to mix their own rations should consider solvent extracted soya bean meal, the best quality vegetable protein.

Growth

Present day growth curves of horses of many light breeds on similar planes of nutrition tend to be remarkably similar in shape. At 6 months, a foal can be expected to have achieved 45% of final adult weight. This rises to 66% at 12 months and 81% at 18 months. Final adult weight is achieved between 3 and 4 years. Ponies tend to be somewhat faster maturing. However, there is no reason to suppose that present growth curves represent the optimum although they do appear to be approaching the maximum. Thirty to forty years ago, growth was 10–15% slower. The difference is probably almost entirely due to the provision of higher quality diets and has little to do with genetic selection. While rapid early growth is clearly necessary for animals destined to race as two-year-olds, its advantage for other categories of horse or pony is by no means clear. On the contrary, so-called overnutrition is frequently associated with the development of skeletal problems in young horses. There is some evidence from equine research and much from farm animal research that skeletal abnormalities are much more common on high planes of nutrition even when diets are apparently perfectly balanced with respect to energy, protein, minerals and vitamins. It is therefore not easy to offer advice on optimum energy intake for growth although one is on more certain ground when suggesting levels for energy protein rations. Even in the latter case, however, it is necessary to assume that the digestible protein will have a reasonable amino acid composition and the importance of the first limiting amino acid, lysine, is becoming better understood as equine research advances.

Table 5.3 summarises the daily digestible energy intakes expressed as a percentage of the maintenance energy intakes at final adult weight of foals and yearlings of riding horse and pony types. These figures would be expected to provide near maximal growth rates provided the diets were balanced. It

Table 5.3. Percentage of adult maintenance energy intake needed for rapid growth at various ages

Age (months)	Horse	Pony
3	75	90
6	92	104
12	100	96
18	103	98

can thus be seen that, from the age of 3 months in ponies and 6 months in horses, youngsters have very nearly the same energy requirements as their parents at maintenance if they are to grow fast.

Foals should receive 8 g of DCP per MJ of DE and yearlings 6.25 g. This presupposes protein of good amino acid balance. For foals, lysine should form 5% of the total crude protein and, for yearlings, at least 3.5%. The lysine levels of cereal and grass proteins are only about 3% while those of animal protein and soya bean meal are 7% or over. When protein supplements of poorer quality, such as peas or field beans are fed, more protein must be provided since their lysine levels are only 5%.

FEEDS, FEED ANALYSIS AND FEED COMPOSITION

Pasture and conserved forages

Pastures may vary in their proportions and species of grasses, legumes and herbs. The most productive pastures in the United Kingdom are predominantly composed of grasses, the nutritional composition of which changes markedly with season. Grass is at its most nutritive in May and June. With advancing maturity, moisture levels decline and the increasing dry matter contains decreasing proportions of digestible energy, protein and minerals. The subjects of pasture establishment and management are beyond the scope of this chapter but have been reviewed by Archer (1980).

Many horses, mature and growing, cope satisfactorily when given free access to spring grass. Some, however, and particularly ponies, tend to overconsume and not limit appetite to energy requirement. Such animals will become excessively fat and may develop laminitis. If growing, they may also suffer bone and joint problems. As an example, one can consider a 12-month-old hunter yearling, weighing 400 kg with an expected adult weight of 600 kg. Reference to Table 5.3 indicates an energy requirement equivalent to 100% of that of adult maintenance. Table 5.1 shows that this is 79 MJ DE/day. Requirement for DCP was stated to be 6.25 g/MJ for yearlings. This represents $6.25 \times 79 = 494$ g. Assume the young grass to contain 10.2 MJ DE and 170 g DCP/kg DM and to have a dry matter content of 22%. The yearling could satisfy its energy requirement by consuming $79/10.2 = 7.745$ kg of grass dry matter. This is equivalent to $7.745/0.22 = 35.2$ kg of fresh grass, providing $7.745 \times 170 = 1317$ g DCP, over 2.5 times the requirement. However, an intake of 7.75 kg represents only 1.9% of body weight while the upper limit may be nearer 3% assuming grazing time and accompanying moisture intake not to be previously limiting. A considerable energy excess, as well as protein excess, is thus possible. This, of course, is even more the case for adult animals. The mature 600 kg hunter, with the same energy requirement as the 400 kg yearling, would only need to consume $7.745/600 = 1.29\%$ of body weight to satisfy its needs, leaving a greater potential for surplus intake. It is for this reason that many horse owners sensibly limit access to pasture when it is in its young early growth stage. In any event there should never be a sudden switch from hay concentrate diets to all grass in the spring. The change over should be gradual to avoid digestive disturbance. This also mitigates against the possibility of hypomagnesaemia as does the introduction of calcined magnesite into salt licks at this time. Clovers are less palatable to horses than are grasses and should not form a high proportion of horse pastures. Compared with grasses of equivalent growth stage, they contain less energy, more protein and considerably more calcium.

Of the conserved forages, artificially dried grass and lucerne meals more or less resemble the original products in nutritive terms. The material may be pressed into cobs and wafers as the meals themselves are dusty. Notwithstanding, they are not always as palatable to horses as one might expect and, under conditions of inappropriate storage, are liable to moulding. They are also expensive. Grass hay is the primary conserved forage used for equines in the United Kingdom. The nutritive value of hay depends upon the quality of the grass from which it was made and upon conservation conditions. Hay from young grass with a high leaf stem ratio that is well made may have DE and DCP levels of 9.5 MJ and 65 g/kg DM. However, hay from more mature grass that is less well made is possibly much more typical of British hay and may only contain 7.5 MJ DE and 28 g DCP/kg DM. It is not easy to evaluate the feeding value of hay by visual examination. Clearly, if it looks mouldy or smells musty, it should be avoided. Those using large quantities of hay should seek to have it analysed by the local Agricultural Development and Advisory Service laboratory. The analysis will provide the Metabolisable Energy (ME) level of the hay for cattle as well as the crude protein level. In the absence of better information, it is reasonable to assume, because horses digest fibre less well than cattle, that the ME level for cattle is the same as the DE level for horses. The crude protein can be converted to DCP for horses by multiplying by a factor of 0.74 and subtracting 25. Thus, with a crude protein of 85 g/kg DM, the DCP figure becomes $(85 \times 0.74) - 25 = 38$ g/kg DM. It is best to assume that hay has no vitamin A or vitamin E activity.

Dry hay may be an inappropriate forage for horses with broken wind or other respiratory problems. Damping the hay may help reduce dust but this may not be enough. Alternative solutions are to feed molassed, chopped straw or high dry matter silage (haylage). The former material can also be used to good effect to mix into the concentrate rations of horses that bolt their feed. Haylage spoils on exposure to air and thus tends to be sold to owners in small bags which are very expensive in relation to their contained nutrients. In the absence of proper analysis being available, one should assume that haylage has a feeding value equivalent to that of good to moderate hay.

There is a tendency among British racehorse trainers, and even breeders, to regard hay as a required source of dietary fibre but to assume that it makes a negligible contribution to energy intake. One can explain this on the fact that British haymaking conditions are often adverse early in the season. Hay least likely to be moulded thus tends to be very stemmy and overmature. Nevertheless, assuming that hay is well made, horsemen would be illogical to opt for anything other than high nutritive value material.

The principal energy concentrates for horses are the cereal grains, oats, barley and maize. Oats are preferred by most horsemen. Of the three, they have the lowest energy and highest fibre levels. Oats are also the most variable of the cereal grains in nutritive value in that the proportion of husk to grain can differ markedly. Thus, good oats should have a high bushel weight since low bushel weights are associated with much husk and shrivelled grains. The horse's stomach is reasonably small and not an important site of bacterial digestion. Nevertheless, it contains a portion that is richly populated with lactobacilli which produce lactic acid and gas by fermenting sugars and starch. A single meal of cereals may remain in the stomach for long periods awaiting the stimulus of further food being chewed in the mouth before being passed to the small intestine. The contents may ferment producing a lactacidosis and, in extreme cases, this can result in colic or even stomach rupture. Oats, with their high fibre, provide a more open textured material in the stomach and are thus the least likely of the cereals to cause this eventuality. Wheat is the most likely because its high gluten content leads to the formation of a doughy lump in the stomach. It should thus not be used except in very small quantities or in low levels in compound feeds. Oat grains are quite soft and easily ruptured during mastication. Whole oats, therefore, are almost as well digested by the majority of horses as are bruised or rolled oats. Once the outer coating of the oat grain has been damaged by crushing, the relatively high levels of unsaturated fat within are exposed to the air and can oxidise. This rancidity makes stored, crushed oats unpalatable and is also associated with destruction of vitamin E. Barley and maize are hard grains and should not be fed whole. Barley may be rolled, flaked or boiled while maize is almost always fed as flaked maize in the United Kingdom. The flaking process involves the use of heat and this gelatinises the starch, possibly rendering it slightly more digestible. Horsemen often mix horse rations on the basis of scoops rather than by weight. It is therefore pertinent to point out that a scoop of barley or maize may easily weigh double that of a scoop of oats. Furthermore, a given weight of either grain, has a higher DE level than of oats. Freshly harvested cereal grains may be associated with digestive disturbance and it is sensible to store them for a couple of months before use.

The commonest alternative energy concentrate to cereals is dried, molassed sugar beet pulp. It is inferior to oats in protein but equivalent in energy. It has a high fibre component and is not associated with digestive disorders if soaked before feeding. Fed dry, the material may cause problems as it can swell in the stomach. Straight beet molasses are also to be regarded as an energy concentrate, equivalent in energy on an as fed basis to the molassed beet pulp but containing very low protein. Molasses are often poured on to a concentrate feed to enhance palatability, as a source of energy and, sometimes, as a vehicle for nutritional supplements. Carrots, too, may be thought of as an energy concentrate with added water. On a dry matter basis, they are equivalent in energy to put poorer in protein than cereals. Carrots are very palatable and are a useful source of vitamin A activity in the winter.

Bran, by tradition a much favoured food for horses, is wheat with most of its starch removed. It thus has less energy and more fibre, protein and minerals than the original product. Bran is a by-product of the flourmilling industry. It used to be very cheap and thus quite a sensible food for horses when used with discretion. Attention has already been drawn to its very poor calcium to phosphorus ratio. In the past, the feeding of high levels of bran was associated with bone abnormalities in horses — so called Big Head or Miller's disease, millers having unlimited access to bran. Nowadays, partly no doubt due to the fashion for high fibre breakfast cereals, bran has become expensive, often more expensive than oats. Its role in equine feeding is thus hard to justify although many horsemen, often by nature traditionalists, continue to use it in surprising amounts. It is frequently used to dilute the energy levels of concentrate feeds if a horse's workload is reduced. A cheaper alternative is to alter the forage concentrate ratio. It is also sometimes mixed with cereals for horses that consume their concentrates too quickly without chewing. However, chopped straw or hay will do as well for this purpose. Bran, because of its high fibre level, draws a lot of water into the gut and thus has a laxative action. When fed as a mash, steeped in boiling water with added salt, it can make a palatable feed for tired horses after hard work such as hunting. The relatively high protein content of bran is not particularly useful because it is not a good source of lysine. Mature, non-breeding animals almost always receive enough protein from oats and hay and breeding and growing animals needing extra protein should receive high quality supplementary protein. Furthermore, very considerable care should be taken with calcium supplements if bran is to be fed to young stock.

Both lucerne and grass meals may be regarded as protein concentrates as well as sources of fibre and energy. Good quality lucerne meal, for example, may contain as much as 130 g DCP/kg DM. However, lysine only represents about 4.25% of the crude protein. Solvent extracted soya bean meal is almost certainly the protein supplement of choice for those wishing to mix their own critical diets such as those for foals. It is not commonly used in Britain by horse owners although it is the most widely used vegetable protein supplement in all farm livestock and is widely used for horses in the USA. It is, of course, used by British horse feed compounders.

Generally, linseed is the only source of concentrated fat that is fed by traditional British horse owners. Linseed is an oilseed. In such seeds, soya being another example, energy is stored as fat rather than as starch. The seeds of linseed are toxic when fed raw and must therefore be boiled in water. This produces mucilage in large quantities and the resulting product is often fed with bran on an occasional basis to give a gloss or bloom to the coat, due to the relatively high level of unsaturated fat. The same effect can be achieved very much more conveniently by occasionally pouring cooking oil on to the feed, be it olive, maize, sunflower or soya bean oil. It has previously been stated that horses are well able to digest high levels of fat and those interested in heavy or fast work performance from their horses concurrent with relatively low dry matter intakes should consider the addition of up to 10% of a product such as maize oil for this purpose. However, a gradual build up to this level would be desirable rather than its sudden introduction.

There remains to be considered the subject of commercially manufactured horse feeds. These may be sold either as coarse mixes or as cubes. There is a huge range available varying from rations designed as complete feeds inclusive of forage to grain balancers. Most commonly, however, the feeds are recommended as replacements for the concentrate part of the ration only. Appropriate feeds are available for all classes of stock at all activities. For this reason, it is extremely difficult to make other than generalised comments. Possibly, therefore, the first thing to realise is that all companies manufacturing such products will be employing their own nutritionists or using nutritional consultants. Thus, if one wishes more information than that stated on the bag, one should consult the company nutritionist. If the enquirer does not receive helpful advice, he should change to another compounder. As an example, energy levels are not declared in compound animal feeds. However, the formulator will have mixed the feed ingredients to a known energy level and should be able to help. Most compound feeds will have been fortified with supplements. These, however, may not be adequate if the feed is mixed with other ingredients of the concentrate ration. This is unlikely to be a serious worry except in the case of young stock and breeding stock provided always that a salt lick is available. As has been previously stated,

Table 5.4. Average composition of common horse feeds[a]

Feed	Dry matter (%)	Nutrient concentration on dry matter basis			
		DE (MJ/kg)	DCP (g/kg)	Ca(g/kg)	P(g/kg)
Pasture grass					
Spring	20	10.5	170	5.0	3.9
Summer	26	9.0	80	4.0	3.1
Grass Hay					
Good	88	9.5	64	3.3	2.6
Average	88	8.5	35	3.3	2.4
Poor	88	7.5	28	3.3	2.1
Lucerne Hay					
Good	88	9.0	110	13.5	2.7
Oats	90	12.9	84	1.1	3.9
Barley	90	15.3	82	0.9	4.7
Maize	90	16.8	53	0.2	2.8
Dried molassed sugar beet pulp	90	12.1	57	8.8	1.1
Carrots	12	15.8	60	3.7	3.2
Beet molasses	77	16.4	50	2.1	0.4
Solvent extracted soya bean meal	90	14.8	447	3.6	7.5
Bran	90	10.8	129	1.6	13.2
Linseed	94	21.2	200	2.4	5.6
Maize oil	98	35.0	0	0	0.0

[a] The forage feeds are extremely variable. These figures are for guidance only. Mineral levels vary with soil type, plant maturity etc.

Table 5.5. The weights of common horse feeds (as fed) which
provide the equivalent digestible energy to 1 kg of oats (as fed)

Barley	0.84
Maize	0.77
Dried molassed sugar beet pulp	1.07
Bran	1.19
Solvent extracted soya bean meal	0.87
Beet molasses	0.92
Linseed	0.58
Maize oil	0.34
Good hay	1.39
Poor hay	1.76
Carrots	6.12

there is a tendency for feed manufacturers to err on the generous side in respect of protein for performance horses. While this may not be a good thing, to an extent, it reflects public demand, however misinformed. The most widely sold and distributed type of compound horse feed is probably the so-called horse and pony cube, usually equivalent in energy to an oat bran mixture and more expensive than either. If readily available in the area, a more specialist feed, either cereal balancer or higher energy cube, often represents better value for money. The great advantage of compound feeds is their convenience of use.

Typical analyses of the feeds mentioned are summarised in Table 5.4. Table 5.5 summarises the energy levels of various feeds relative to the digestible energy levels of oats on an as fed basis. For example, either 0.8 kg barley or 6.12 kg carrots would replace 1 kg of oats.

THE NEED FOR VITAMIN AND MINERAL SUPPLEMENTS

Vitamins

There is little evidence that supplementary B vitamins are ever beneficial to any type of horse fed other than most peculiar rations. However, since tiny improvements in performance in, for example, racehorses are not possible to measure reliably by scientific means, it seems not unreasonable that trainers may wish to consider the use of B vitamin supplements, either alone or as part of a multivitamin trace mineral supplement. B vitamin excesses are not harmful and the vitamins themselves are generally cheap relative to total ration cost although there are many rather overpriced proprietary preparations on the market. Those not specifically formulated for horses generally give better value for money. It is considered likely that folic acid and thiamin are the most important of the B group to supplement. It would be sensible to supplement foal feeds and most horse feed compounders supplement their full range of feeds.

All horses not receiving green feed are likely to become deficient in vitamin A after reserves built up during grazing have become exhausted. Daily requirements are in the region of 2500 i.u./100 kg liveweight for mature non-breeding animals and double that for brood mares and young stock. It is recommended that non-grazing horses be supplemented with four times their requirement levels. Lack of exposure to sunlight will prevent the horse's ability to produce vitamin D by solar irradiation of the skin. However, adult horses on rations well balanced in calcium and phosphorus are likely to have minimal requirements. Growing stock will have much higher needs. Daily supplementation at the level of 1000 i.u. vitamin D/100 kg liveweight should be ample for all types of equines not receiving solar irradiation. There are many devotees of high levels of vitamin E supplementation who believe they will improve early season breeding and athletic performances. There is conflicting evidence on these matters. Vitamin E is destroyed in rancid diets and is needed at greater levels in its antioxidant role in diets high in unsaturated fat. Daily supplementation levels of between 50 and 200 i.u./100 kg liveweight have been suggested.

Trace minerals

Probably the great majority of British horses do not receive trace mineral supplementation and are not obviously the worse for this lack. Under most circumstances, natural diets contain sufficient. However, trace minerals are cheap and it is known that deficiencies of copper, iodine and selenium have been identified in horses. It would therefore be reasonable for horse owners to insure against risk by routine supplementation, particularly of young stock, brood mares and high performance animals. It should be remembered that iodine and selenium are both toxic in excess, the margin between requirement and excess in each case being in the order of 1:20 with requirements being extremely low at approximately 0.1 ppm in the diet. The risks of trace mineral deficiencies are enhanced if total reliance is placed on home grown feeds since trace mineral levels in plants tend to reflect those of soil. Owners should be able to ascertain whether any trace minerals are likely to be deficient in their own areas by reference to the regional Agricultural Development and Advisory Service laboratory.

Major minerals

While many horses thrive with no vitamin or trace mineral supplementation, the same is not true in the case of major minerals.

Herbage and cereals are poor sources of sodium. A 600 kg hunter may lose 25 g sodium in sweat and 12 g in urine per day when at work while only receiving 15 g from oats and hay. Compound feeds are normally supplemented with sodium to provide 0.2% (equivalent to 0.5% salt).

Table 5.6. Phosphorus requirements of horses (g/day)

	Adult weight (kg)		
	200	400	600
Maintenance	6.0	12.0	18.0
Work	7.0	13.0	19.0
Late Pregnancy	8.0	15.0	21.0
Peak Lactation	23.4	3.56	39.0
Growth (months)			
3	10.9	16.4	32.2
6	10.4	21.9	32.0
12	7.5	14.8	20.6
18	6.5	13.8	19.6

However, even when using compound feeds, one should also provide a free access salt lick. Horses can adjust intake to their own requirements in this way. Horses permanently at grass should have access to salt in their paddocks.

The other major area of concern in respect of equine supplementation relates to calcium and phosphorus. It is probable that, in Britain, lack of calcium is the commonest of all horse feed problems. Summarised in Table 5.6 are the daily phosphorus requirements of horses and ponies of various ages and in various physiological states. Calcium requirements should be regarded as 1.5 times phosphorus requirements. When diets contain levels of phosphorus in excess of minimum requirement, which is often the case when cereals are fed, then dietary calcium levels should be increased such that they remain 1.5 times greater than those of phosphorus. This will be demonstrated with the examples below.

A working hunter of 600 kg has a phosphorus requirement of 19.0 g. Suppose that it is being fed 6 kg each of oats and average hay on a dry matter basis. Reference to Table 5.4 enables one to determine that it is receiving $(6 \times 3.9) + (6 \times 2.4) = 37.8$ g phosphorus and $(6 \times 1.1) + (6 \times 3.3) = 26.4$ g calcium. Had the dietary phosphorus requirement been exactly met, calcium requirement would only have been $(19 \times 1.5) = 28.5$ g. However, because the diet supplies surplus phosphorus, the horse should receive $(37.8 \times 1.5) = 56.7$ g calcium. There is thus a deficit of $(56.7 - 26.4) = 30.3$ g calcium. It is therefore recommended that supplementary limestone be fed. This contains approximately 35% calcium. The quantity to feed is thus $30.3/0.35 = 87$ g.

An example will now be taken of a 600 kg mare at peak lactation receiving, on a dry matter basis, a diet comprising 5 kg oats, 1 kg soya bean meal and 8 kg hay. In the same manner as above, it can be determined that the diet would provide 46.2 g phosphorus. Minimum phosphorus requirement

(Table 5.6) is 39.0 g and is thus satisfied by diet. Calcium requirement is going to be $(46.2 \times 1.5) = 69.3$ g. The diet only provides 35.5 g. The deficit is thus 33.8 g which can be supplied with 97 g of limestone. In both of the examples above, only supplementary calcium is needed. However, in the case of young stock, phosphorus may also be limiting as illustrated below.

Consider a 6-month-old hunter foal with an anticipated adult weight of 600 kg. Its energy and protein needs can be satisfied with a daily feed intake of 3 kg hay, 3 kg oats and 0.6 kg soya bean meal on a dry matter basis. The diet will provide 23.4 g phosphorus and 15.4 g calcium. The requirement is for 32.0 g phosphorus and $(32.0 \times 1.5) = 48$ g calcium. There is thus a deficit of 8.6 g of phosphorus and 32.6 g of calcium. Phosphorus can be supplemented in the form of steamed bone flour (24% calcium: 12% phosphorus) or of dicalcium phosphate (20% calcium: 18.5% phosphorus). In the former case, $8.6/0.12 = 72$ g would be needed. This would also supply $2 \times 8.6 = 17.2$ g of calcium. The remaining calcium deficit of $32.6 - 17.2 = 15.4$ g would be met with 44 g of limestone supplementation. If one opted for dicalcium phosphate, one would need 46 g plus 67 g of limestone.

The above examples are based on the assumption that phosphorus from vegetable concentrates has the same availability as that from other sources such as forage. In fact, one would be more accurate only to count two thirds of the phosphorus in cereals and soya when making the calculation. However, for practical purposes, the simple method described, while tending to over supply calcium, is convenient and will do no harm since both growing and mature horses are capable of dealing with calcium to phosphorus ratios well in excess of 1.5: 1 provided that the diet supplies adequate phosphorus.

It should be remembered that legume hays contain approximately four times more calcium than do grass hays. Therefore, if they constitute the principal forage, limestone supplementation probably will not be necessary in adult animals. For those people who do not wish to indulge in calculations, a rule of thumb guide for mature working horses would be to provide 15 g of limestone per 100 kg liveweight per day when grass hay is the main forage and a cereal is the principal concentrate. Diets for young stock need more attention and owners who are unprepared to design their own balanced rations should either seek help or buy already balanced compound feeds in the manner instructed by the compounder.

REFERENCES

Archer, M. (1980). Grassland management for horses. *Vet. Rec.* **107**, 171–174.
Hintz, H. F. (1985). Recent advances in equine nutrition research. *Equine Practice* **7** (6), 11–21.

National Research Council (1978). Nutrient Requirements of Domestic Animals, No. 6. Nutrient Requirements of Horses, 4th ed. Washington, D.C., National Academy of Sciences—National Research Council.

Ott, E. A. (1982). Recent Advances in Equine Nutrition. Proc. 30th Annual Pfizer Research Conference, pp. 101–109.

Ross, R. F. (1986). Sweet itch. *Vet. Rec.* **118** (6), 163.

6   Equipment for the Care, Protection and Restraint of Horses

I. M. WRIGHT

INTRODUCTION

Successful equine husbandry is entirely dependent on an ability to utilise the available equipment and in general the greater the demands made on the horse the greater the care and attention necessary. This chapter is based on keeping the adult riding horse although much of the information can be equally well applied to young stock, the breeding animal or driving horses.

It is hoped that this chapter will form a bridge between the traditional texts and the plethora of new ideas which have paralleled the resurgence of the horse's popularity. It is not aimed at the professional horseman nor is it intended to be a technical manual but rather the aim is to supply the salient facts concerning equipment which is encountered in the day to day care of horses.

Clothing

Clothing, usually in the form of rugs, is used for keeping horses clean and preventing heat loss. Specialised rugs are available for various occasions depending upon the horse's activity, whether it is stabled, working, in transit or at grass, whether the horse is clipped and the weather conditions. Until recently man-made fibres were not commonly employed in the manufacture of equine clothing because they lack the absorbency of natural materials. However, when suitably lined the advantages of these materials (cost, durability, lightness and ease of maintenance) may be reaped.

Rugs are marketed in specific lengths measured from the centre of the horse's brisket to approximately 10 cm from the base of the dock. For each length a range of depths (vertical height) is usually available.

Horse Management 2nd edition
ISBN: 0-12-347218-0 case

In addition to the care of rugs discussed with each type particular caution is necessary to prevent infections being carried by these items. The most important of these so called fomite transmissions are the ringworms. These are fungal infections, of which the two most common species are *Trichophyton Equinum* and *Microsporum Equinum*.

Confining rugs to an individual is a sensible precaution in controlling ringworm and, in the face of outbreaks, treatment of an affected horse's rugs is important to prevent reinfection. Washing in iodine based preparations is useful in this respect but perhaps the most efficacious treatments are applications of natamycin (Mycophyt: Mycofarm Ltd.) or sodium benzuldazate (Defungit: Hoechst U.K. Ltd.). Contaminated clothing may be sprayed, sponged, brushed or washed with these preparations.

New Zealand rugs   New Zealand rugs (Fig. 6.1) are constructed of waterproof material to provide protection against wind and rain, and are principally used on clipped or partially clipped horses at grass during the winter months.

New Zealand rugs consist of two separate layers. The outer waterproof layer is usually made of cotton or flax canvas although man-made fibres such as nylon are sometimes utilised. Flax canvas is said to be stronger than cotton and while the man-made fibres are lighter their durability has been questioned and in general they are more expensive. Either chemical or wax (oil) proofing may be employed.

Traditionally the rugs are half, three quarter or fully lined with wool. However, polyester quilt and cotton have become increasingly popular and recently materials with high thermal (tog) values such as polyvinyl chloride (PVC) coated aluminium filaments enclosed in cotton quilts have become available. These reflect, and therefore conserve, body heat.

Rot resistant chrome leather is the material most frequently used for leg straps but soft nylon and PVC coated nylon are easier to keep supple.

A number of individual design features are found in New Zealand rugs. Some manufacturers incorporate sheepskin wither and shoulder pads or smooth nylon patches inside the breast flaps to prevent rubbing. The neck/breast fastenings may have two or three alternative strap and buckle positions for accurate fit. Leather reinforcement of these attachments, which are prone to wear and tear, is a commendable option. Additional strength in this area is obtained by incorporating a neck rope. However, increased rigidity is to be avoided and so some rugs have a soft binding, such as leather, at the neck to improve flexibility. The rear edges may also be bound.

Those areas of the rug which are subject to the greatest movement may accommodate this by the inclusion of tailored areas (darts) over the shoulders and/or quarters. A still closer fit can be obtained by the use of draw strings at the rear of the rug.

Figure 6.1. New Zealand rug with surcingle passing through the rug and Australian leg straps.

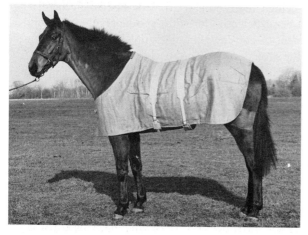

Figure 6.2. Traditional jute night rug secured by double surcingles.

Figure 6.3. Quilted stable rug with integral cross surcingles.

An integral tail flap provides additional protection at the rear and detachable neck covers can be anchored to the front of the rug. The latter also attach to the headcollar and beneath the horse's neck with quick release buckles. Damage to the shoulders of the rug and horse from fences, particularly barbed wire, can be reduced by leather plates sewn over this area.

Two additional designs merit description. The first involves an inner quilted rug which may be used alone in the stable or covered by a detachable cotton canvas for outside use. The second incorporates a semi-permeable membrane which the manufacturers claim has pores sufficiently small to prevent penetration of water drops but large enough to allow water vapour (from sweat) to escape. Both rugs are of lightweight design.

Particular care in fitting is necessary for safe and efficient use since these rugs must cope with extremes of movement, galloping, rolling etc., and resist elevation by the wind or being caught on gates, fences, hedges etc. Rollers, surcingles and straps must be sufficiently close fitting to satisfy these demands without impeding movement or causing skin abrasions and pressure sores. Hair loss at the points of the shoulders or over the withers, which are pressure points, is a common sequel to prolonged use of a New Zealand rug. The most satisfactory rugs have adjustable neck/breast fitting positions which allow the rug to accommodate both narrow and wide bodied horses. Even so it may be necessary to provide additional protection by sewing sheepskin pads onto the rug lining over the pressure points.

Some New Zealand rugs are made with an integral surcingle which passes through the sides of the rug, others require no surcingle and are therefore less likely to cause a sore back. The use of padding beneath surcingles is rarely satisfactory and frequently becomes dislodged.

Traditionally these rugs have Australian style leg straps which pass from a point in front of the hindlimbs to the rear of the rug. These are crossed between the hindlegs to avoid chafing, and should be fastened sufficiently closely to prevent the horse putting its leg on or through them. Additional protection against rubbing may be necessary and purpose-made covers of aerated wool/latex combination are available but an efficient economical alternative is to thread the straps through the soft rubber of a bicycle tyre innertube. More intricate strap systems are often included in those rugs which require no surcingles.

The top of a New Zealand rug should extend from just in front of the withers to the base of the tail. Rugs which are not tailored should be fitted full over the shoulders such that contact with the rug is avoided even when the limbs are maximally protracted. Although manufacturers vary in the recommended depth of New Zealand rugs they should always extend beyond the bodyline. If fitted below the knee or hock the risk of tearing is increased by the horse standing on the rug.

Daily removal and refitting is mandatory. At this time it is important to inspect those areas which are particularly susceptible to damage; the points of the shoulders, withers, girth, inner-thigh and groin. At the first sign of skin thinning, abrasion or tenderness the rug must be changed or refitted. The application of emollients such as vaseline or uddercream may be of additional benefit but necessitate increased rug care.

Any mud or dried sweat found on the horse's coat or rug lining should be removed daily with a stiff dry brush since both will cause the lining to become rough and abrasive. If either the horse or rug lining become wet then thorough drying is necessary before re-application, and for this reason a spare (dry) rug is frequently invaluable. Washing New Zealand rugs with soaps or detergents should be avoided since this destroys the proofing. At the end of winter gross debris should be brushed off before the rug is washed down with water only.

The effects of inclement weather necessitate preservation of supple leg straps and oiling of fastening clips to prevent rusting. To ensure continued rug efficiency any tears in the outer waterproof layer must be promptly repaired.

Night or stable rugs Night or stable rugs offer an insulating layer to the horse when necessitated by low ambient temperatures and/or clipped coat. They may be used alone or with an underblanket.

There are two broad types. The traditional stable rugs are made of twilled jute, hemp or flax, and hence are frequently referred to simply as "jute rugs" (Fig. 6.2). They are half or fully lined with woollen blanket, are resistant to rotting and tearing and are washable and hardwearing. Weight is a good guide both to their quality and insulation properties.

Since the late 1960s rugs of quilted man-made fibre have been available and are now the most popular stable rugs (Fig. 6.3). These too are rot proof and long lasting and in addition, due to their increased ability to trap air, are more efficient insulators than the traditional night rugs. They are also lighter (weight is not an effective guide to their thermal properties) and more easily washable but they do tear. The outerlayer may be nylon or polyester/cotton mixture, which are relatively hardwearing, while the lining is usually of cotton for its superior absorbancy. The thermal properties of these rugs are obtained by polyester fibre filling of the quilts. These may be of varying thickness and provide "duvet" type insulation.

A leather strap and metal buckle are found at the breast of most jute rugs whereas the most popular breast fasteners of quilted rugs are nylon straps and quick release plastic buckles. Although these rugs are not required to accommodate the degree of movement demanded of a New Zealand rug the front should be sufficiently loose fitting to enable the horse to lie down without pinching or chafing. Adjustable positions for breast straps are found only occasionally.

Both traditional and quilted rugs may have integral or detachable surcingles of conventional design. However these are generally considered undesirable due to the danger of producing pressure sores from continual use. When necessary these rugs are best secured with a padded roller. More satisfactory are the adjustable nylon cross surcingles with plastic or light metallic clips found on many quilted rugs. These pass from upper shoulder to the opposite quarter and are both secure in preventing rug movement and avoid pressure problems. Some designs of both types of stable rug are fitted with adjustable nylon leg straps, as found on New Zealand rugs, which may negate the need for any fasteners across the trunk. Fillet strings may be employed on quilted rugs which are devoid of leg straps or cross-over surcingles.

Jute stable rugs should be brushed daily on both sides to remove bedding and dried droppings. Quilted rugs require little more than daily shaking. However, manure stains can be removed with a damp sponge and fillet strings washed as necessary.

Both types of stable rug should be washed before storing. Jute rugs are more difficult due to their bulk but may be spread on a clean surface and scrubbed. Detergents should be thoroughly rinsed out before the rug is drip dried. Quilted rugs can be machine washed and dried. Repairs should be carried out before storage in a mouse, moth and damp-proof container. Traditionally chests with secure lids are employed and the addition of moth balls is a sensible precaution.

Blankets Blankets are fitted beneath the stable rug to provide an additional insulating layer for the clipped horse during the colder winter months. This practice is infinitely preferable to attempts at increasing the ambient temperature by closing stable doors or air ducts which severely compromise ventilation.

Horse blankets may be made of wool or man-made fibres. The most traditional is a woollen Witney blanket which is recognised by its characteristic golden colour with red and navy blue stripes. Acrylic versions are much cheaper but of inferior thermal quality and a better second choice, although less aesthetically pleasing, are woollen army blankets. Weight is a reasonable guide to the durability and insulating quality of these fabrics but cannot be applied to materials such as dacron (polyethylene phthalate). This is a highly resilient helical coil and is available enclosed in a fleece fabric giving, it is claimed, superior thermal properties to conventional blankets.

The traditional horse blanket is oblong in shape but fitted underblankets of both conventional materials and synthetic fleece are now increasingly popular. These are shaped to fit beneath a rug and are either provided with their own breast fasteners or loops through which those of the rug may pass. Leg straps are also fitted by some manufacturers.

A conventional blanket of the correct size should extend from the horses' poll to its croup (Fig. 6.4(a)). Although the depth is largely a matter of personal choice, for maximal efficiency it should be similar to the stable rug. Once the blanket is positioned in this manner both front corners are folded back to the withers (Fig. 6.4(b)). The rug is then applied as usual (Fig. 6.4(c)) before the free apex of the blanket, produced by the first manoeuvre, is folded back over the outside of the rug and secured beneath a roller (Fig. 6.4(d)).

When in use blankets require little more than daily shaking but if heavily soiled they should be washed. A large domestic or industrial washing machine is suitable and this should be repeated before storage in similar conditions to other rugs.

Figure 6.4. Fitting a blanket.

(a) Blanket extends from poll to croup.

(b) Front corners folded back to withers.

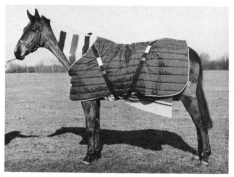

(c) Rug applied and fastened.

(d) Blanket folded back and secured beneath roller; here an anti-cast roller with a breast girth is employed.

Day rugs   Day rugs were originally used to replace the night rug of a stabled horse after exercise but are now most frequently employed for travelling and use at shows/events etc.

The traditional day rug is made of wool and is either a single colour bound with contrasting braid (Fig. 6.5) or may be of 'Melton' or 'Witney' design. Synthetic quilts are however now seen with increased frequency.

A double layer of wool may be utilised as padding at the withers and shoulders. Some rugs are tailored at the quarters, and like summer, paddock and exercise sheets, owner, trainer or sponsor's initials or logos are often displayed on the rear corners.

Day rugs are fitted with the same criteria as stable rugs. A cotton fillet string is standard and supplements the breaststrap with either a matching surcingle or wool web roller completing the fittings. Occasionally cotton leg straps are employed and a 'D' ring may be present for attachment of a tail guard.

While in use the day rug requires only shaking on removal although fillet strings usually need to be washed on a regular basis. Washing in a detergent suitable for wool is indicated prior to storage.

Summer sheets or fly sheets   Summer sheets replace day rugs in warm weather. They are used, as the name implies, to protect the horse from irritation by flies and also to keep the coat free from dust and stable stains. Light finely woven cotton or linen are the most common materials and may either be double or single layered.

Summer sheets are frequently made in a single colour with contrasting braided edges or alternatively may be of Tattersalls' check. Some are tailored although this is, as yet, uncommon.

All sheets should be fitted as for day rugs. Surcingles may or may not be integral but fillet strings are usual. An alternative design incorporates leg straps. The lightweight nature of these rugs facilitates their regular washing and drying during use.

Paddock sheets or quarter sheets   As the name implies these are used while walking horses in the paddock prior to racing. According to the weather conditions paddock sheets may be made from linen, wool or waxed cotton. They are devoid of front fastenings and cut off at the shoulder/withers but are secured by a breastplate or paddock roller with a breastplate. Fillet strings are useful to prevent the rug being lifted by the wind.

Exercise sheets   These are also referred to as galloping sheets, and are of a similar pattern to paddock sheets. They are placed under the saddle during work to keep the loins warm in cold weather.

Figure 6.5. Woollen day rug secured by a web roller.

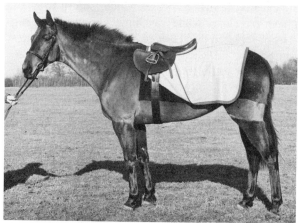

Figure 6.6. Fitted exercise sheet; note absence of fillet string.

Figure 6.7. Anti-sweat rug.

Exercise sheets are usually woollen but more recently waterproof materials have been introduced such as waxed cotton and rubberised man-made fibres. The majority completely cover the horse's back but some have a cut out area for the saddle. During fast work secure fitting beneath the saddle is essential and this is achieved by folding the front corners upward between the girth and the saddle panels (Fig. 6.6). A fillet string is also necessary to prevent elevation of the rug.

Frequent washing is necessary to avoid accumulations of dried sweat and, to minimise the risk of spread of contagious skin diseases such as ringworm, individual rug use is recommended.

Anti-sweat rugs   Anti-sweat or anti-chill rugs may be used alone or beneath a natural fibre rug on a horse that is wet, whether from sweat or extraneous water. They are also used beneath the rugs of horses which tend to sweat or "break-out" after return from exercise or while travelling. The anti-sweat rug can be both supplemented by or replaced by a layer of straw. In both instances the rationale is to create air pockets adjacent to the horse which provide an insulating layer but in which there is sufficient air movement for evaporation.

Cotton is exclusively employed in the body of anti-sweat rugs but they may be bound either with cotton or nylon. The latter is generally more resilient and minimises stretching of the rugs.

An open mesh design is most commonly utilised for both anti-sweat (Fig. 6.7) and anti-chill rugs but the weave of the mesh is closer in the latter. Recently reschel knitted rugs have been introduced.

Breast fasteners are the only consistent fittings and are usually plastic. Fillet strings can be useful when the rug is used alone but the security of a roller may be necessary if horses are loose in the stable.

Repeated soaking in sweat necessitates frequent washing but it is important to thoroughly rinse detergents from these rugs to minimise skin irritation.

Surcingles and rollers   Surcingles and rollers are used to keep rugs in place. The former are single strapped and unpadded while the latter are usually double strapped with integral padding on each side of the vertebral column.

Single or double surcingles sewn into rugs used to be popular. When only a single surcingle is employed this is fitted at the level of a girth and ideally has a pad sewn in place on each side of the withers to avoid direct pressure on the summit. When used, the second surcingle is fitted loosely beneath the horse's abdomen, and therefore is sewn flat to the top of the rug. The second surcingle, although not tightly fitting, adds significantly to the security of the rug particularly when the animal lies down or rolls. However, surcingles are inferior to rollers for most purposes because the latter produce superior pressure distribution.

Rollers are indicated when a rug is not fitted with a surcingle or leg straps (Fig. 6.5). They may be made from leather, which is easiest to maintain and hardest wearing, hemp, flax or wool, and are fitted with padding on each side of the summit with a clear passage at the pressure point between them. Straps may be of leather or nylon. Rollers with inadequate padding whether through design or wear should be used with a thick conforming pad, such as foam rubber, over the spine.

One type of roller which merits individual attention is the anticast or arch roller (Fig. 6.4(d)). The wither pads are joined by a metal arch which prevents the horse rolling onto its back and becoming cast. In some cases the arch is hinged on each side which permits a more accurate accommodation of the individual horse's contours.

Breast girths are a useful adjunct to rollers to prevent them from slipping backward. These are attached to "Ds" on the front of the roller.

Fillet strings   Fillet strings or tail strings are fitted to rugs via eyelets in the rear border and pass around the animal's hamstrings. They are usually made of coloured braided cotton, but can be of thin leather. Fillet strings prevent rugs sliding forward or flapping up and are used on rugs without leg straps.

Tail bandages   The two main indications for tail bandages are to provide protection against injury or rubbing particularly during transport and to improve the appearance of the tail and keep the hair of a pulled tail in position (Fig. 6.8). They may also be used to keep the perineal area free from hair during examination or service and by folding the free hair of the tail up over the dock keep this out of wet muddy conditions (Fig.6.9).

Tail bandages should be 6.5 to 7.5 cm wide and 2.5 to 3 metres long. Crepe, cotton or felted cotton are the most suitable materials. Stockinette (tubular woven cotton) has a low coefficient of friction and tends to slip more easily and, since the blood supply of the dock is particularly sensitive to pressure, elasticated bandages should be avoided. Tail bandages are secured by tapes of non-stretch material. They are rolled tape side inward and applied with the roll to the outside. To set the tail hair it is first dampened with a water brush (wetting the bandage is not recommended as the material may shrink on drying) and the bandage then begun at the base of the tail. The first 10 cm is left as a free end and after the first turn around the tail is folded down and covered by subsequent folds of bandage, each overlapping its predecessor by approximately 50% of its width. At the end of the dock the bandage is redirected to ascend the dock before tying the tapes on the outside near the tail base.

Care is necessary to avoid excessive pressure during application or tying. White marks ringing the dock are a commonly encountered sign of previous local pressure necrosis. For similar reasons the practice of prolonged use,

Figure 6.8

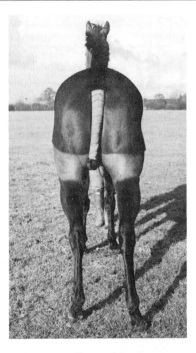

Figure 6.9

Figure 6.10

Figure 6.8. Tail bandage fitted in conventional manner.

Figure 6.9. Free hair of tail plaited, folded up and enclosed by tail bandage.

Figure 6.10. Blanket tail guard applied over a bandage and secured to day rug roller.

such as overnight, should be discouraged. The tapered contour of the dock means that following release of the tapes a tail bandage can be removed by gently but firmly pulling from top to bottom with the hair line.

Tail guards   Tail guards may be made of felt, blanket or leather with straps or ties at the side and a long strap or tie at the top. It encloses the dock preventing the horse rubbing hair from its tail and is most frequently employed during travelling over a tail bandage (Fig. 6.10). The top strap or tape is secured to the horse's roller preventing the guard from slipping down and the dock is then enclosed only with sufficient tension to prevent it opening since the circulation in this area is susceptible to interference from external pressure.

Poll guards   Poll guards are used to protect the top of the head during transit. They are usually made of leather and can be full fitting with holes for the ears or a single strip over the apex of the poll (Fig. 6.11). Both are usually attached to the headpiece and browband of the headcollar and are heavily padded with felt, foam or similar materials.

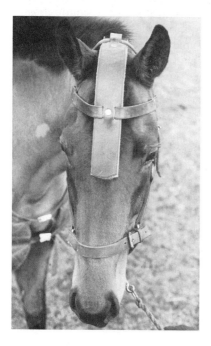

Figure 6.11. Padded leather poll guard attached to headpiece and browband of headcollar.

Figure 6.12. Fly fringe fitted to headcollar.

Fly fringes A fly fringe (Fig. 6.12) acts as a long artificial forelock to discourage insects irritating the eyes. It consists of light cord tassels which attach to the browband of a bridle or headcollar.

Boots and bandages

The majority of horses are kept as athletes. This applies equally well to the animal used as a hack at weekends as to the racehorse, the difference is merely one of degree. The most common limiting factor to performance is lameness and therefore considerable effort is directed to reducing the incidence of trauma to susceptible areas of the horse's limbs. External trauma related lameness varies from small cuts and contusions to fractures and can largely be controlled by fitting protective boots and bandages. By contrast there is little evidence to support the prevention of stress related injuries, sprained joints and ligaments, strained tendons, stress fractures etc. by the application of external devices to the limb.

Brushing occurs when the supporting limb is struck by its advancing partner. Strictly speaking this injury is confined to the region of the fetlock and lower cannon whereas speedycutting injuries, so called because they most frequently occur during fast work, are found in the upper cannon area. A cut produced by the toe of an advancing hind foot on the back of the forelimb of the same side is called an over-reach. Typically these injuries occur at the bulbs of the heel and consist of a flap of skin attached at its lower margin. They are caused by the inside edge of the shoe when the horse is moving at speed or jumping particularly in soft (holding) ground. High over-reach injuries to the fetlock or flexor tendons are less common.

Brushing boots As the name implies brushing boots are employed to prevent interference injuries to the inner aspect of the cannon, splint bone and fetlock, rarely they are referred to as splint boots (Fig. 6.13). The traditional materials are box cloth, kersey, felt and leather which may or may not be lined with felt, rubber or synthetic materials. However, due to their durability and ease of maintenance, boots made entirely of synthetic materials are now widely used. The most popular of these has a firm plastic outer layer lined with soft expanded foam which overlaps the upper edge to prevent skin chafing (Fig. 6.14).

The basic rectangular shape has a rounded protruberance on the lower border which covers the inside of the fetlock. This is common to all brushing boots although individual designs may incorporate other features. The padding of these boots is reinforced to varying degrees on the inner aspect of the limb by adding a padded cap or increasing the thickness of the lining.

Traditional forelimb brushing boots are secured with three or four straps and buckles, either of which may have elasticated attachments, and the larger

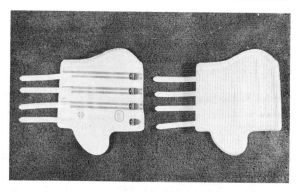

Figure 6.13 (left). Leather forelimb brushing boot with padded cap.
Figure 6.14 (above). Plastic brushing boots with expanded foam lining and ski-type clip fasteners.

hindlimb boots with five. Straps are fastened on the outside of the limb with ends pointing toward the rear of the horse. During application the centre strap should be secured first followed by those above it and then those below. The lowest strap should not be tight since this will limit joint movement. When removing, boots straps are undone from bottom to top to avoid the boot falling down and flapping about the horse's leg. The use of keepers is important as loose straps may lead to buckles becoming undone. On recent designs single or multiple self-adhesive velcro type fastenings, some with overlying elasticated straps, or straps with ski-type clips have been employed.

All boots require cleaning after use. Leather is cared for as other pieces of harness to keep it supple. This avoids cracks and breakages, difficulty in fastening and unfastening and the production of skin abrasions. Cloths are best dried and brushed clean daily with periodic washing as necessary while the plastics are well suited to daily washing.

Heel boots   Heel boots are a type of brushing boot which also provide protection over the back of the fetlock against injury from ground contact or advancing hindfeet. This is achieved by a flap with varying degrees of mobility (Fig. 6.15).

Speedycutting boots   These are also termed speedicut or speedycut boots. Speedycutting boots are essentially similar to brushing boots and are fitted in a similar manner but provide protection only above the fetlock (Fig. 6.16).

Figure 6.15

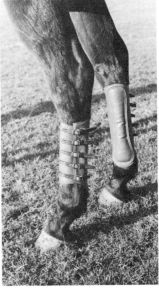

Figure 6.16

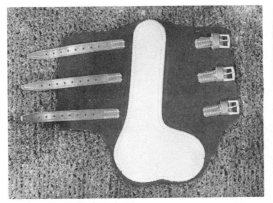

Figure 6.17

Figure 6.15. Heel boot demonstrating extended leather cap.
Figure 6.16. Leather speedycutting boots.
Figure 6.17. Open fronted plastic tendon boots; velcro fastener reinforced by overlying elastic strap.

The conventional boots were of box cloth or leather but have now been largely replaced by the more comprehensive brushing boot.

Tendon boots   Tendon boots protect the forelimb digital flexor tendons and their sheath from high over-reach injuries and for this reason are sometimes referred to as showjumping boots. There is no evidence for manufacturers' claims that these boots provide significant support.

The true tendon boot (Fig. 6.17) is fitted from knee to fetlock and is open fronted but it may be combined with a brushing boot or extend to cover the fetlock. The body of the boot is shaped to the caudal aspect of the limb and is constructed of impact resistant material. Boots of leather and kersey, the original materials employed, have largely been superseded by boots with a touch plastic outer shell lined by soft expanded foam, rubber or acrylic. In these instances the lining should overlap the edges of the shell to avoid chafing particularly at the upper edge.

As tendon boots are used during strenuous exercise security of fitting is paramount. The original boots were fastened by four or five leather or leather and elastic straps and buckles, while the modern designs have plastic straps with ski-type clips or elasticated velcro fasteners. On some boots the latter is reinforced by an overlying elastic strap and clip.

Fetlock boots   Fetlock boots are sometimes referred to as ankle boots and are used to prevent injuries from interference particularly in hindlimbs. Nevertheless they may be single or double sided i.e. cover the inner aspect of the fetlock or both inside and outside (Fig. 6.19). Due to the range of movement of this joint the latter are notched in the midline.

The original fetlock boots were leather, either blocked and unlined, or with kersey or rubber lining, but more popular now are wear resistant plastic boots with expanded foam or aerated wool/latex linings. In both types a single fastener is employed above the fetlock, a leather strap and buckle on the former and plastic strap with ski-type clip on the latter.

Yorkshire boot   A Yorkshire boot is a rectangle of felt with a cotton tape sewn along the centre. This is tied on the outside of the leg just above the fetlock (Fig. 6.18(a)) and the top half of the boot is then folded down over the tape to provide a double layer of felt over the fetlock (Fig. 6.18(b)). Yorkshire boots are used on hindlimbs for the same reasons as fetlock boots. Although they provide less protection against impact, are more difficult to maintain and are less hard wearing, they are conforming and can be of use as a precautionary measure on horses which do not normally interfere but which move close behind.

Polo boots   Polo boots are made in a variety of shapes designed to protect the distal limb not only from interference injuries such as brushing, speedycutting and treads but also against blows from balls and sticks. Perhaps the most widely used is the Californian pattern which are leather with felt lining. These extend from below the knee to the coronary band, although they are secured by elasticated leather straps only over the cannon thereby avoiding impedement of movement.

Fetlock rings   A rubber ring fitted above the fetlock is called a fetlock ring boot or brushing ring (Fig. 6.20). These are secured by a leather strap running

Figure 6.18. Yorkshire boot.

(a) Tapes tied on the outside of the limb;   (b) Top of boot folded down over tapes.

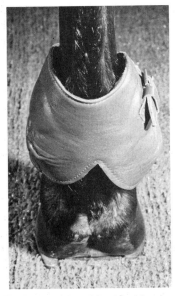 

Figure 6.19. Double sided leather fetlock boot viewed from the rear.   Figure 6.20. Fetlock ring used in the prevention of brushing injuries.

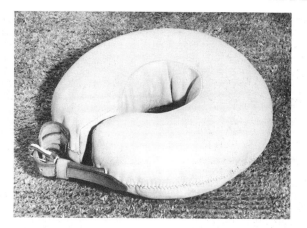

Figure 6.21. Sausage boot used in the prevention of capped elbows.

through the rubber and are used to prevent brushing injuries to this area without encompassing the limb in a full brushing boot. Variations of the fetlock ring are a scalper boot which has a leather flap covering the inside of the fetlock and a quarter boot which is similar but with a more extensive area of protection around the whole joint.

Sausage boots   Sausage boots or doughnuts are thick padded rings of leather fitted around the pasterns (Fig. 6.21). They prevent the heel of the shoe traumatising the elbow, when the horse is lying down, which causes a capped elbow or shoe boil.

Coronet boots   Coronet boots are designed to prevent tread injuries to the coronary band whether self inflicted, e.g. during travelling, or from other horses, e.g. playing polo (Fig. 6.22). They should cover as much of the pastern, heel and coronary band as possible and are usually made of thick padding such as felt with a leather covering. Coronet boots are fitted loosely around the pastern and are normally secured by two straps.

Over-reach boots   Over-reach boots fit around the pasterns and cover the coronary band and bulbs of the heels. In addition to preventing over-reach injuries they also protect the coronary band from tread injuries. The vast majority are made of rubber and because of their shape are often referred to as bell boots. The most popular designs have no fastenings (Fig. 6.23). These are applied inside out and pulled over the hoof inverted before being correctly orientated, and are removed by stretching the upper margin over the hoof. Others are divided and secured by laces, velcro or a vertical strap

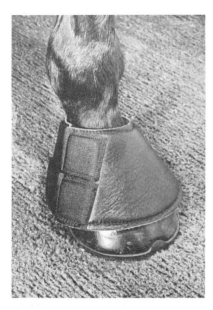

Figure 6.22

Figure 6.23

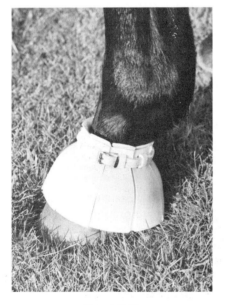

Figure 6.24

Figure 6.22. Coronet boot with velcro fastenings.
Figure 6.23. Ribbed over-reach boot.
Figure 6.24. Segmented over-reach boot.

running through "Ds". Despite the ease of application and removal of the latter designs current consensus of opinion favours the former for safety and security. The body of both types of boot may be plain or vertically ribbed and of various weights.

A novel design has recently been introduced which it is claimed overcomes the principal problem associated with conventional over-reach boots — that of inversion during use. It is constructed of individual replaceable petals connected and fitted by a single strap around the pastern (Fig. 6.24). This also enables one size of boot to fit all horses by removal or addition of individual segments.

Serving boots   Serving boots are fitted to mares' hind feet to protect the stallion from the impact of kicks (Fig. 6.25). They are made of felt or soft leather, have a thick padded base and are secured around the pastern.

Knee boots   Knee boots are used when transporting horses and during exercise on roads. Penetrating wounds of the joints or tendon sheaths are a particular risk when horses are worked on roads. These so called "broken knees" usually result from the horse stumbling and are frequently contaminated with gravel, etc. Knee boots are secured by straps at the top and bottom. The top strap should be tight enough to prevent the boot slipping down while the bottom strap must be loose enough to allow movement and is merely used to prevent the boot flapping about.

Two types of knee boot are available, plain and skeleton. The former (Fig. 6.26) consists of a padded leather top secured by a leather strap which is usually set on elastic. The body of the boot is a large pad of felt, leather or wool rugging (the latter frequently in a colour to match the day/travelling rug) reinforced in the centre with a blocked and stuffed leather cap. The boot is completed by a lower strap.

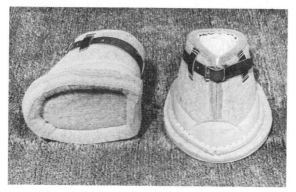

Figure 6.25. Pair of felt serving boots illustrating thick padded base.

Figure 6.26. Plain knee boot; secure upper fitting and loose lower strap.

Figure 6.27. Skeleton knee boot.

The skeleton boot (Fig. 6.27) consists of upper and lower straps but has only the reinforced cap between them. In consequence it is lighter and often preferred for working horses.

Hock boots   Hock boots may be employed during transit or on the stabled horse to reduce trauma to the point of the hock which results in bursal enlargements—so-called capped hocks. They are shaped to fit the hock and are normally made of thick felt, wool or leather with a padded blocked leather cap over the point of the hock (Fig. 6.28). The boots are fastened by straps at the top and bottom and like knee boots the upper is firmly fastened and the lower loosely.

Travelling boots   Some authorities consider travelling boots inferior to bandages since they provide less support. It is questionable whether either functions as such and, indeed, it is doubtful whether support is as important during transport as providing protection from external trauma. In addition they are quicker to put on and remove than bandages. Several designs are available and range from simple tubes which encase the lower limb as far as the knee and hock (Fig. 6.29), to tailored boots which include the knee and hock and overlap the coronet. A number of materials have been employed

Figure 6.28

Figure 6.29

Figure 6.28. Cloth hock boot secured by elasticated upper strap with velcro fastening and loosely fitting lower strap and buckle.
Figure 6.29. PVC travelling boot with synthetic fur lining and velcro fasteners.
Figure 6.30. Forelimb stable bandage with tapes tied and loose ends tucked away.

Figure 6.30

in their construction: thick felt or foam with leather or synthetic backing, cotton with polyester fibre filling and acrylic fur lining with an outer layer of PVC or cotton canvas. Those designs which are easy to wash have a distinct advantage. Conventional straps and buckles or ties may be used but velcro fastenings are the most popular method of securing travelling boots.

Stable and travelling bandages   Almost all horse bandages require some form of conforming material between them and the leg to ensure even pressure distribution and stable and travelling bandages are no exception. Their function is to provide protection against trauma and support to the limb particularly after strenuous exercise when some control of synovial effusions such as windgalls can be obtained. Applied over soft straw or hay, stable bandages help to dry wet legs and they also play a greatly underrated role in conserving body heat.

Stable bandages should be 10 to 12.5 cm wide and 2.1 to 2.4 metres long. They are usually made of wool, wool/synthetic combinations or other inelastic materials such as stockinette which is cheaper but not as warm or secure as woollen bandages. Sandown bandages consist of an initially applied length of fleecy wool followed by stockinette which is placed over this and as such can be employed without any under material.

Following application of a consistent thickness of under material the stable bandage is commenced below the knee or hock and passed around the leg in even turns, each with a uniform overlap of one half to two thirds of the previous turn. This should be performed without wrinkles and with even pressure on the bandage from top to bottom. On reaching the coronary band the bandage takes a natural turn upward and is continued in a similar fashion to finish near the top of the cannon. Two principal means of securing the bandage are available: velcro type fasteners and tapes. The latter should be of inelastic material such as cotton and lie flat around the limb before being tied in a bow against the outside of the leg at the same tension as the rest of the bandage. The loose ends are then tucked beneath adjacent folds (Fig. 6.30). It is important that knots are not placed on the front or back of the limb where pressure points may be created.

The bandage is removed by untying the tapes or releasing the velcro and passing the accumulating ball from hand to hand around the limb. It should then be re-rolled firmly and accurately with the fasteners on the inside of the first turn. This is essential if the bandage is to be used correctly on the next occasion.

Exercise bandages   Exercise bandages are used primarily to protect the digital flexor tendons from injuries such as over-reaches, etc. since trauma of this nature can be extensive. They are not as impact resistant as boots but are conforming, which is important in horses in fast work. In addition it is widely believed that exercise bandages provide better support to the flexor tendons than that which is obtained with boots. This view is however difficult to justify since the mechanical load applied to these tendons is axial, i.e. in the long axis of the limb, whereas pressure from bandages is circumferential, i.e. at right angles to this axis.

Most exercise bandages are approximately 7.5 cm wide and 2 metres long. Crepe and cotton (Newmarket bandages) are the most common materials and have a certain inherent elasticity. Similarly elasticated cotton produces the bandage security necessary for fast work but means that there is a danger of applying excessive pressure. In addition if wetted these materials tighten on drying. Plain stockinette and Sandown bandages are less likely to overtighten but lack security; nevertheless, the latter are popular in polo work.

Exercise bandages are fitted from knee or hock to fetlock in a manner which avoids interference with movement. They should be applied over a conforming material which must lie flat. The bandage is started at the top of the cannon leaving a 7.5 to 10 cm free end on the outside (Fig. 6.31(a)). The bandage is applied in the same direction as the padding overlap with equal tension throughout its length and each turn covering approximately two thirds of the width of the bandage. The cross-over from descending to ascending turns takes place on the front of the limb just above the fetlock producing a shallow inverted V. The bandage is completed by folding over the original free end and covering this with the ascending bandage (Fig. 6.31(b)). It is secured by tapes which should lie flat and are tied on the outside of the leg before tucking in the ends. For added security the knot may be covered with adhesive tape or sewn to the bandage (Fig. 6.31(c)). Any excess protruding under the material should then be trimmed.

While it is important that exercise bandages are under sufficient tension to avoid slipping during strenuous exercise, of at least equal importance is that the bandage is not so tight as to compromise the circulation of either the skin or underlying flexor tendons.

Support bandages These have been called pressure bandages or dressings but this is a misleading term. Support bandages are used on injured limbs to minimise swelling and on uninjured partners which have to withstand increased weightbearing loads. Such bandages can be constructed from basic bandaging materials although one-piece elasticated dressings are commercially available in various sizes for the fetlock, knee and hock. They are made of lycra and nylon and are secured by zip fasteners with velcro reinforcement at the extremeties. These are lightweight, re-usable, contoured to minimise slippage and avoid pressure points and are particularly useful for the hock since many horses do not readily tolerate bulky dressings in this area (Fig. 6.32).

Figure 6.31. Application of an exercise bandage.

(a) Free end left at the beginning of bandaging.

(b) Free portion folded down between the first (descending) and second (ascending) layers of bandage.

(c) Completed bandage with sewn tapes; note inverted 'V' produced at front of fetlock.

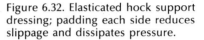

Figure 6.32. Elasticated hock support dressing; padding each side reduces slippage and dissipates pressure.

Support bandages for the distal limb are applied in a similar manner to stable bandages but with increased pressure. This may impede venous return and therefore care should be taken to avoid an unbandaged gap at the coronary band which will result in swelling in this area. Whenever possible support bandages should be applied with the limb weightbearing. A generous comforming layer is applied first to eliminate irregularities in the limb contour. This is followed by a compressive layer which must be fitted in a manner which discourages slippage, particularly in a tapering region of the leg, yet avoids potential pressure points. The bandage is completed by an elasticated layer which adds a dynamic component to the support. Susceptible pressure points over the knee are the bony eminences each side of the limb at the lower border of the radius and most importantly the accessory carpal bone which is the hindmost protruberance of the knee. The point of the hock and tendons immediately above it are the pressure points of most concern in the hindlimb. If these points cannot be excluded as the bandage is constructed then its outer layers can be split by a single sharp cut over the protruberance on completion of the dressing.

Undermaterial A variety of conforming materials are available for use beneath bandages and once the advantages and disadvantages of each have been evaluated there is usually still a choice which can then be based on subjective data and personal preference.

Gamgee, the archetype material, is a strong white tissue produced by enclosing cotton wool in gauze. It is cut to the required size and is re-usable only in the short term since it is not washable. Rayon wool bandages (Velband®; Johnson and Johnson) are a softer more conforming alternative and cotton wool is a cheaper substitute but both lack the strength of gamgee. Rectangles of foam rubber and synthetic fleece are becoming increasingly popular as undermaterial since both are washable and resilient with re-use. As a temporary measure, particularly on wet legs and for keeping horses warm, a useful insulating layer can be produced by soft straw or hay.

Grooming equipment

Grooming is performed not only for aesthetic reasons but for the general health of the horse, particularly the coat, skin, mane, tail and feet. The benefits of a regular grooming regime to the domesticated horse have long been recognised. The Greek cavalry general Xenophon in 400 BC had his horses groomed with wooden body combs and this was followed by hand massage and use of a date palm rubber. Throughout the centuries of equine domestication there has been little change in equipment or techniques. By the early seventeenth century the use of brushes, currycombs, wisps and stable rubbers is well documented but the use of grooms' wet hands recorded at this time appears to have fallen into disuse. The promise that electric grooming machines would be the next, long awaited, evolutionary step in grooming has not been fulfilled.

Most horses require some form of daily grooming. Horses at grass require the minimum of attention as a routine, i.e. picking out hooves and the removal of dried mud before exercise and dried sweat after exercise. The stabled horse, by contrast, is deprived of environmental influences and demands a more thorough regime for three reasons; in the first place to remove dirt, sweat and exfoliating skin from the coat; second, to encourage sebaceous activity; and third, to provide a massage effect.

Traditionally grooming is performed twice daily and this practice is maintained in many professional yards. The first grooming, called quartering, is carried out in the morning before exercise and consists of picking out the hooves, brushing out the mane and tail and removing stable stains. The name is derived from the practice of folding the rug of the clipped horse up over each quarter in turn. The exposed area is then brushed and the rug replaced to keep the horse warm on cold winter mornings.

Strapping is the main grooming period and is usually carried out in the afternoon once the horse has recovered from exercise. It is important that the coat is completely dry before this begins. With gross contamination or the presence of abundant loose hair a plastic or rubber currycomb may be used first but the first implement is usually the dandy brush. This is begun at the top of the neck and continued down the whole of one side of the horse including the legs and then repeated on the other side. The body brush follows and includes the head before being used on the mane and tail. The latter can then be layed with a slightly damp waterbrush. If considered desirable the horse can then be massaged with a wisp. or massager, and the coat finished by a lightly dampened stable rubber. The hairless areas are cleaned with a sponge and the hooves picked out and/or brushed. If necessary the feet can be washed with a waterbrush; hoof oil is then applied to both wall and sole.

Hygienic care of grooming equipment is important; all items should be regularly cleaned and each kit confined to one horse. In the advent of infectious skin disease such as ringworm then grooming instruments should be managed in a similar manner to rugs.

Plastic/rubber currycombs Plastic and rubber currycombs are used to remove dried mud and also dead hair when horses are shedding their winter coats. The plastic variety (Fig. 6.33(a)), which are also available with hose connectors, have an oval body from which protrude numerous rows of firm plastic teeth. These are usually separated by channels in which debris collects during use. Rubber currycombs (Fig. 6.33(b)) consist of three or more

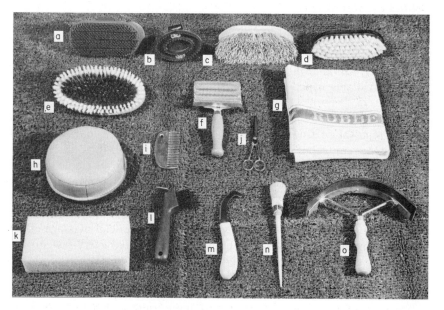

Figure 6.33. Grooming kit. (a) Plastic currycomb; (b) rubber currycomb; (c) dandy brush; (d) water brush; (e) body brush; (f) metal currycomb; (g) stable rubber; (h) leather massage pad; (i) mane/tail comb; (j) thinning scissors; (k) sponge; (l) hoof pick/brush; (m) bot egg knife; (n) hoof oil brush; (o) sweat scraper.

concentric ovals of soft rubber serrations with a similar base. Both are fitted with a handstrap which may be adjustable and are available in various sizes to fit the operator's hand. Short curved or circular movements are the most effective.

An alternative for use on thin skinned or clipped horses is a rubber grooming glove or mitten. These may be double sided with small rubber pegs on one side and bristles on the other.

Dandy brushes   Dandy brushes (Fig. 6.33(c)) are designed to remove heavy dirt from the horse's coat e.g. caked mud, stable dirt and dried sweat. They are used with the hairline in short strokes flicked away from the horse with a wrist action. It is generally suggested that dandy brushes should not be employed on manes and tails as hair may be pulled out or broken and use on clipped areas is usually resented. In this instance a soft rubber currycomb offers a suitable alternative.

Lacquered wooden stocks are universal but grooved edges for a secure grip are optional. The best brushes have the stock made in two parts with the tufts hand drawn and anchored between them. In cheaper although less hardwearing machine-made brushes the stock is a single unit into which the bristles are forced.

The tufts are 40–45 mm long and although available in varying degrees of rigidity all are relatively stiff. Brushes containing synthetic goose quill are very stiff, natural fibre bristles are stiff, while monofilament nylon and plastic are relatively soft and white Mexican fibre is very soft. Combinations of these are also available which give a spectrum of bristle characteristics.

Body brushes   Body brushes (Fig. 6.33(e)) are used on stabled horses for the removal of fine dust, scurf and grease from the coat, mane and tail once the removal of gross debris has been completed. In addition to the inefficiency of use on horses living at grass in winter it is advisable to avoid their use since the skin debris and sebaceous secretions assist in waterproofing the skin. On the body they are used in short arc or circular strokes in the direction of the coat with the weight of the groom's body behind them. Firm handling of this nature is also thought to have a beneficial massage effect to underlying muscle masses. The body brush is cleaned every few strokes by drawing the face of the brush across the teeth of a metal currycomb. As debris accumulates in the currycomb this too is cleaned by tapping it against the floor.

The stock of body brushes is usually a flat oval, although it may be shaped to fit the hand, and may be rigid or flexible. The former are usually glossed or waxed wood and the latter leather. Both have a loose band of material which fits over the back of the hand for secure use. Stock size varies from about 70 mm × 150 mm to 90 mm × 215 mm since good hand/brush congruency is essential for efficient use.

Body brushes may be hand drawn or machine made and contain short (approximately 20 to 25 mm) fine, closely set bristles designed to penetrate the coat. A selection of natural and synthetic materials are employed, horse hair, pig bristles, Mexican fibre, nylon and PVC. There are also combinations of these either with mixed materials throughout or with outer rows of, frequently slightly larger, relatively stiff material and centre of softer tufts. In general the synthetic materials are somewhat stiffer than those of natural origin.

Metal currycombs Metal currycombs (Fig. 6.33(f)) are used for cleaning the bodybrush. These are metal plates bearing between 4 and 8 rows of serrations separated by channels in which debris removed from the bodybrush accumulates. Currycombs are either fitted with wooden handles or hand straps (military style) and sometimes with metal buffers for tapping on the floor to clear the debris. Metal parts are usually zinc plated to minimise corrosion.

Water brushes Water brushes (Fig. 6.33(d)) are damped for use on the mane and tail and used wet for feet. The mane and tail are layed with a water brush by brushing from root to tip. When washing feet it is generally recommended that the thumb of the hand holding the horse's foot is firmly pressed into the cleft between the bulbs of the heels. This prevents soaking the area which is difficult to dry and when devoid of its natural secretions is liable to secondary infection (cracked heels).

The waxed wooden stock of a water brush is narrower than the body brush. The sides may be grooved for grip and bristles hand drawn or inserted by machine. These are normally longer than those of a body brush (up to 45 mm) and are traditionally white. Mexican fibre, tampico bristles and PVC are the most commonly used materials either alone or in combinations to provide a variety of bristle characteristics.

Following use water brushes should be thoroughly washed and then stood bristle side down to dry which significantly prolongs their useful life.

Wisps A wisp is made by fashioning a damped hay or soft straw rope into a firm hand size pad and is used as a form of massage beneficial to both skin and, in appropriate areas, to underlying muscles. They are now infrequently used outside of racing yards, having largely been replaced by leather covered felt massage pads (Fig. 6.33(h)) which have a handstrap. Both are used with energy and to avoid contusions palpable bony eminences should be avoided.

Stable rubbers   Stable rubbers (Fig. 6.33(g)) are used as pads to produce a polishing effect at the end of grooming. They should be slightly damped and used with some force over muscle masses in the direction of the hair lines. The traditional materials are twill linen and cactus cloth but a sheepskin or synthetic grooming mitten may be used in a similar manner.

Sponges   A damp sponge (Fig. 6.33(k)) is necessary for cleaning the eyes, nostrils, lips, dock, anus, vulva, sheath and udder. It is a sensible hygienic precaution to include two sponges in the grooming kit, one for the head and the other for the dock and groin areas. The importance of sponging out nostrils and beneath the dock is often over emphasised. However, regular cleansing of the sheath is extremely important as sebum accumulation not only leads to the formation of concretions but is also thought to be a carcinogen involved in the production of penile tumours (squamous cell carcinoma). Sponges should be thoroughly washed after each use and then allowed to dry before being returned to the grooming kit. Whether synthetic or natural sponge is employed is unimportant.

Hoof picks   Hoof picks are essentially blunt hooks of metal or plastic with a variety of handle designs some of which are folding. Double headed instruments with a hoof pick on one side and small wire brush on the other are popular (Fig.6.33(l)), and used in this order permit a more thorough cleansing than when a hoof pick is used alone.

Generations of children have been taught to use hoof picks from heel toward the toe to avoid penetration of the soft horn of the frog and thus the sensitive structures of the foot. However, in the absence of gross pathological change the possibility of this is remote. Particular attention should be paid to the lateral clefts and central sulcus of the frog where accumulation of detritus is a potent predisposing factor to thrush. Exposure of these areas to air is a reliable means of controlling this opportunist infection by anaerobic organisms. In addition the state of the foot and wear and security of shoes may be usefully assessed at this time.

Hoof oil   In addition to the cosmetic effect, hoof oil provides an impermeable barrier to regulate water loss from the hoof. Over the wall this function is normally carried out by the periople and so oil is of particular benefit when this has been destroyed or is deficient resulting in excessive moisture loss and hence brittle/cracked hooves. Excessive use can however be detrimental in preventing all evaporation from the hoof thereby producing soft crumbly horn. A small round-headed brush is conventionally used for application (Fig. 6.33(n)) but a suitably sized paint brush is perfectly adequate.

Mane and tail combs   Mane and tail combs (Fig. 6.33(i)) are used for pulling rather than for combing out manes and tails to avoid hair breakage. Pulling of manes and tails should be performed regularly on a little and often basis rather than as an occasional single event. It is most readily performed with the minimum of resentment when the horse is warm after exercise or on a warm day. The bases of the hairs to be pulled are isolated a few at a time by back combing their neighbours before being plucked from the crest, forelock or dock by a short sharp movement. This is designed to remove the hair roots and therefore thin the mane and tail, whereas pulling at the apex breaks the hairs and fails to produce any thinning.

Although frequently used as dual purpose tools, strictly speaking mane combs have deeper teeth than tail combs and plaiting combs have four large deep teeth for separating the mane prior to braiding. They are usually held at the base and most are aluminium. However, plastic combs are becoming increasingly popular.

Sweat scrapers   Sweat scrapers are employed to remove excess water after a horse has been bathed or washed down. Horses which are soaked with sweat dry quicker and have a cleaner coat if washed down with water. There are two basic designs. The most common is a curved metal band with a protruding rubber edge and a single centrally attached handle (Fig. 6.33(o)). The rubber side is applied to the horse and used in long firm strokes with the hairline towards the dependent parts of the body. This design is also available in plastic. The other is a thin flat metal band, such as brass, which is malleable and fitted with a leather (or similar) handle at each end. This type of sweat scraper may or may not have rubber on one edge. A further variation has a finely toothed serrated edge on one side for removing old matted hair and is referred to as a shedding blade. An American sweat scraper is a curved or flat metal blade with a shallow V cross-section which is both effective and cheap.

Scissors   Three types of scissors are frequent additions to grooming kits. A large flat pair are used for banging tails, that is cutting them off square just below the hock, cutting gamgee etc. A small pointed pair are useful for plaiting and trimming coronary band hair of show horses. While a pair of thinning scissors (Fig. 6.33(j)) are most satisfactory for trimming lower limb feather should this be desirable.

Bot egg knives   Bot egg knives are not included in all grooming kits. They consist of a handle and a curved, usually stainless steel, blade with a serrated edge on the outer surface and over the tip (Fig. 6.33(m)). Three species of bots are encountered in the United Kingdom but of principal concern is the species *Gastrophilus intestinalis*. The adult flies are active during the summer months and lay individual yellow coloured eggs mainly on the horse's forequarters. These are securely stuck to the tip of hairs and are not removed by ordinary grooming. Use of a bot knife with the lay of the hair gives a reasonable cosmetic effect and helps to reduce the subsequent larval load in the horse's stomach.

Electric grooming machines   Electric grooming machines are of two basic types, revolving brush and vacuum. The first consists of a motor driven brush and polisher on a moveable arm. The brush revolves at high speed and in either direction. Caution must be observed in the use of these lest the mane or tail become entangled. The second consists of a rubber currycomb handpiece connected via a long flexible tube to a vacuum unit. Neither remove grease from the horse's coat but can be labour saving in the removal of gross debris. In view of their limitations it is unlikely that grooming machines are economically viable in all but the larger yards.

Clippers   The principal reasons for clipping are (i) to enable the horse to carry out fast work without impedance to efficient heat loss from a dense coat; (ii) to facilitate drying off or washing down after work; and (iii) to save labour in grooming. It therefore follows that most animals are only clipped in winter. For the types or patterns of clips which have evolved as suitable for various work/management regimes the reader is referred to the bibliography. Shaving and singeing of coats are now rarely performed; however, facial whiskers of show animals are sometimes removed in this manner.

Clippers may be hand-operated, wheel machine or electric. The former, operated by squeezing the handles together, are slow and laborious. Although they may still have a place for use around the head of nervous animals, they have largely been replaced in this role by small low noise electrical apparatus. Wheel machines which require one person to power the machine while another clips, are little more than museum pieces.

Electric clippers are now almost universally used and three basic types are available. Hanging electric clippers consist of a mains operated motor unit suspended in a case away from the clipper head. This means that these machines are relatively immobile but the blades do not become as hot as hand operated electric clippers. The latter have the motor contained in the hand unit and are now the most popular (Fig.6.34). They are compact and can be operated wherever there is an electricity supply. The final variation also consists of a totally enclosed hand held unit but are cordless and rechargeable. These clippers are usually quiet and the small sizes are eminently suitable for use around the head. All contain two serrated blades, one fixed and the

(a)   (b)

Figure 6.34. Hand held electric clippers. (a) Suitable for general use. (b) Small pair for use around head.

other movable, of varying coarseness. The closeness of apposition of these blades or "tension" is usually adjustable by a screw fitted to the top of the cutting head.

The available equipment, individual horse's temperament and skill of the operator determine the degree of restraint necessary for clipping. Noise is usually the animal's main objection and their anxiety can frequently be reduced by placing cotton wool plugs in the ears. With more fractious individuals chemical tranquillisation may be necessary.

The coat must be clean and dry for clipping since few machines will remove heavily soiled or wet hair. Clippers are used against the hairline in long steady sweeps. Efficient progress is dependent on the maintenance of clean, sharp. cool blades used at the correct tension. Poorly cared for apparatus is noisy, pulls the hair and the heat generated can be sufficient to burn the horse. Despite the presence of cooling devices clippers may become hot during a clip. If this occurs then clipping is halted and the machine cleaned and given a light oiling. Following use machines should be stripped, cleaned and oiled before storage. Regular blade sharpening is also necessary.

The time and frequency of clipping is determined by the demands made of the horse, its environment and ultimately the individual coat characteristics. However, it is traditional not to clip the lower limbs to preserve the waterproofing role of the hair and not to clip horses after January to avoid damaging the subsequent summer coat.

Restraint

In discussing means of restraint the principle of paramount importance is that this should always be minimal, conducive to the efficacy of the procedure to be carried out and safety of the participants. Some understanding of equine psychology is desirable since a trial of physical strength is futile. Control of the horse's head is vital and diversionary activities may be useful but there is a wide individual variation in the efficacy of the response to various means of restraint. Use should also be made of the immediate surroundings since as a rule of thumb most horses will tolerate interference of any nature better when confined than allowed freedom of movement. Simple procedures such as backing a horse into the corner of a stable can be extremely useful.

Harness is not within the scope of this chapter, however a bridle is a useful means of restraint. Some authorities suggest that this should be avoided other than for control of a led horse but it can be employed without injuring the mouth and a significant number of horses tolerate minor interferences while bridled which they would not in a headcollar. Whether this is diversionary or "respect" for aids conveyed via the bit is not entirely clear.

Headcollar   The most common means of restraint is the headcollar which may be of webbing, nylon or leather. The best leather headcollars are made from the shoulder section of cowhide which is coarser and heavier, and thus harder wearing than butt or belly leather. Sometimes specialised leathers such as rawhide and helvetia are employed for increased strength and soft calf skin linings are occasionally added.

The most basic and popular leather headcollar is the Newmarket pattern which is fitted by an open headpiece which may buckle either on the left or on both sides. Neither the throatlatch, nosepiece, back stay or cheekpieces are adjustable (Fig. 6.35). The rings or stop squares which connect these sections and buckles are usually made of brass and it is both fashionable and practical to have a brass nameplate riveted to a cheekpiece. Lead reins may be attached to the stop square at the back of the noseband while pillar reins can be secured to the two side stop squares at the nosepiece/cheekpiece junction. A French headcollar is similar but includes a "D" ring at the front

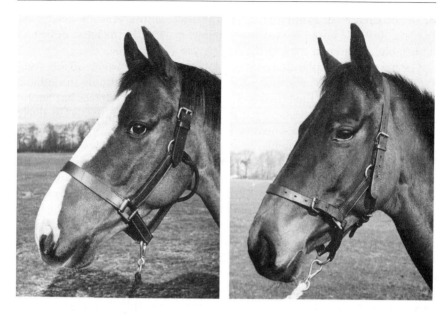

Figure 6.35

Figure 6.36

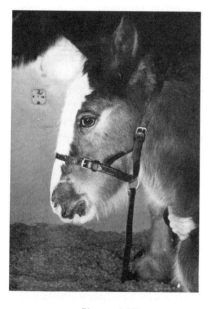

Figure 6.35. Newmarket headcollar with brass fittings.

Figure 6.36. Yearling headcollar with adjustable nosepiece and throatlatch; lead rope attached by spring loaded hook.

Figure 6.37. Foal slip or Dutch slip.

Figure 6.37

of the nosepiece for leading from this position, and a yearling headcollar has an adjustable nosepiece to accommodate growth. In the latter design an adjustable throatlatch may also be employed (Fig. 6.36).

The next most popular design is the Albert headcollar which differs from the Newmarket by the addition of a browband and adjustable throatlatch which passes through the browband slots and over the headpiece but has no rings at the top of the cheek pieces. It is claimed that this design is less likely to break if a horse pulls back because the strain is taken at the poll and inhibits the horse. A Queen's headcollar is a variation of the Albert design in which the throatlatch is attached to metal rings at the point where the browband also fastens. Both cheekpieces and throatlatch are adjustable but it is not considered to be as strong as the Albert headcollar. The Salisbury headcollar is similar to the Queen's but lacks a browband.

A number of minor variations of these basic designs are also encountered such as clip release attachments from the throatlatch to the headpiece but two other specialised headcollars merit inclusion. The foal slip or Dutch slip is the most popular headcollar for foals (Fig. 6.37). This consists of a continuous head and cheekpiece which passes through a ring beneath the foal's lower jaw and a nosepiece which also passes through this ring. Both are adjustable on the left side and are connected by an oblique strap on each side. A short free strap is usually attached to the metal ring to aid in catching the foal.

The headcollar bridle is used for showing bitted animals ''in hand'' and is therefore frequently fancy-stitched. The bit is attached via adjustable buckles on the nosepiece stop squares. Headcollar bridles may be of Newmarket or Albert designs.

All designs are available in nylon which is strong, heardwearing, easily washable and cheap. These are made either with conventional or rolling buckles and either metal or plastic fittings.

Lead rein   Lead reins or shanks provide a means of leading or securing the horse via a headcollar. A variety of materials may be employed. Cotton rope leads are the most popular as they are soft and easily tied into quick release knots. However, they are not very durable and their life expectancy is limited. They are 1.5 to 1.8 metres long and 12 to 20 mm in diameter. Similar nylon ropes are a more durable alternative but are harder on the hands and produce less secure knots.

The most common form of attachment to the headcollar is a spring loaded hook. This provides a quick and convenient method of securing a horse, but must always be attached with the open side facing outward to reduce the risk of facial lacerations. For similar reasons such clips should be maintained in good working order. More recently safety clips have appeared on the market which are blunt ended clasps with a quick release mechanism.

Tubular cotton web lead reins are used for show purposes and may attach to the headcollar via a buckle and strap or short length of brass chain and spring loaded clip. Leather lead reins are occasionally employed, chiefly on stallions, and are attached in a similar manner. Stabled horses which chew rope lead reins may be secured by chain leads or rack chains attached by spring loaded clips but these are not suitable for leading horses.

Halter   A halter is a one piece noseband, headpiece and lead rein. They are now less popular but may be of use in emergency situations since one size will fit virtually any horse. In addition it is customary for some heavy horses to be shown in halters. Rope halters may be of cotton, synthetic yarn or polypropylene all of which are washable (Fig. 6.38). The rope is usually 12 to 14 mm in diameter for most horses, but 16 to 18 mm rope is used for Shire halters. Head shanks are generally longer than lead reins for headcollars, being 3.0 to 4.5 metres long and, once fitted, a knot is tied at the base of the lead to prevent the halter from tightening or loosening.

A Yorkshire halter has a ribbed hemp headpiece and nosepiece, a cord throatlatch which is tied for additional security, and a rope lead attached to the rear of the nosepiece (Fig. 6.39). These are much more size specific than rope halters and outside the show ring have few routine uses.

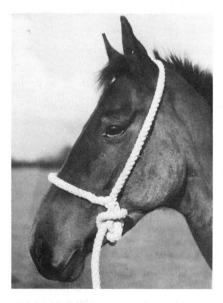

Figure 6.38. Cotton rope halter with knot tied at base of lead to prevent tightening or loosening.

Figure 6.39. Yorkshire halter.

Figure 6.40. Chifney anti-rearing bit.

An American halter is essentially a rope headcollar and is used widely elsewhere in the world. They are usually made of polypropylene which is strong, durable and washable.

Chifney A chifney anti-rearing bit (Fig. 6.40) is particularly useful for leading stallions but can be applied, with varying success, to restrain any animal which persists in rearing when being handled. It is fitted with three rings, two for cheekpieces and one for the lead rein, and has a shallow inverted port mouthpiece.

Twitch A twitch is a device for gripping and exerting pressure on the horse's upper lip. It has long been recognised that some animals are effectively restrained from fractious behaviour by a device of this nature. Its mode of action is controversial; originally it was believed that the discomfort associated with the application of a twitch was merely a diversionary action that discouraged movement and thus further pain. Recently however it has been suggested that pressure on the horse's lip may stimulate the production of endorphines which have a centrally acting sedative effect.

It must be appreciated that an individual animal's response to application of a twitch is idiosyncratic and while many horses will be restrained by this instrument violent behaviour will be precipitated in others.

Various types of twitch are encountered. The most common is a wooden pole, either short for one handed use or long for double handed grip, with a loop of string through one end (Fig. 6.41(a)). This is passed over a hand

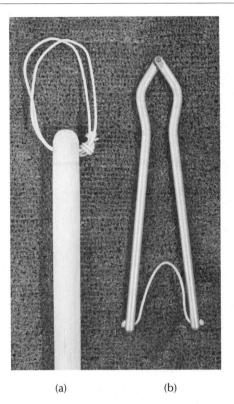

(a)   (b)

Figure 6.41 (a) Conventional twitch. (b) Hand held or humane twitch.

gripping the upper lip and tightened by twisting the pole. Care is necessary in the use of a twitch since excessive pressure or prolonged application will result in ischaemic necrosis of the underlying tissue. Following initial tissue loss a ring of depigmented skin is produced. A wooden handle with a chain loop is said to be more humane since it cannot be twisted as tightly as string. A hand held or humane twitch (Fig.6.41(b)) is a pair of shaped metal handles hinged at the apex which are used to squeeze the upper lip. These may have a string to hold the end of the handles together thus eliminating the need for continual manual pressure.

A twitch should only be applied to the upper lip and sites such as the ear and tongue which contain more pressure sensitive tissues avoided.

REFERENCES

The British Horse Society and the Pony Club (1983). *The Manual of Horsemanship* 8th Edition (Marabel Hadfield, ed.). Threshold Books, London.

Edwards, E. Hartley (1963). *Saddlery*. J. A. Allen and Co., London.
Ferguson, G. S. (1984). *Stable Management in Horse Management* (John Hickman, ed.). Academic Press, London.
Fitzwygram, F. (1869). *Horses and Stables*. Longmans, London.
McBane, Susan (1984). *Keeping a Horse Outdoors*. David and Charles, Newton Abbot.
Richardson, Julie (Ed.) (1982). *Horse Tack: the complete equipment guide for riding and driving*. Pelham Books, London.
Watson, Valerie (1986) *Manes and Tails*. Threshold Books, London.
Watson, Valerie (1986). *Trimming and Clipping*. Threshold Books, London.

7 The Foot and Shoeing

J. HICKMAN and M. HUMPHREY

INTRODUCTION

Horseshoeing was unknown to the Ancient Greeks. It started with the Gauls and Celts from whom it passed to the Romans, although they were not an equestrian nation. Several hundred years of writing, discussion and experimentation in farriery reached a peak in the early part of this century when the veterinary profession was very close to the farriery trade.

Farriery went into a decline with the advent of the motor car. However, with the increasing popularity of the horse in the last thirty years farriery has become again an essential and important trade. Most of the theory and much of the practice in shoeing comes from a bygone age when the vast majority of horses were a means of transport, doing a lot of road work at a slow pace, and not athletes as they are today.

The Farriers (Registration) Act 1975 and the Farriers (Registration) (Amendment) Act 1977 were passed to prevent horses suffering from unskilled shoeing, to promote the training of farriers, and to establish the Farriers Registration Council to register farriers.

A four year apprenticeship with an approved training farrier is the only way to become a farrier except through the Royal Army Veterinary Corps. During this time an apprentice will spend a total of six months at the farriery school in Hereford. The qualification Dip.WCF (Diploma of the Worshipful Company of Farriers) which is taken at the end of the apprenticeship entitles the farrier to registration. Higher examinations are the AWCF (Associate) and the FWCF (Fellowship) which are attainable by showing a higher understanding of theory than for the Diploma and an ability to make a variety of surgical shoes.

Horse Management 2nd edition
ISBN: 0-12-347218-0 case

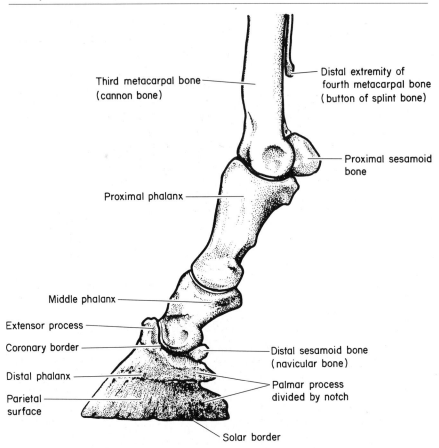

Third metacarpal bone
(cannon bone)

Distal extremity of
fourth metacarpal bone
(button of splint bone)

Proximal sesamoid
bone

Proximal phalanx

Middle phalanx

Extensor process

Coronary border

Distal phalanx

Parietal
surface

Distal sesamoid bone
(navicular bone)

Palmar process
divided by notch

Solar border

Figure 7.1. Digit of the horse. Front leg, lateral aspect.

ANATOMY AND FUNCTION OF THE FOOT

The digit of the horse comprises three main bones, called the phalanges, and a sesamoid bone, the distal sesamoid bone (navicular bone) (Fig. 7.1).

The proximal phalanx (long pastern) occupies an oblique position, being directed downwards and forwards at an angle of about 55°, between the third metacarpal bone (cannon bone) and the middle phalanx with which it forms the proximal interphalangeal joint (pastern joint).

The middle phalanx (short pastern) is situated obliquely between the proximal and distal phalanx, with which it forms the distal interphalangeal joint (corono-pedal joint).

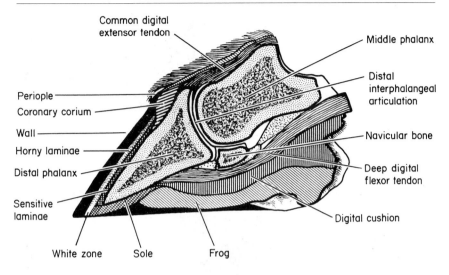

Figure 7.2. Saggital section of the foot. Note the distal phalanx is enclosed within the hoof, its solar surface is not parallel with the bearing surface of the foot, and the position of the navicular bone.

The distal phalanx (pedal bone) is entirely enclosed within the hoof (Fig. 7.2), to which it has some resemblance in shape, and is composed of very compact bone. Attached to the upper borders of each palmar process (wing) is a plate of cartilage which extends sufficiently above the hoof to be palpable. They lose their elasticity with age and may be converted into bone when they are called side bones.

The distal sesamoid (navicular bone) is a small shuttle shaped bone, situated at the back of and forming part of the distal interphalangeal joint. It has two surfaces, an articular surface which articulates with the distal articular surface of the middle phalanx and a flexor or tendon surface which is covered by fibro-cartilage and over which passes the tendon of the deep digital flexor muscle.

The foot receives a very good blood supply from the digital arteries. The majority of the veins are valveless, which form plexuses communicating with each other and are drained by the digital veins. It is interesting to reflect on the part the function of the foot plays in maintaining its venous circulation. When the foot takes weight the expansion at the heels results in a rise of blood pressure in the veins and they empty. When the foot is raised off the ground the pressure is reduced and the veins fill. The nerves are branches of the digital nerve and accompany the distribution of the arteries and veins. (Fig. 7.3).

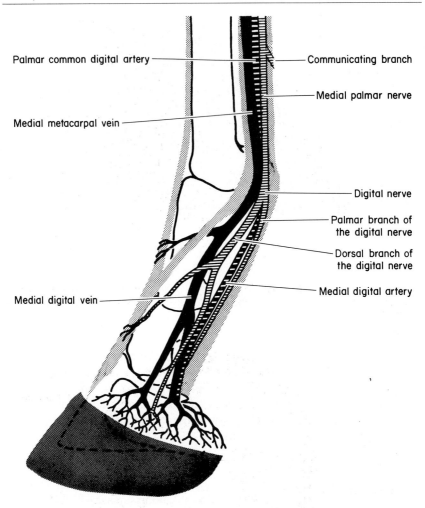

Palmar common digital artery

Communicating branch

Medial palmar nerve

Medial metacarpal vein

Digital nerve

Palmar branch of
the digital nerve

Dorsal branch of
the digital nerve

Medial digital artery

Medial digital vein

Figure 7.3. Arteries, veins and nerves, front leg and foot, medial aspect.

External foot or hoof

The hoof is the covering of horn at the end of the digit and is non-vascular and insensitive. It comprises the wall, sole and frog (Figs 7.4 and 7.5). The junction of skin and hoof is called the coronet.

Wall The wall is that part of the hoof which is seen when the foot is on the ground, and for description is divided into toe, quarters and heels (Fig. 7.4). It does not form a complete circle but is reflected inwards and

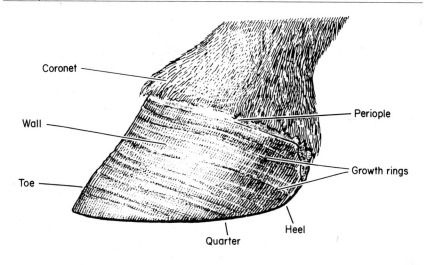

Figure 7.4. The hoof, side view. Note the wavy growth of the horn which appears as a number of rings parallel with the coronet and indicates alterations in the rate of growth.

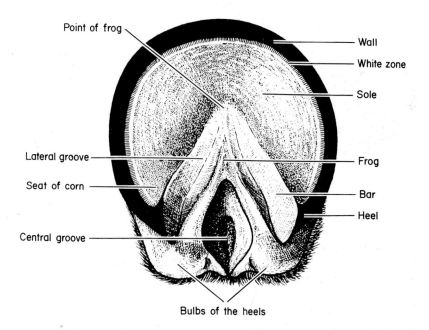

Figure 7.5. Front foot, ground surface.

forwards at an acute angle at the heels to form the bars. The area of the sole between the wall and the bars is called the "seat of corn". The wall is thicker at the toe than at the heels. The surface of the wall is smooth, but is crossed by a number of waves or rings of horn which run parallel to the coronet and indicate variations in the rate of growth.

The only part of the wall which is subjected to wear is its bearing surface. The wall grows evenly all round the coronet at the rate of about 2.5 cm (1 in) in three months. Therefore it takes, on average, from 9–12 months for the horn to grow from the coronet to the toe, and to the heels in about six months. The rate of growth of the wall of horses in the wild keeps pace with the rate of wear but when they are domesticated and put to work on roads then the rate of growth cannot keep pace with the rate of wear and they soon become foot sore. Therefore wear has to be controlled by protecting the bearing surface with a rim of iron, a shoe, which in reality is only an extension of the wall.

Bridging the junction between the skin and the wall is a thin layer of soft horn, the periople. It is thickest proximally, and gradually disintegrates as it grows down the wall. When the hoof is dry the periople is not obvious, but is most conspicuous when the foot gets wet. The periople prevents undue evaporation from the underlying horn, and if removed continuously by rasping it can lead to brittle feet.

The internal surface of the wall is traversed by some 500 to 600 horny laminae (leaves) which extend from the proximal border of the hoof to the bearing surface and onto the bars (Fig. 7.6). These insensitive laminae dovetail with the corresponding sensitive laminae of the foot to establish a very strong union. It is interesting to note that the weight of the horse is supported by this union, the distal phalanx being suspended by the laminae from the hoof wall.

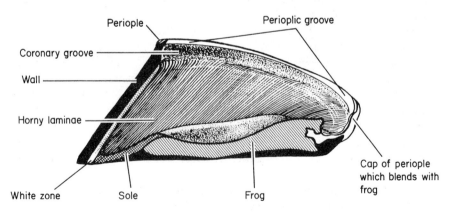

Figure 7.6. Saggital section of hoof—internal surface.

The proximal or coronary border is thin and hollowed out to form the large coronary groove which lodges the coronary corium, and immediately above it is the perioplic groove which lodges the corium of the periople. These grooves merge at the heels.

The distal (bearing) border comes into contact with the ground and here its inner surface is united with the periphery of the sole. This union is indicated on the bearing surface of the foot by a well defined white zone. This is a most important anatomical feature and a most helpful guide for the farrier as it indicates the thickness of the wall and hence the position of the sensitive structures. This enables the farrier to drive the nails to secure a good hold without pricking or causing pressure on the sensitive structures, a so-called nail bind.

Sole   The sole constitutes the greater part of the ground surface of the hoof and is designed to protect the sensitive structures above it and to take weight around its border. The sole is arched or vaulted, a feature which is more pronounced in the hind than in the front feet. The thickness of soles varies considerably; in some horses they are firm and rigid, whereas in others they are thin and yield to pressure. Growth of the sole differs from that of the wall as it exfoliates or flakes off when its fibres have attained sufficient length. This is a necessary feature if it is to maintain its natural vaulted configuration as it is not exposed to or worn by friction.

Frog   The frog is a wedge shaped mass of soft elastic horn, having the character of india-rubber, which occupies the angle between the bars and the sole. It is an integral part of the expansion mechanism of the foot and to function properly it must be in contact with the ground and take weight. This results in compression which is followed by expansion: if this does not occur then atrophy results.

The horn of the wall and sole, which is almost completely keratinised, is tunnelled with numerous small tubules which contain fluid absorbed from either the sensitive foot or from the surface of the hoof. The horn of the frog also contains tubules, but the horn is not so keratinised and therefore relatively soft and elastic. The water content of the hoof is an important factor, and ranges from approximately 25% in the wall to 40% in the frog. Constant evaporation is taking place from the surface of the wall and sole and if it is excessive the hoof becomes brittle and may crack. On the other hand, if evaporation is checked the fluid accumulates and the horn becomes soft.

Horn is a bad conductor of heat and hence proves a very effective protection to the underlying sensitive structures against the effects of cold and the application of a hot shoe. It should not be overlooked that horn

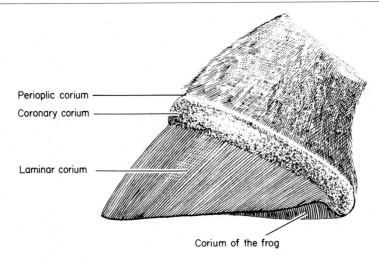

Perioplic corium

Coronary corium

Laminar corium

Corium of the frog

Figure 7.7. The hoof removed to show the sensitive foot, side view.

cells are dissolved by alkalis and therefore the harmful effects of ammonia present in decomposing manure and urine are an important factor in stable management.

Internal or sensitive foot

The sensitive structures of the foot are a modified part of the deeper layers of skin, which contain the blood vessels and nerves, and provide the nutrition to the corresponding parts of the hoof (Fig.7.7).

The perioplic corium lies in the perioplic groove and supplies nutrition to the periople. The coronary corium lies in the coronary groove and supplies nutrition to the wall. The sensitive laminae are attached to the parietal surface of the distal phalanx. They extend from the coronary corium to the solar border of the bone and dovetail with the horny laminae to which they supply nutrition. The sensitive sole and sensitive frog blend with the solar surface of the distal phalanx and the digital cushion respectively and supply nutrition to the sole and frog. The digital cushion is a wedge shaped fibro-elastic pad which occupies the posterior part of the foot and fills in the hollow of the heels. It is moulded to the sensitive frog below and it plays an important part in reducing concussion by expanding when the foot takes weight.

Function of the foot

The function of the foot is first to reduce concussion and second to prevent slipping. The effects of concussion are minimised by the angular structure of the limbs and the expansion of the foot when it comes to the ground and the weight of the horse passes over it. When considering how the concussion of weight bearing is reduced it must be appreciated that the various structures involved are interdependent. Since the heels come to the ground first the most important anti-concussion structures are located towards the heels. They are of great importance because at stages of gallop the whole of the horse's weight is supported by one foot.

The foot is protected against concussion by the yielding articulation of the distal interphalangeal joint (corono-pedal joint), a slight descent of the distal phalanx and sole, the elasticity of the frog and digital cushion, the flexibility of the cartilages, and expansion of the heels when the foot takes weight (Fig. 7.8). The amount of expansion depends on the degree of compression which in turn is related to the development of the frog, the height of the heels and the type of ground surface. This compression is followed by pressure on the digital cushion and bars, which in turn press on the cartilages and force the heels apart. When weight is taken off the foot all these structures return to their resting positions. If the degree of expansion in a front foot is measured by lifting the opposite leg it will be found to be in the order of 1.0 mm. Obviously, when the horse is in motion the degree of expansion will be correspondingly increased.

The concave or saucer-like shape of the sole and the wedge shaped frog which grips firmly when pressed into the ground combine to assist the horse to maintain its balance and prevent slipping when it pulls up or turns sharply.

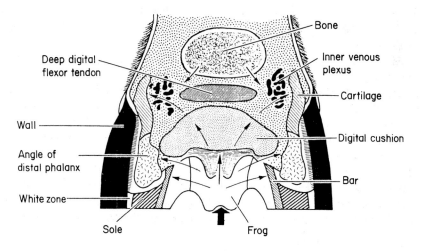

Figure 7.8. Frontal section of the foot. When the foot takes weight the frog is compressed. This results in pressure on the bars and digital cushion. These in turn press on the cartilages which yield and is followed by expansion of the foot at the heels.

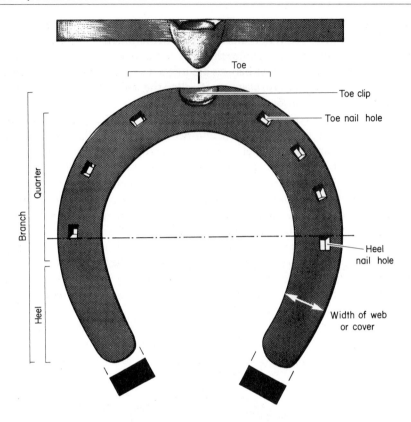

Figure 7.9. Standard horseshoe with a plain foot and ground surface and correctly placed nail holes. In each branch three nail holes are stamped in front of an imaginary line which divides the shoe into two equal parts, with the heel nail hole in the outer branch placed just behind this line.

HORSESHOE AND HORSESHOE NAILS

Horseshoes have been made from a variety of materials, but steel is the most suitable. The type of shoe fitted varies according to the horse and the work it has to perform, but a number of basic design features are common to all shoes.

A horseshoe, like the hoof, is described as having a toe, quarters and heels (Fig. 7.9). From the toe to heel on each side is called a branch, the width and thickness of the branch is referred to as the web, and the width of the web is called the cover. The shoe has two surfaces and two edges. The foot surface is in contact with the hoof and the ground surface in contact with the ground and there is an inner and outer edge.

The width of the web is related to the natural bearing surface of the foot and should cover the wall, white zone and outer border of the sole. If it is too wide it predisposes to grit lodging under it and is liable to be sucked off in mud. A shoe which is too thick raises the foot too far off the ground which reduces normal frog pressure and excessively large nails are necessary to retain it which tend to split the hoof.

The foot surface of the shoe supports the hoof and a flat surface provides a firm base for the foot to rest on and is suitable for all normal feet. To relieve pressure on the sole of horses with flat or dropped soles the inner edge of the shoe is sloped and is referred to as a seated-out shoe (Fig. 7.35).

The ground surface of the shoe has to be considered in relation to the foothold it provides, its wearing properties and its support for the nails. There are (i) a plain stamped shoe (Fig.7.9) which has a flat ground surface and wears well; (ii) a fullered shoe which has a groove around the ground surface—this improves the foothold, helps to prevent slipping but does not wear as well as a plain foot surface, and (iii) a concave shoe which has the inner ground surface sloped to conform with the natural concavity of the sole.

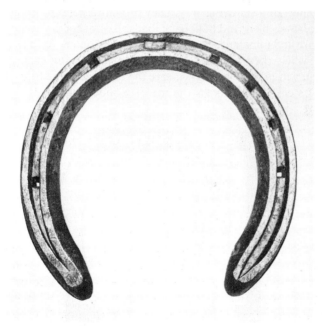

Figure 7.10. Concave fullered front shoe.

For all general purposes, a combination of fullering and concaving, the concave fullered shoe (Fig. 7.10) which combines the advantages of both types is the most popular.

The lighter the shoe the better but obviously this must be related to wear. It should be appreciated that the extra weight of shoes increase the expenditure of effort necessary for a horse to perform a day's work and the heavier the shoes the greater the contribution to stresses on the limbs and to fatigue.

Machine made shoes have regularity of form and a true foot surface. They save the farrier valuable time by reducing manual labour. A good machine made shoe is not as good as a good hand made shoe but because machine made shoes are of consistent quality they are preferable to all but the most carefully made hand made shoes.

Features of horseshoes additional to the basic designs

Clips   Thin triangular projections drawn from the outer edge of the shoe at the toe and quarters. They rest against the wall, resist shearing forces and thus support the nails. For riding horses it is traditional for front shoes to have a single toe clip and hind shoes two quarter clips.

Calkins   Projections formed by either turning down or welding a piece of steel to the heel of a shoe (Fig. 11). They provide a good foothold and assist draught horses to back and hold back loads. By raising the heels they reduce frog pressure and unbalance the foot. It is not customary to fit them on front shoes but when they are used, on draught horses for example, it is in conjunction with a toe-piece to preserve the balance of the foot. A *toe-piece* is a rod of steel welded across the toe of a shoe (Fig. 7.11).

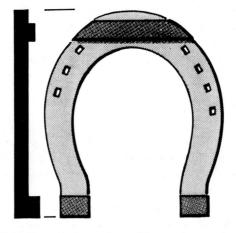

Figure 7.11. Draught horseshoe with toe-piece and calkins.

Figure 7.12. Bar shoe. An ordinary shoe which is joined at the heels by a bar that presses on the frog.

Bars   A bar joins a shoe across the heels (Fig. 7.12) and is designed either to press on the frog, thus taking weight from the hoof wall at the heels, or to clear the frog, thus relieving it completely of weight bearing.

Horseshoe nails   are made from the best mild steel, smooth and polished. They are strong enough to be driven through the hoof without buckling yet sufficiently ductile not to snap when the point is turned over.

There are two basic types, the European which has a long tapering head (Fig. 7.13) and is designed for thick shoes, and the USA type which has a shorter head and is more suited for thin shoes. The point of all horseshoe nails is bevelled on the inner face. Therefore the nail must always be driven with the bevel to the inside so that the point will turn away from the sensitive foot and emerge on the outside of the wall.

STYLES OF HORSESHOES

Racehorses require the lightest shoes possible that are strong enough not to twist or break. Concave fullered shoes made of an aluminium alloy and called racing plates (Fig. 7.14) are used. For training, light concave fullered shoes made of a mild steel are used.

There are a number of varieties of racing plates. Some have two grooves supposedly to give a better foothold, some have a steel insert at the toe to increase wear and others have a synthetic rim pad bonded to the foot surface

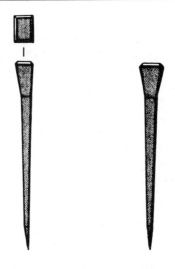

Figure 7.13. Horseshoe nail, European type.

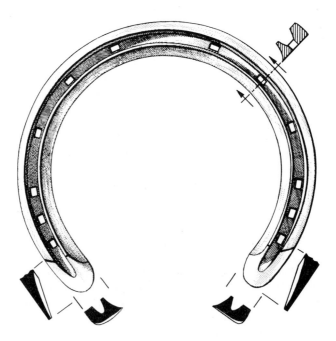

Figure 7.14. Racing plate, aluminium alloy, machine made. front shoe with sloping or "pencilled" heels.

to reduce concussion. Hind shoes are with or without calkins at the outside heel and may be full sized or with a short inside branch to reduce the risk of brushing. A set of racing plates weigh from 170 to 280 g. (6–8 oz).

Hunters require a shoe which provides a good foothold. Concave fullered shoes (Fig. 7.10) are the most satisfactory. Front shoes have a toe clip and the inner branch may be fitted close to prevent brushing. Hind shoes have the toe squared slightly and two quarter clips with the toe set back to prevent over-reach injuries. It is common practice to have a wedge heel and calkin but these are being replaced by studs.

Showjumpers have to twist and turn at speed and to pull up sharply. These problems are met by fitting concave fullered shoes with the heels tapped to take studs.

Hacks do a lot of road work and therefore require a reasonably heavy shoe. Shoes can be three quartered fullered or concave fullered which are often fitted in front with three quartered fullered behind. To prevent slipping on the roads tungsten carbide plugs or nails (Fig. 7.15) are used in one or both heels of each shoe. These are preferable to calkins and studs which by raising the heels unbalance the feet. A set of shoes for a hunter, showjumper or hack weigh about 1.8 kg (4 lb).

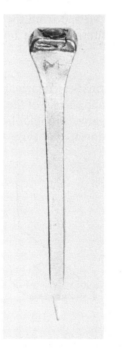

Figure 7.15. Horseshoe nail with tungsten carbide plug.

CONFORMATION AND THE PRACTICE OF SHOEING

Conformation is the make and shape of a horse. This has a direct bearing on the distribution of a horse's weight and its action. Deficits of conformation cannot be corrected, but the effects can be controlled and alleviated by attention to the horse's feet and to shoeing.

Deficits — front limb

Both limbs should be straight and parallel. If the feet are placed close together (base narrow) the hoof tends to have the medial side longer and more sloping than the lateral side. Placing wide apart (base wide) results in the hoof on the lateral side being longer and more sloping than the medial side.

When viewed from the side if the knee is deviated backward it is referred to as calf knee and if bowed a goat or bucked knee.

Deficits — hind limb

Feet placed close behind (base narrow) invariably result in the hocks being turned out which is called bow legs, and if the feet are placed too far apart the hocks are brought close together which is called cow hocks.

The most common defect viewed from the side is an excessively curved hock joint which is referred to as a sickle hock.

Pastern foot axis

When viewed from the front this is an imaginary straight line from the centre of the fetlock joint to the ground surface of the foot which divides the pastern and the foot into two equal parts. A similar line when viewed from the side is at the same angle as the dorsal wall (Fig. 7.16).

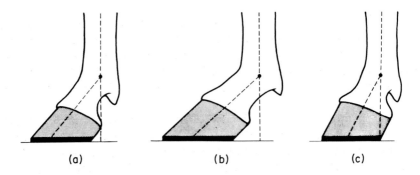

(a)   (b)   (c)

Figure 7.16. Side view of a normal pastern foot axis: (a) a normal foot; (b) a sloping foot; (c) an upright foot.

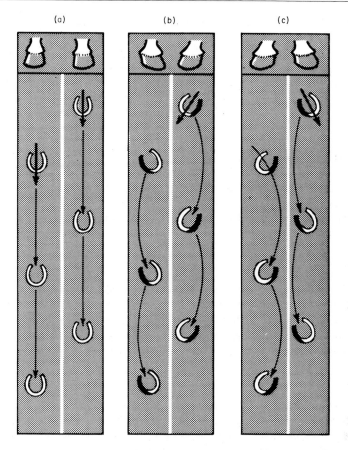

Figure 7.17. Flight of the front feet: (a) normal flight; (b) winging out, due to toe-in conformation; (c) winging in, due to toe-out conformation.

Front view   Toe-in conformation is when viewed from the front, the pastern foot axis slopes inwards (Fig. 7.17(b)), and results in the medial side of the hoof being longer and more sloping than the lateral side. Whereas in a *toe-out conformation* the pastern foot axis slopes outwards (Fig. 7.17(c)), and the lateral side of the hoof is longer and more sloping than the medial side.

Side view   It is essential to differentiate a natural abnormality of the pastern foot axis from one that has developed due to an excessive growth of horn at either the toe or the heels. A horse with a naturally sloping foot (Fig. 7.16(b)), that is one which is long at the toe and low at the heels, has a straight pastern foot axis, whereas a similar conformation due to excessive growth of the horn at the toe has a pastern foot axis which is broken back (Fig. 7.18(a)).

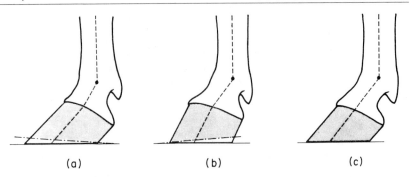

Figure 7.18. Side view of a foot with an abnormal pastern foot axis due to: (a) excess horn at the toe, pastern foot axis broken back; (b) excess horn at the heels, pastern foot axis broken forwards; (c) pastern foot axis restored to normal by either lowering the toe or the heels.

Similarly, a horse with a naturally upright foot (Fig. 7.16(c)), that is one with a short toe and high heels, also has a straight pastern foot axis, but if it is due to excessive growth of horn at the heels then the pastern foot axis is broken forwards (Fig.7.18(b)).

When a sloping or upright foot has resulted from an excessive growth of horn at the toe or the heels, as opposed to the natural state, then the pastern foot axis can be corrected by shortening the toe in the former and lowering the heels in the latter case (Fig. 7.18).

The hoof

A normal front hoof is rounder at the toe than a hind hoof. The medial and lateral hoof wall are of equal length and slope. The wall is thickest at the toe. The sole is slightly concave and the frog is large, elastic and with a shallow central and deep medial and lateral grooves.

The surface of the wall is not absolutely flat, but broken by a wavy growth of horn which appears as a number of rings parallel with the coronet. These rings are a normal feature and indicate alterations in the rate of growth due to either changes in the diet or some febrile disease. It is important to differentiate between these rings and those associated with chronic laminitis which are characterised by being widely spaced at the heels and converging towards the toe.

The foot axis and the angle of the wall at the toe and the heels should correspond and be between 50° and 55° (Fig. 7.19(b) and (c)).

A normal hind foot is oval at the toe and widest towards its posterior third. In comparison with a front foot the sole is more concave, the frog smaller, the slope of the wall at both the inner and outer quarters more upright and the foot axis is at an angle between 50° and 55° as for a front foot.

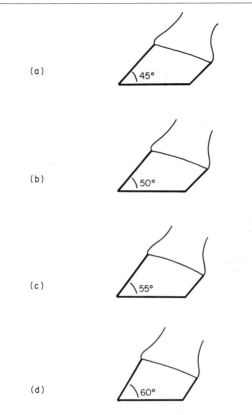

Figure 7.19. Types of hoof: (a) sloping hoof; (b) and (c) normal hooves; (d) upright hoof.

Abnormal hooves   A hoof may be abnormal in shape due to a natural defect or the result of irregular wear, a defect of conformation.

Sloping hooves. The hoof has a long toe and is low at the heels. The angle of the foot axis is decreased and the angle at the toe is less than 50° (Fig. 19(a)).

Upright hooves. The foot has a short toe and is high at the heels. The angle of the foot axis is increased and the angle at the toe is more than 55° (Fig. 19(d)).

Club foot. An accentuated upright foot. The angle of the foot axis is greatly increased and the angle at the toe is in excess of 60°. As a rule the condition is bilateral when it is congenital in origin, but unilateral cases are seen associated with chronic contraction of the digital flexor tendons in young stock.

Flat feet. The sole lacks its normal concavity and obviously the condition is conducive to bruising.

Dropped sole. The sole is convex, below the bearing surface of the wall, is subject to bruising and is a feature of chronic laminitis.

Thin sole. An inherited condition which is usually accompanied by a thin wall. The conformation of the foot is normal but the sole being thin yields under pressure. Horses with this conformation are subject to bruised soles.

Brittle feet. May be an inherited defect but the majority of cases are encountered in dry weather due to loss of moisture from the hoof. White feet are more frequently affected. Brittle feet are very liable to crack and therefore create a shoeing problem as they are easily broken and split by the nails. The condition can be improved by soaking the feet in water or by the application of vaseline or oil to limit evaporation. A crack can be prevented from extending by cutting a groove across its apex.

The horse's gait

To assess a horse's gait it should be observed at the walk and trot, coming towards, going away from and past the observer with the horse on a loose rein.

Front view   A sound horse moves its legs in alignment with its body, its toe pointed forward and the foot set down flat.

Winging-out (Fig 7.20(a)). The flight of the foot is outwards and then inwards in a circular movement. The foot lands and breaks over on the outer branch of the shoe which results in excessive wear of the outside quarter.

Winging-in (Fig. 7.20(b)). The flight of the foot is inwards and then outwards in a circular movement. The foot lands on and breaks over on the inner branch of the shoe which results in excessive wear of the inside quarter.

Plaiting (Fig. 7.20(c)). The foot moves inwards to land more or less in front of the other foot. The foot comes down flat and so the shoe is not abnormally worn.

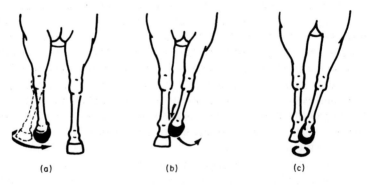

(a)   (b)   (c)

Figure 7.20. Abnormalities of gait: (a) winging-out; (b) winging-in; (c) plaiting.

Figure 7.21. Flight of a front foot. The foot breaks over at the toe with the foot following a smooth arc which reaches its peak as it passes the opposite leg.

Side view   The normal action of a front leg is characterised by the foot breaking over at the toe with the flight of the foot following a smooth arc which reaches its peak as it passes the opposite leg. (Fig. 7.21).

Factors affecting gait   Apart from lameness the two most important factors are conformation and the hoof and shoe balance.

A horse with a toe-in conformation tends to wing-out and with a toe-out conformation tends to wing-in, resulting in excessive wear on the outer and inner branch of the shoe respectively.

With a sloping pastern foot axis the break over of the foot is delayed which results in the horse having an increased stride and keeping its feet close to the ground, whereas with an upright pastern foot axis the break over of the foot is quick resulting in the horse having a short stride and bringing its feet to the ground sharply increasing concussion and being an uncomfortable ride.

An excessively long hoof results in increased flexion of all the joints of the limb and break over is delayed. If a hoof is not balanced with the limb conformation it will be twisted towards the longer side when on the ground and tend to wing in the other direction when in flight.

Wear of the shoe   The most important defects of conformation, which result in abnormal wear of the shoe, are those of the pastern foot axis. Therefore it is essential to study the wear of a shoe when it is removed so that these defects and any faults in the preparation of the foot can be taken into account when fitting a new shoe.

Even wear of a shoe indicates that the foot has been reduced to its correct proportions and the type of shoe is suitable. If horses with a normal action wear out their shoes more quickly than average, then a wider webbed shoe rather than a thicker shoe should be fitted.

Unevenness of wear results from a variety of causes. It may be due to faulty preparation of the foot, abnormal conformation of the limbs or pastern foot axis, or the horse's gait whereby it goes on its toes or heels.

Excessive wear of the toe of the shoe may be because the toe is too long, the shoe has excessively high calkins, or the horse goes on its toe to alleviate pain, as is seen in navicular and bone spavin disease. This can be corrected by turning up the toe of the shoe, out of the line of wear, to resemble the worn surface of the old shoe. Excessive wear should never be compensated for by increasing the thickness of the shoe.

Excessive wear at the quarter of the shoe is due to an uneven bearing surface or an abnormal pastern foot axis with a toe-in or toe-out conformation. If the wear is due to the former then all that is required is to lower the foot on the appropriate side to its normal proportions.

When the wear is due to a toe-in conformation it can be rectified by the margin of the wall at the inside toe, back almost to the quarters, being reduced and rounded off and the shoe fitted close to the inside toe and quarters; but on the outside, from the quarter to the heel, it should be fitted a little wider than normal. For a toe-out conformation the reverse procedure should be adopted. Excessive wear at the heels of the shoe is generally due to either chronic laminitis or degenerative joint disease of the distal interphalangeal joint (low ring bone disease). Increasing the wear of the shoe in the former case is not by increasing the thickness of the heels of the shoe, but rather by fitting a shoe a little long at the heels.

Removing a shoe and dressing the foot

A shoe is removed by first cutting off the clenches, by placing the blade of the buffer under each clench and giving it a sharp blow with the shoeing hammer (Fig. 7.22). Next, using the farriers' pincers, the inside heel of the shoe is eased (Fig. 7.23). Then the outside heel is similarly eased. This is continued alternately along each branch until it is loose, when the shoe is grasped at the toe and pulled backwards across the foot and off.

After removing the shoe the farrier must study the shape and proportions of the foot in relation to the pastern foot axis, with a view to making the necessary corrections to balance the foot, and then pick up the foot to examine the bearing surface for irregularities of outline and any unevenness.

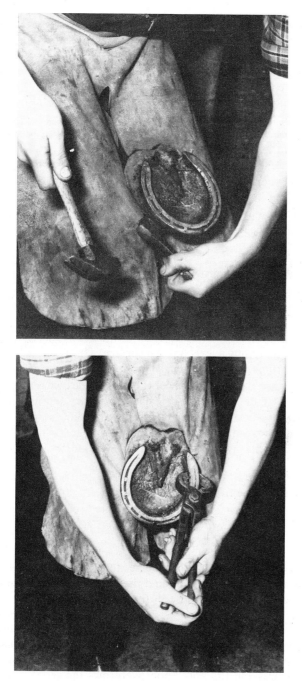

Figure 7.22. To remove a shoe. The clenches are cut off or straightened using a buffer with the blade held close against the wall to prevent cutting into it.

Figure 7.23. The inside heel of the shoe is raised by closing the jaws of the pincers under it and prising downwards towards the toe.

Attention to the frog and sole For the frog to play its important role in the anti-concussion mechanism of the foot and in preventing slipping it must be large, firm and prominent, yet not project beyond the bearing surface at the heels by more than the thickness of the shoe.

The frog must be pared sparingly—only to remove any ragged or loose tags of horn. On no account should it be pared for neatness or in the mistaken ideas that it should not be in contact with the ground. If it is kept pared down so that it does not make contact with the ground it cannot function normally, atrophies and leads to a contracted foot.

The sole protects the foot against injury and excessive paring is a bad and unnecessary practice. Only flakes of sole that have failed to shed should be removed. It is a mistaken idea that the descent of the sole is fundamental to expansion of the foot. Paring out the sole until it yields under thumb pressure not only deprives the foot of its normal protection, but also reduces the bearing surface provided by its margin.

Trimming the wall A slight overgrowth of horn is easily reduced with a drawing knife, but if excessive a hoof cutter is required. Hoof cutters must be used with the handles held perpendicular to the bearing surface to control the depth of cut.

The final preparation of the foot for the shoe comprises reducing it to its normal proportions by rasping a level bearing surface. This is carried out in four stages by a continuous circular movement of the rasp around the hoof (Fig. 7.24(a–d)). First, the outer aspect of the bearing surface is rasped along the line from heel to toe. Second, it is rasped across the quarters and toe; third, on the inner aspect along the line from toe to heel; and fourth, the edge of wall is lightly rasped around, using the file surface of the rasp, to remove its sharp edge. This prevents the hoof from breaking or splitting.

Faults in using the rasp At all times the rasp must be held level. If not, more horn will be removed from one side than the other and result in an unlevel bearing surface (Fig. 7.25), or the toe and opposite heel may be over-dressed leading to an uneven pressure distribution (Fig. 7.26).

Fitting the shoe

When fitting a shoe the foot surface of the shoe has to adapt exactly to the bearing surface of the hoof (surface fitting) and the outer edge of the shoe has to adapt to the circumference of the wall (outline fitting).

Close fitting Infers that the edge of the shoe is brought within the circumference of the wall and the projecting horn is rasped away. It is a bad practice because bearing surface is lost, less wall remains to secure the nails, it contributes to brittleness and the hoof soon overgrows the shoe.

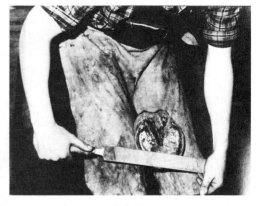

(b)

(a)

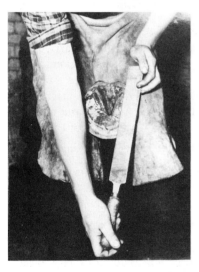

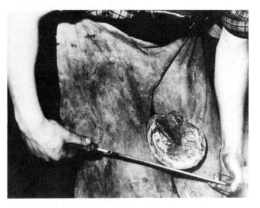

(d)

(c)

Figure 7.24. Rasping the foot to prepare a level bearing surface: (a) outer aspect of bearing surface is rasped along the line from heel to toe; (b) bearing surface is rasped across quarters and toe; (c) inner aspect of bearing surface is rasped along the line from toe to heel; (d) edge of wall is lightly rasped around to remove sharp edges and to prevent the hoof from breaking.

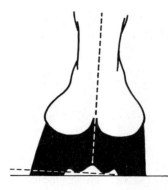

Figure 7.25. Uneven bearing surface due to one side of the foot being over-dressed.

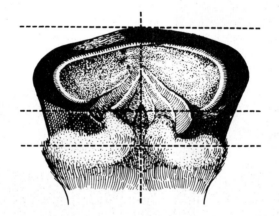

Figure 7.26. Uneven bearing surface due to the toe and opposite heel being over-dressed.

Wide fitting Implies that the edge of the shoe projects beyond the circumference of the wall with the risk that the protruding edge will be stepped on and the shoe lost. It also predisposes to brushing injuries. It is customary for heavy horses to be fitted with shoes wide at the heels to provide a firm base for support. It is important that the heels of a shoe extend back to the end of the bearing surface to cover the wall and the bars so that the weight of the horse is spread over the whole hoof wall.

Hot shoeing Horn is a poor conductor of heat and therefore a shoe can be applied to the horse's hoof without causing pain or injuring the underlying sensitive foot. The shoe is applied at a dull red heat and the horn is charred at the areas of contact, thus revealing any irregularities between the surface of the shoe and hoof. The charred areas can be taken down with a rasp to obtain a perfect fit.

Cold shoeing As the name implies this comprises fitting a shoe cold. However, there are a number of limitations when compared with hot shoeing. The outline of the shoe can only be altered slightly by opening or closing a toe or heel (Fig. 7.27(a) and (b)). They are more liable to become loose because such a good fit cannot be obtained, and the nails cannot always be driven through sound horn as the nail holes are in standard positions, and not specifically punched to meet the requirements of each individual hoof.

Nailing on and finishing off To nail on a shoe, the shoe is held in the fitted position and the first nail is driven (Fig. 7.28(a)). It is matter of personal preference whether the first nail should be driven at the toe or at the heel.

The nail selected must be proportional to the size of the shoe and the foot, and driven with the straight side of the nail towards the outer edge of the shoe which ensures that, due to the bevelled point of the nail, the point will

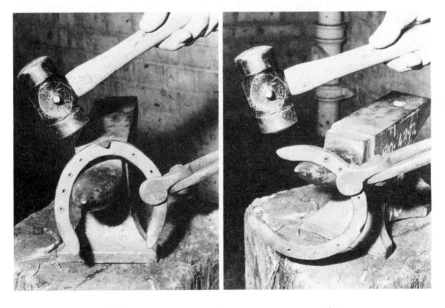

(a)   (b)

Figure 7.27. Cold shoeing: (a) opening a toe over the beak of the anvil; (b) closing a heel over the beak of the anvil.

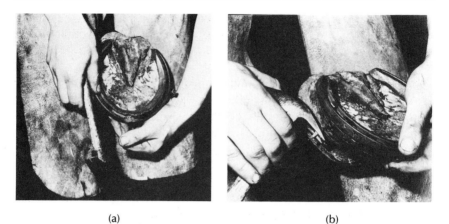

(a)   (b)

Figure 7.28. Nailing on a shoe; (a) shoe is held in the fitted position and the first nail is driven with the straight side of the nail outside; (b) immediately each nail is hammered home the protruding point is twisted off, using the claw of the hammer, leaving enough of the shank to form the clench.

turn away from the sensitive structures. If the nail is too large it will split the horn and if too small the shoe will not be secured. Nails should enter the wall at the white zone and be pitched to emerge about one third of the way up the wall. No nail should be driven to enter or cross a crack and if the feet are broken then the nail holes should be stamped so that the nails are driven through sound horn.

The remaining nails are then driven on each side alternately. Only the minimum number of nails necessary to secure the shoe should be used. Most shoes can be secured with six nails, three in each branch though seven is the traditional number, three in the inside and four in the outside branch.

As each nail is hammered home the protruding point is immediately twisted off, using the claw of the hammer, making sure that enough of the shank remains to form the clench (Fig. 7.28(b)). If a nail is pitched too high it can press on the sensitive structures resulting in pain and lameness, a so-called "nail bind", or it may even prick the sensitive structures resulting in the escape of blood and immediate pain and lameness. On the other hand if the nail is pitched too low, a secure hold is not obtained.

When all the nails have been driven and the points wrung off, the closed jaws of the pincers are pressed firmly upwards against the stubs and the heads driven home by repeated blows of the hammer. This draws up the nail to tighten the shoe on the foot and at the same time turns over the protruding stub to form a clench.

Next, using the file-edge of the rasp, the split horn underlying the clench is smoothed off and the clench shaped so that it is equal in length to the width of the shank. The clench is now turned and tapped with the hammer until it lies flat against the wall. At the same time the closed jaws of the pincers are pressed firmly upwards against the head of each nail in turn to prevent it being driven back.

Finally, the clenches are smoothed off using the file side of the rasp, care being taken not to weaken them by excessive rasping, the wall below the clenches is lightly rasped and the edge of the wall is underrasped to give a final finish.

Points to be observed when examining a newly shod horse

Foot on the ground (Fig. 7.29(a))

1. Both front and hind feet should be pairs, the same size and shape and with the same correctly aligned pastern foot axis.
2. Clenches even, flat and broad. Nails pitched higher at the toe than at the heels and not driven into cracks.
3. No rasping of the wall unless there has been flaring and a little below the clenches when they are rasped smooth.
4. No shortening of the toe of the front feet to conceal a badly fitted shoe.

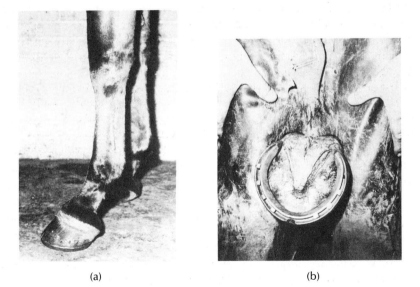

(a)   (b)

Figure 7.29. Examining a newly-shod horse; (a) the clenches should be flat and broad and no rasping of the wall above the clenches or dumping of the toe. Clips low and broad and toe clip centred. Shoes fit the outline of the foot, with heels of correct length; (b) no unnecessary paring of the frog or sole, toe clip centred, and nails driven home with the heads fitting the countersinks. The shoe fits the foot and the heels do not interfere with the frog.

5. Clips low and broad and the toe clips centred.
6. Shoe fits the outline of the foot and heels are of the correct length.

Foot lifted off the ground (Fig. 7.29(b))

1. Nails driven home, heads fit their holes and protrude slightly.
2. Heels not opened up (opening up the heels means cutting away the bars to make the foot appear wider).
3. Toe clip centred and in line with the point of the frog.
4. Frog and sole not excessively pared.
5. Sole eased at the seat of corn.
6. No daylight between the foot and the shoe, which would indicate an unevenness of either the bearing surface of the foot or the foot surface of the shoe.
7. Shoe fits the foot and the heels do not interfere with the function of the frog.

Finally, the horse should be walked and trotted up to reassess its gait.

CARE OF THE FEET

Stable hygiene is most important. Horn cells are dissolved by alkalis and the ammonia present in decomposing urine and manure softens the horn. For feet to be kept in a healthy state they require to be picked out at least twice daily, morning and evening, and each time the horse returns from work. Care must be taken to keep the space between the shoe and the sole, and the grooves of the frog, free from all dirt and grit.

Shoeing   Shoes require to be removed every 4 to 6 weeks, any loose flakes of the frog or sole removed and the feet trimmed to balance them. Only by regular shoeing can the feet be maintained in the best possible condition.

Moisture   If horses are stabled for long periods and during spells of dry weather the hoof is unable to make good the moisture lost by evaporation and becomes dry, brittle and cracks. If the feet are washed daily, using a soft brush, sufficient moisture should be absorbed to make good this loss.

Hoof oil   The question of applying impervious materials such as hoof oil merit careful consideration. Under ordinary conditions the continuous application of hoof oil prevents evaporation and the hoof becomes soft and crumbly. In wet conditions it has the opposite effect and by controlling the excessive absorption of moisture it prevents the horn becoming too soft. On the other hand in dry conditions it can be used to reduce loss of moisture and if dry and brittle hooves are soaked in water this can be followed by an application to help retain it.

Thrush This is a disease of the frog which is characterised by an offensive and dark coloured discharge from the central and sometimes the medial and lateral grooves, associated with the disintegration of the horn.

It does not cause lameness and the cause is not known but it is invariably associated with poor stable management and lack of exercise. Stalls and loose-boxes not regularly mucked out, wet standings, failure to pick out the feet and irregular shoeing favour its development. It is a preventable disease and in the first instance improving stable management and general hygiene, the regular use of the hoof pick and regular shoeing are necessary.

Local treatment comprises paring away all loose and diseased horn, keeping the grooves of the frog clean and dry and the application of an astringent powder, such as a mixture of 1 part powdered copper sulphate to 2 parts of boric acid powder every time the hooves are picked out.

The unshod horse

It is customary to turn out all types of working horses for a few months during the summer. This not only provides them with a rest but permits their shoes to be removed which gives the hooves a chance to grow without being damaged by nails.

When a horse is out at grass its feet require attention every 4 to 6 weeks as normal growth of the hoof is not always controlled by friction. The toe will require to be shortened, the heels lowered to balance the foot, and the edge of the wall rounded off to prevent it from splitting. To control excessive wear and the horse becoming foot sore, a little more of the wall should be left than when dressing the foot to fit a shoe.

Young stock

Foals should have their feet examined and kept trimmed from one month of age and then monthly, or every one or two weeks if a problem exists that requires correction.

Yearlings should have their feet attended to at regular intervals of four to six weeks and given adequate exercise on hard ground, especially if reared under cover, because the hoof will grow too long, especially at the toe, the heels curl inwards and the foot become unbalanced.

Angular deviations

It is essential that defects of conformation are recognised and dealt with as soon as possible. Deviation of the limb is usually caused by uneven growth at the growth plate.

Deviation at the knee Generally originates from the growth plate at the distal radius. The forelegs are deviated laterally or medially from the carpus. If laterally the medial side of the hoof is worn excessively and therefore is corrected by lowering the lateral side of the hoof. The converse holds good for medial deviation.

Deviation of the hock   Generally originates from the growth plate at the distal tibia. Foals with *bow legs* wear the hoof on the lateral side and are corrected by lowering the medial side of the hoof. Foals with *cow hocks* are corrected by lowering the lateral side of the hoof.

Deviation at the fetlock   Generally originates from the growth plate at the distal end of the metacarpal and metatarsal bones. Toes turning in from the fetlock are corrected by lowering the medial side of the hoof and in turning out by lowering the lateral side of the hoof. The longer attempts to correct deviations of the fetlock are delayed after 3 months of age the less likely they are to be successful.

It is important that in all cases of corrective trimming care is taken not to lower excessively the appropriate side of the hoof in the hope of obtaining a quicker and better correction as this will only result in abnormal and excessive strains on the joints which may cause the onset of degenerative joint disease.

Contraction of the flexor tendons

Contraction of the flexor tendons occurs most frequently during the first summer and the spring and summer of the second year of life. It is a problem associated with overfeeding and rapid growth and it is important that the rate of growth is slowed down by reducing the food intake. With suckling foals the mare is fed less to reduce her milk yield and weaned foals are fed only hay.

Contraction of the superficial digital flexor tendon results in the fetlock knuckling with the hoof remaining flat on the ground. This condition is most frequently seen in yearlings and the farrier can only assist by keeping the hoof trimmed and balanced.

Contraction of the deep digital flexor tendon results in flexion of the distal interphalangeal joint with the heels raised off the ground. Treatment comprises keeping the heels lowered and encouraging the foal to take plenty of exercise which keeps the tendons under constant tension and prevents their progressive shortening.

If angular deviation and contraction of the flexor tendons do not respond to corrective shoeing or gradually deteriorate then veterinary advice should be sought as surgical procedures may be required.

SPECIAL HORSE SHOES

Farriers are frequently called upon to make special shoes to prevent injuries, to improve defective feet and to alleviate lameness.

Injuries caused by the shoe

Capped elbow is the name given to any swelling at the point of the elbow. In most cases it is an acquired bursa due to slight but repeated injury resulting from contact with the ground. It is frequently said to be due to the inner heel when the horse is lying down. If the position a horse adopts when it is lying down is observed it will be seen the heel of the shoe lies near the elbow but does not come in contact with it unless the heel of the shoe is very long and especially if with calkins.

To determine the cause the horse should be observed lying down and if it is due to the heel of the shoe striking the point of the elbow then a shoe should be fitted with the inner heel shortened and fitted close, or hot rasped off obliquely so that it resembles the back of the bowl of a spoon (Fig. 7.30).

Injuries caused by abnormalities of gait

Brushing This is an injury caused by the horse striking the inside of one leg, in the region of the fetlock, with the shoe of the opposite foot. It may be due to fitting the shoe too wide on the inside or a toe-out conformation. It is not always easy to decide which part of the shoe is responsible for the injury, but this can be ascertained by chalking or greasing the shoe and noting where it is rubbed off.

Slight cases can be prevented by such measures as fitting a flat and lighter shoe, shoeing behind without calkins, replacing a calkin with a wedge heel and fitting the inner branch of the shoe close and rasping off the overhanging wall.

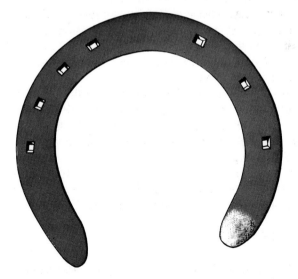

Figure 7.30. Shoe for capped elbow made from flat bar.

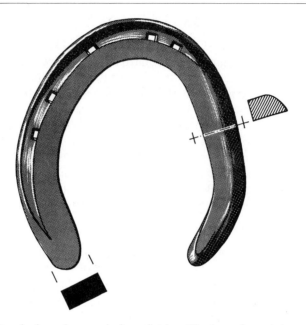

Figure 7.31. Knocked-up shoe made from flat bar. The inner branch is narrow with the ground surface sloped downwards and inwards and rounded off.

There are two types of shoes frequently fitted to prevent brushing. First, the knocked-up shoe (Fig. 7.31), which has a narrow inner branch, thereby reducing the ground surface which is sloped downwards and inwards, and rounded off. The branch is fitted close, any projecting wall is rounded off with the rasp, and it is blind except for one or two nails at the toe. It is recommended to prevent brushing injuries caused by the toe or mid quarter of the toe of a shoe.

Second, the knocked-down shoe (Fig. 7.32), which has both branches of the same width, but the inner branch where it strikes the opposite leg, is knocked down and rounded off. This shoe is recommended for horses with a toe-out conformation which results in injuries being caused by the posterior quarter of the branch and the heel of the shoe.

Forging. Forging is due to faulty action at the trot when the heel or inside of the toe of a front shoe is struck with the corresponding hind shoe (Fig. 7.33). It can also occur as the hind foot comes past the outside of the front foot and the inside of the toe of the shoe strikes the outside heel of the front shoe.

Forging occurs in young and unfit horses and it gradually disappears as the horse gets fit. The actual contact between the shoes takes place when

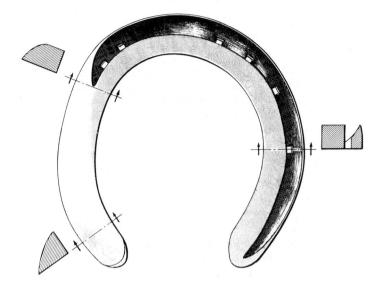

Figure 7.32. Knocked-down shoe. The inner branch is knocked-down and rounded off.

the hind foot is reaching the end of its flight, just as the front foot leaves the ground. Therefore, to prevent forging a shoe has to be fitted which hastens the breakover of the front feet and delays that of the hind feet. This can be attained in front by fitting a concave shoe and increasing the break-over by rolling the toe and slightly raising the heels. If this does not prove satisfactory then set the hind shoe well back, and if necessary, square the toe. To delay break-over of the hind feet the heels of the shoe should be lowered and left a little long; this acts as a brake when the foot comes to the ground.

The diamond-toed shoe, which is frequently illustrated as an anti-forging shoe, is only required on very rare occasions for horses which forge by carrying a hind foot outside a front foot.

Figure 7.33. Forging. The inside of the toe of a front shoe is struck by the toe of the corresponding hind shoe.

Figure 7.34. Over-reaching. An over-reach occurs when a front leg is overtaken and struck by the inner edge of the corresponding hind shoe, usually on the bulb of the heel.

Over-reaching   This is the name given to an injury at the back of the leg caused by the inner edge of the hind shoe striking the front leg when the horse is moving at speed (Fig. 7.34).

To prevent this injury the horse is shod to hasten the break-over of the front feet and delay the break-over of the hind foot so that it does not overtake and strike the front leg. This is achieved by fitting a rolled toe shoe with raised heels in front and hind shoes with the heels half the thickness of the shoe at the toe, and left a little long to delay breaking-over of the foot. In addition, the shoe should have the inner border of the toe concave and well rounded out, and be set well back.

Stumbling   A horse will stumble if it catches or digs its toes into the ground. This is most liable to occur when the animal is tired, has long overgrown feet, goes on its toe due to lameness, or has reduced flexion of its fetlock joint. A horse is always more liable to stumble when recently shod and before the shoe has worn to conform with the horse's action.

To prevent stumbling the horse must be shod to prevent the toe from coming into premature contact with the ground. This is attained by rasping the toe short and fitting a rolled-toe shoe with raised heels to increase break-over. An idea of the extent to which the toe of the shoe should be "rolled" or "set up" can be obtained by studying the wear of the worn shoe.

Injuries, which are frequently referred to as "interferences", may often be due to working an unfit or tired horse or to inappropriate foot dressing or shoeing as well as to poor conformation and action. Therefore, it is essential to seek and identify the cause for each horse individually as horses with a similar injury need not necessarily require to be shod with the same type of shoe or in the same manner.

Shoeing defective feet

Flat feet   Care must be taken to differentiate flat feet which are congenital and due to their conformation and flat feet which are acquired as the result of disease such as laminitis.

A flat foot is one which is large in circumference, has an excessively sloping wall, low heels, prominent frog and a sole which lacks the normal concavity. Pressure on the sole can be relieved by fitting a seated-out shoe (Fig. 7.35). It has a wide web and the seating is carried back to the nail holes, except at the heels which are left flat to allow normal weight bearing. Due to the suction created in heavy going these shoes can be easily pulled off.

Upright feet   Upright feet in adult horses do not require any special shoes. A naturally upright foot should not be altered but an unnaturally upright foot should be correctly balanced by trimming the heels.

Weak and unnaturally low heels   Weak heels, which are usually also low heels, curve forwards and the weight is taken on the outside of the wall.

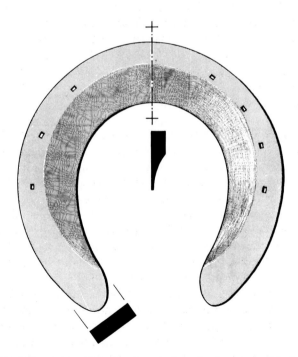

Figure 7.35. Seated-out shoe. The shoe is seated-out back to the nail holes and around the shoe except at the heels.

The best results are obtained by shoeing frequently with wide fitting of wide webbed shoes so that the hoof has ample bearing surface at all times. If there is no improvement it may be best to raise the heels artificially with a raised heel shoe or plastic wedge.

Contracted feet   When one foot is smaller than normal, narrow at the quarters and heels, with an excessively concave sole and atrophied frog, it is termed a contracted foot. The method of treatment depends on the cause. If associated with disease or injury which responds to treatment, then as normal function returns so gradually the foot regains its normal shape. If it is associated with an incurable condition then there is no useful purpose in trying to effect expansion. Turning a horse out unshod so that the foot can take weight, obtain frog pressure and function normally and grooving the heels to obtain expansion are helpful in some cases.

Surgical shoes

Surgical shoes are designed to assist in the treatment of disease and injuries of limb and foot by providing protection and relieving pressure.

Corns   A corn is a bruise of the sole in the angle between the wall and the bar. They are most frequently seen involving the inner heel of front feet and are rarely seen in hind feet. They are especially common in horses with wide flat feet, and although they may result from a stone getting under the shoe, they more often are due to pressure from the heels of shoes which are left on for too long or from shoes which are fitted narrow and close and with short heels.

A shoe is required which will relieve pressure on and protect the seat of corn. In the majority of cases all that is required is to remove or correct the cause, pare away the discoloured horn to below the level of the wall, thereby relieving pressure, and then fit an ordinary shoe making sure the heel rests on the wall and the bar. The only special shoe that needs to be considered is the "set" heel shoe (Fig. 7.36) which is the opposite to the "dropped" heel shoe. It has the ground surface of the heel lowered by about 6 mm (0.25 in) and so makes no contact with the ground. It is a satisfactory shoe in that it relieves pressure and protects the area whilst providing normal contact between the shoe and bearing surface. The loss of some ground surface at the heel is of little practical significance.

Chronic laminitis   Laminitis is an aseptic inflammation of the sensitive laminae. Usually both front feet are affected but it can develop in one foot. Chronic laminitis is a sequel to the acute form and is characterised by changes in the shape of the foot and the horse's action. Due to the rotation of the distal phalanx and irregularities of horn growth the foot becomes elongated

Figure 7.36. Shoe with "set" heel. The ground surface of the heel is lowered by 6 mm (0.25 in), thereby relieving pressure and at the same time protecting the seat of corn.

and narrow with a flat or convex sole (dropped sole). The wall at the heels is longer than normal and at the toe is concave in profile. The hoof wall is characterised by the so-called laminitic rings which are widely spaced at the heels but close together at the toe.

Treatment comprises trying to reduce weight on the wall at the toe thereby relieving pain and encouraging new wall to grow down close to and parallel with the dorsal surface of the distal phalanx. This is attained by trimming the lower part of the wall back at the toe to the white zone and lowering the heels (Fig. 7.37) to tip the distal phalanx backwards to assume a more normal angle with the bearing surface of the foot.

A shoe is required which will relieve pain, protect the dropped sole from pressure and improve the horse's gait. In many cases an ordinary shoe, with quarter clips instead of a toe clip and seated out to relieve pressure on the sole, is quite satisfactory. A rocker bar shoe which is twice the thickness of a standard shoe at the quarters and gradually gets thinner towards the toe and the heels, is seated out and may have a rolled toe, is popular for horses used at the walk. As a horse with chronic laminitis lands on its heels and then rolls its foot over onto its toe, the shoe reduces concussion and assists the action of the foot.

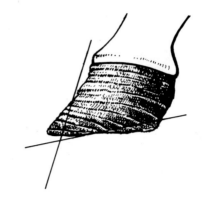

Figure 7.37. Chronic laminitis. The wall at the toe is trimmed back to the white zone and the heels lowered to encourage the distal phalanx to assume a more normal angle with the bearing surface of the foot.

Navicular disease   Navicular disease is the name given to a pathological lesion of the navicular bone which is characterised by erosion of the fibro-cartilage on its tendon surface with destruction of the underlying bone. It is an incurable and progressive disease and all that can be achieved by shoeing, in the early stages, is to alleviate pain by taking weight off the posterior aspect of the foot and to prevent stumbling. The best guide to this is to fit a shoe which conforms to the worn shoe, and to this end a rolled toe shoe with raised heels improves the horse's action.

Sandcrack   A sandcrack is a fissure of the wall which commences at the coronet and extends a variable distance down the hoof. Treatment is directed to removing pressure from the extremity of the crack and immobilising its edges. This can be done, in incomplete sandcracks by cutting two grooves, 5 mm (one fifth of an inch) wide and down to the white zone, in the form of a "V" running from the coronet to the tip of the crack (Fig. 7.38(a)). In complete sandcracks when the foot takes weight and the crack opens up, a plain flat shoe with toe or quarter clips, eased under the crack, is required to supplement the grooves cut parallel to the crack (Fig. 7.38(b)).

This condition should not be confused with a false sandcrack which is a simple crack that commences at the ground surface and extends a variable distance up the hoof. All that is required in these cases to prevent the crack from extending is to cut a horizontal groove, about 12 mm in length, across the top and down to the white zone, and relieve pressure at the extremity by either easing the bearing surface of the wall or the foot surface of the shoe.

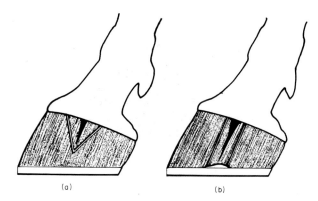

Figure 7.38. Sandcrack; (a) incomplete. Two grooves are cut, to the depth of the white zone, in the form of a "V" from the coronet to meet at the lower limit of the crack; (b) complete. Two parallel grooves are cut, one on each side of the crack, from the coronet to the bearing surface. Note that the hoof has been eased under the extremity of the crack.

Seedy toe This is the name given to a cavity at the toe which results from a separation of the horny and sensitive laminae at the white zone and is filled with a crumbly type of horn. Treatment comprises scooping out the degenerate horn and packing the cavity with cotton wool and Stockholm tar. To protect the bearing surface and retain the dressing it only requires a plain shoe to be fitted with a wide web at the toe. In extensive cases it may be advisable to provide extra protection by fitting a pad. It is important that no clips are placed or nails driven near the cavity.

Sidebone This is an ossification of the cartilage of the distal phalanx and is particularly common in draught horses. The condition only needs to be treated with a special shoe if the horse is lame and the shoe abnormally worn. The most satisfactory shoe to fit is one which corresponds to the worn shoe (Fig. 7.39). The branch of the shoe on the affected side requires gradual thinning down from the nail hole at the toe to the end of the heel, and the width of the shoe gradually increased until its outer edge corresponds with the outline of the overhanging coronet. The branch extends a little beyond the heel and to allow for expansion no nails are driven except for two at the toe.

Spavin Bone spavin disease is an osteoarthritis which results in a lameness characterised by imperfect flexion of the hock, and in consequence the horse drags its toe.

To overcome the horse dragging its toe and to facilitate its action the toe is shortened and a shoe fitted with a rolled toe and raised heels. As calkins

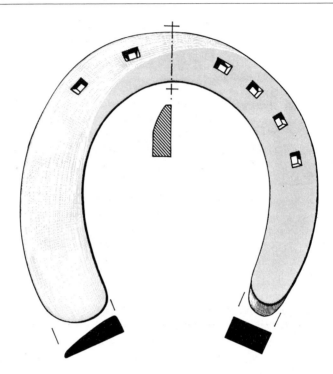

Figure 7.39. A shoe for unilateral sidebone disease. The branch on the affected side is thinned down and gradually increased in width from the toe to the heel and is blind except for the two nail holes at the toe.

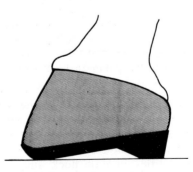

Figure 7.40. Shoe for bone spavin disease. The toe of the shoe is rolled and the wedge heels are sloped so as not to catch and allow the foot to slide.

are inclined to catch in the ground they are replaced by sloping wedge heels (Fig. 7.40). This shoe improves the horse's action but it must be appreciated that it does not have any direct influence on the disease process.

Strained tendon   A horse with a strain of the superficial and/or deep digital flexor tendon of a front leg is lame and pain can be relieved by reducing pressure in the tendons by fitting a shoe with raised heels such as a Patten shoe. The height the heels are raised depends on the severity of the injury, between 2.5 cm and 7.5 cm (1 and 3 in) but it cannot be over-emphasised that sprained tendons heal by fibrosis which is followed by contraction. Therefore it is essential to lower the heels to their normal height without undue delay because otherwise a chronic and irreversible shortening of the tendons will result with the horse unable to get his heels to the ground.

It is doubtful if raising the heels of a shoe to relieve pain and tension on the strained tendons has a place in their treatment.

ACKNOWLEDGEMENTS

The authors wish to acknowledge the reproduction of the illustrations from *Farring* by John Hickman and published by J. A. Allen; also Neil Chalmers for drawing Fig. 7.21.

REFERENCES

Adams, O. R. (1974). *Lameness in Horses*, 3rd Edn. Lea and Febiger, Philadelphia.
Bridges, J. (1752). No Foot, No Horse; an Essay on the Anatomy of the Foot of that Noble and Useful Animal a Horse. J. Brindlay and R. Baldwin, London. (Quoted in Fleming, 1869.)
Clark, B. (1809). A Series of original experiments on the foot of the living horse, exhibiting the changes produced by shoeing, and the causes of the apparent mystery of this art. B. Clark, London.
Coleman, E. (1798–1802). Observations on the structure, economy and diseases of the foot of the horse, and on the principles and practice of shoeing. Edward Coleman, London.
Dollar, J. A. W. and Wheatley, A. (1898). *A Handbook of Horseshoeing*. David Douglas, Edinburgh.
Fiaschi, C. (1564). Traite de la maniere de bien emboucher, manier, et ferrer les chevaux. Dedie au Roi Henri II. C. Perrier, Paris. (Quoted in Fleming, 1869.)
Farriers (Registration) Act 1975. Her Majesty's Stationery Office.
Farriers (Registration) (Amendment) Act 1977. Her Majesty's Stationery Office.
Fitzwygram, F. (1863). *Notes on Shoeing Horses*, 2nd Edn. Smith, Elder and Company, London.
Fleming, G. (1869). *Horse-shoes and Horse-shoeing*. Chapman and Hall, London.
Fleming, G. (1878). *Practical Horse-shoeing*, 3rd Edn. Chapman and Hall, London.
Hickman, J. (1977). *Farriery, a Complete Illustrated Guide*. J. A. Allen, London and New York.

Holmes, C. M. (1949). *The Principles and Practice of Horse-shoeing.* The Farrier's Journal Publishing Company, Leeds.

Hunting, W. (1922). *The Art of Horse-shoeing*, 4th Edn. (Revised and edited A. B. Mattinson). Bailliere, Tindall and Cox, London.

Lafosse, E. G. (1754). *Observations et Decouvertes Faites sur des Chevaux, avec une Nouvelle Practique sur la Ferrure.* Hochereau le jeune, Paris.

Lungwitz, A. and Adams, J. W. (1966). *A Textbook of Horseshoeing for Horseshoers and Veterinarians*, 11th Edn. Oregon State University Press, Oregon.

Macqueen, J. (1921). *Fleming's Practical Horse-shoeing*, 11th Edn. Bailliere, Tindall and Cox, London.

Miller, W. C. and Robertson, E. D. S. (1947). *Practical Animal Husbandry*, 5th Edn. Oliver and Boyd, Edinburgh.

Reeks, H. C. (1906). *Diseases of the Horse's Foot.* Bailliere, Tindall and Cox, London.

Smith, F. (1921). *A Manual of Veterinary Physiology*, 5th Edn. Bailliere, Tindall and Cox, London.

de Solleysel, J. L. (1706). *The Complete Horseman or, Perfect Farrier.* (Transl. Sir William Hope, 2nd Edn. R. Bonwicke). T. Goodwin, London.

Sparkes, I. G. (1976). *Old Horseshoes.* Shire Publications, Aylesbury.

8   Basic Training

P. J. DAVISON

The procedure in training must be determined by the end to be attained, by the time available for training, by the horse to be trained and by the abilities, knowledge and temper of the instructors.

XENOPHON 435–354 BC
Historian, statesman,
philosopher, essayist,
military commander
and horseman.

INTRODUCTION

Man has been training horses for well over 2000 years and during that time many different techniques of teaching have been used and no single method has evolved. My professional qualifications have enabled me to work with horses in many countries and on every continent, and I have seen horses tamed and schooled by widely contrasting systems. Some practices are gentle and some are cruel; some are calm and some are violent; some methods start when the foal is a few hours old, others wait until he is four or five years of age.

This chapter offers for consideration my own opinions which rely greatly on the earliest possible start to training, the calmest possible approach, and are based on the maximum use of leading the young horse in hand through every conceivable situation before he is backed.

As long ago as the 4th century BC the writings of the great Greek general, philosopher and horseman, Xenophon, described that a gentle approach, and an understanding of the nature of the horse, were intrinsic factors in good horsemanship. Even though Xenophon's horses were used primarily in the frenzy and turmoil of battle, he appreciated that calmness was essential if training was to be effective.

Horse Management 2nd edition
ISBN: 0-12-347218-0 case

When calmness and patience give way to violence and loss of temper then the horse will develop bad habits, resist, and eventually submit through fear of punishment. Almost all equine vices such as rearing, running backwards, biting and kicking can be traced to rough, impatient or thoughtless handling during early training. There are probably very few horses which cannot be cowed into submission by force and intimidation, but the willing submission of successful training should always be achieved by confidence and not by fear.

"Breaking" or "breaking in" are the words most commonly used in the English language to describe the early handling and training of the horse. They are words which few horsemen like, but many use. "Breaking" implies subjecting, overcoming, shouting, and beating into submission, it suggests gadgets, twitches and restraint. The result is a tractable animal with a broken spirit, an unwilling machine that only performs to avoid further punishment. Horses are still "broken" by the gauchos of Argentina, the cowboys of North America, the Romany gypsies of Central Europe, the Cossacks of South Eastern Russia, and the stockmen of Australia, all of whom are seeking absolute submission in the shortest possible time.

The very word "breaking" conjures up an exciting and dangerous image, calling for a strong and fearless master with ropes and whips, dust and noise, intimidation and terror. To obtain a willing riding horse, however, equine education should be a calm and enjoyable experience for both pupil and trainer. Basic training is often best carried out by women who on average are more patient and sensitive than men, and who frequently establish far more trusting relationships with the horses they are teaching.

The French use the word "Débourrage" to describe elementary training to a level where the horse may be ridden quietly at the walk, trot and canter, will respond to basic aids, and is physically fit to progress to more advanced training. In this chapter I prefer to use the term "basic training" to describe the education of the horse to that same level.

In all fields of equitation there are widely differing convictions, there is frequent controversy, and both written and spoken views are much debated. The process of training horses is still evolving, and the opinions that follow are once again only influenced by the observations and experiences of one man. If in my travels I have learned one thing it is that there is no single way to train a young horse, and perhaps also that there are no short cuts. What follows is my own humble contribution to this complex enigma, and I hope that in combination with the writings and teachings of others, it may help to form a foundation for basic training.

AIMS OF BASIC TRAINING

The aim of basic training is to produce a horse which is calm and pleasant to ride, which goes forward freely and obediently, and in response to simple instructions from the rider becomes a useable animal. The result should be a friendly companion, which is a pleasure to handle and ride, is happy, willing and cooperative, well behaved in field and stable, and well mannered when ridden either alone or in company. There should be no force on the part of the rider or the horse. The final result is a platform on which the advanced skills of equitation can be built.

Only when this solid foundation has been established, can a horse with inherited physical and mental ability proceed to higher levels of training. A Badminton event horse, a high goal polo pony and an Aachen showjumper must all have had a sound basic training before their natural talents are directed to their individual disciplines.

Naturally there are always exceptions to any generalisation, as for example the very early backing and racing of juvenile thoroughbreds which is a specialised subject falling beyond the scope of this book. However, even in this difficult sphere of training, where high finance dictates that skeletally immature colts and fillies should compete at the highest levels, patience, calmness and an early start are essential to ensure success.

At the conclusion of his basic training, the young horse must respond to simple positive aids. These are the signals by which the rider conveys his intentions to the horse. The aids will feature strongly throughout this chapter as they are concerned with every mounted and dismounted aspect of training. Aids may be natural such as the sound of the trainer's voice, the touch of his hands, legs and seat, and the visual example set by an experienced "schoolmaster" who leads by demonstration; or they may be artificial such as whips, spurs and special harness. Although the number of aids is comparatively few the ways in which they may be applied are many. The tone and volume of the voice, the pressure and position of the legs, and the contact of the hands can all be varied, and it is the skill and understanding with which the aids are applied that differentiates between good and bad trainers.

The horse that has completed his basic training must be well mannered at all times. He should stand still whilst being mounted, but when asked to do so should move forward freely and willingly. He should walk, trot and canter calmly and under full control of the rider, he should perform balanced turns in response to the appropriate aids. He should progress smoothly from one pace to another, and similarly his downward transitions should be without resistance or loss of balance.

Horses, like people, differ in their natural ability and in their capacity for learning. Some horses will have reached the limit of their natural ability at the completion of basic training, others will show a potential that may be encouraged even at this early stage. The experienced instructor will quickly appreciate during basic training if he is dealing with an honest plodder or a potential equine paragon. The élite horse will be able to accept and benefit from a more comprehensive syllabus which will result in his natural paces having more impulsion, balance and cadence, and the time he takes to master each new lesson will be much shorter than his less talented colleagues.

HORSE PSYCHOLOGY

Psychology is the study of behaviour, attitudes and mind. If we can understand the workings of the equine mind, then training becomes more logical and less complicated. A good school teacher requires a thorough understanding of a child's mind and a sympathetic attitude to its workings. Similarly the mind of the horse, and the natural instincts which govern his behaviour must be fully understood by the equine trainer.

In his natural state the horse evolved on open grasslands. He was a creature of habit with a regular routine of grazing, drinking and resting. He had an accurate built-in metabolic clock and his weeks were not separated by a day of rest. Life was governed by a desire for food and a fear of the unknown. He had a remarkable memory for places and events and would unfailingly return to places with good grazing, and avoid places where predators had previously attacked. The wild horse was gregarious, possessive and governed by a leader, the leader often being more of a herdsman than a blazer of trails. It was the possessive and jealous nature that enabled the leader to emerge from the herd to protect them from predators and other stallions, and which enabled the mare to protect her foal.

Once domesticated the horse became docile and willing, and was naturally friendly towards man. The wild characteristics did not die, and the domesticated horse is still gregarious, he is unwilling to leave a group of other horses, and is keen to return to company. He will trust and follow a leader, and an old schoolmaster is often used as a training aid to give him confidence. He remains timid and will still wheel round to avoid the unknown, and he may even bolt when terrified. He still responds best to a life of habit and repetition, and he has not lost his acute senses. By watching eye movements, ear movements, and changes in head posture, an experienced trainer can anticipate what the next move will be; he can encourage the move if it would be of advantage, or he can discourage or prevent it should it be of disadvantage.

We have an animal which although domesticated for thousands of years still retains strong natural instincts which govern its behaviour. He responds to contact with other horses or to the patting and stroking of man. He acquires habits which if good are to be encouraged, but which if bad may be difficult to break. He retains an uncanny and almost proverbial memory for people and places and knows at once if a person or place has pleasant or unpleasant connotations. He remains a great observer, and misses little especially if it alters his routine. He remains cautious and is easily frightened by the unexpected, by violence, or by pain. He has not lost his ability to trust others and will readily transfer his trust in his mother or leader, to a schoolmaster or trainer.

Is the horse intelligent? People have argued this point since horses were first domesticated. Intelligence is difficult to define in man, and even more difficult to evaluate in animals. If by intelligence we mean the rational powers of the mind, the ability to reason, and the principle of thinking, then the horse has little intellect. He has no faculty for conceptual thinking and his power of reasoning is minimal. He is unable to link cause and effect unless they happen simultaneously, and for this reason reward and reprimand are useless unless administered at the exact instant of the success or disobedience.

All too often we endeavour to put ourselves in the horse's place and try to imagine what we would do in such a position. This is a complete waste of time as the horse has no facility to think out or solve a problem as we would, in fact anatomically he does not even possess that part of the brain with which to think. If a horse does not appear to understand it is usually the fault of the trainer for failing to express himself in a manner that the horse can understand.

EARLY HANDLING

Basic training should start from the day a foal is born. His future is entirely in the hands of those who are to look after him and train him, and first impressions can remain throughout his life influencing his temperament and character.

Like most animals born on the plains a foal in the wild has no nest or home and is surprisingly well developed at birth. The newborn foal is expected to be able to follow its mother within a few hours of birth, and already his senses and physical development are such that he can detect danger and escape from it.

The fact that the foal has acute hearing, good eyesight (except for a blindspot behind him) and is sensitive to touch, must all be appreciated in the trainer's approach to early handling. If confronted by unexpected physical contact, loud noises, or sudden movements, he will resort to his wild instincts and either try to run away, or if cornered, will kick and bite.

If properly handled from birth the foal will soon become accustomed to people and will quickly learn to trust them. For the first few days every advantage should be taken of the young foal's natural inquisitiveness. Whilst dealing with the routine management of the mare, such as checking her udder and teats, picking out her feet, grooming and feeding her, the foal should also be touched and spoken to. Always approach him at his shoulder, never from behind, and for the first few days try not to come between him and his mother otherwise he may feel isolated, and panic.

Occasionally a jealous mare can be so protective that she will physically prevent any approach to her offspring. In these rare cases it is imperative to have a competent assistant to handle the mare whilst a second person handles the foal. Usually within a few days both mare and foal will become manageable.

Adult horses are physically stronger than their handlers and riders, and could easily overpower or escape from their human masters if they wanted to channel their strength into evasion. It is important therefore during the first few weeks of life when the trainer is temporarily stronger than the foal, to make the foal realise this dominance and to imprint on his mind a feeling of respect and submission. This does not mean that the foal should be bullied, but he should be moved gently and firmly, caught safely and without chase, and carried in a firm and confident manner without fear, force or panic.

It is useful to be able to carry the young foal without a struggle. The youngster may have to be brought in out of foul weather, and the easiest method of weighing a foal is to carry him onto a bathroom scale and then subtract the weight of the handler. It is also easier for the veterinary surgeon to deal with retained meconium if the foal is held with his feet off the ground.

A foal should never be carried by putting an arm under his belly as this could damage his sternum or ribs, and might even tear his stomach, intestines or diaphragm. To carry a foal correctly one arm should be placed behind his hindquarters, and the other in front of his chest (Fig. 8.1). This is also the best way to steady him for examination or treatment.

As soon as the foal is used to being approached and touched he should be fitted with his first headcollar which is called a foal slip. A good foal slip is made of soft lightweight leather which can be readily adjusted as the foal grows. It is best made of relatively weak leather so that it will break should it get caught on a protrusion, or should the foal accidentally get his own foot caught in it as he gambols or scratches himself. Modern slips made of tough synthetic materials should never be left on when the foal is unattended, and it is better in the long run to buy expensive leather slips and pay to have them stitched if they break. It is cheaper to stitch leather than foals!

The first fitting of the foal slip must not involve any fuss or fear on the part of the foal, and no force on the part of the handler. By stroking his neck, tickling his chin and rubbing his nose and ears, it is possible both to ease the slip into place and to adjust it, without him noticing. Talk to him reassuringly the whole time, and reward him with a molasses cube and a pat when the job is done.

Figure 8.1. How to carry a young foal.

If possible the slip should be put on and taken off several times during the first few days, and not left on for more than about ten minutes at a time unless under supervision. After a few days it may be left on for longer periods both in the stable and the paddock. Care must be taken to ensure that the buckles of the foal slip do not rub the mare during suckling.

The handling of the ears is most important, for horses which do not allow their ears to be touched may prove difficult to manage in later life. Most horses enjoy having their ears rubbed especially after hard work or exercise in wet weather. There is nothing worse than a headshy horse; he is hard to catch, difficult to tack up, it is awkward to adjust his bridle in situ, he may refuse to have his head groomed or his eyes and ears cleaned, he will be impossible to plait, and clipping becomes a traumatic nightmare.

In connection with headshy horses it is worth mentioning here the deplorable habit which has crept into many yards of people waving their arms, hats or even handkerchiefs at horses to make them stand back in the box in order to give the owner, buyer or visitor a better view. If you want to show off a horse then put a headcollar or bridle on, take him out of the box, and present him properly in a professional manner. The horse is naturally inquisitive and trusting. We normally want him to approach us confidently at all times, and the last thing we should do is to send him away with a raised arm.

The next stage of the foal's education is to teach him to be led. Despite being a vital part of his early training this is usually quite a simple process. The young foal will never stray far from his mother and will instinctively follow her from stable to paddock, and this close affinity is used to accustom him to being led. The important rule here is to never pull an unwilling foal otherwise he will learn to resist by running back and rearing up. Right from the start he must move forward freely and willingly. Three aids are used to teach the foal to walk in hand; first, the mother walking in front; second, the voice with a quiet but firm "Walk on"; and third, an arm around his hindquarters to push, guide and steady him. For the first few days of being led it is advisable always to leave one hand on his hindquarters to counteract any tendency to slow down or stop. As soon as he moves forward make much of him with voice, a pat and a titbit — if he hesitates repeat the "Walk on" and encourage with the hand behind the hindquarters.

Teach him to be led from both sides and gradually encourage him to walk close to his mother or further away, as you dictate. Eventually after a week or two he will be walking totally at your command, and not just aimlessly following his mother.

In difficult cases where the foal does not learn to be led as easily as has been described, the trainer must use patience, firmness and common sense. If the foal tries to throw himself down then teach him in an indoor school or a paddock with soft ground. If he tries to run back then have an assistant to push him forward. If he refuses to move at all then lead his mother further away. Always make sure that wheelbarrows, buckets, bicycles and other such hazards are removed from the training area.

Use the voice all the time to reassure the foal whilst he is learning to be led, and gradually introduce him to new words of command. "Walk on" has already been mentioned. Second, when slowing down and stopping, a longer, calmer and more drawn out "Whoa". Other words of command to be introduced at this stage are "Stand" and "Steady". Obviously the foal has no comprehension of spelling, and it is the tone and phonetics that give the meaning which he learns to associate with moving off, stopping, standing still and being calm. Always be conscious of the tone of the instruction from the sharp "Walk on" to the lengthened "Whoa".

The next part of the training of the young foal is a gradual and continual education which takes place every day, and concerns his stable manners. The details of the various procedures are mentioned elsewhere in this book under stable management, and are concerned with tying up or short racking, hoof care and farriery, grooming and clipping, deworming and vaccination, loading and transport. With all these routine tasks, calmness and patience are essential and there should never be any struggling or shouting. Always use the mare, as long as she sets a good example, to instil confidence. When the mare is being groomed, shod or treated, let the foal be present and allow his inquisitive nature to explore, and discover that there is nothing to fear. Never shoo him away, and only move him back if his curiosity should endanger him.

The voice should always be used if the foal has to be moved in the box. When moving him back say "Back" and when moving him sideways say "Over", similarly whenever his feet are picked up say "Foot" as you run your hand down his leg and he will soon learn to pick up his feet on command.

If the young foal is alarmed by the sound of the clippers then they should be left running outside his stable to accustom him to them gradually. I know of one yard that has a recording of noisy clippers which is played endlessly to nervous horses. If the foal is worried by the smell of burning hooves then always allow him to be present when the farrier is at work. Reward each step towards confidence with voice, touch and after special success, a titbit.

Releasing and catching the foal in the paddock requires special mention. It gives great pleasure and is a wonderful conserver of patience and temper if when one calls across to grazing horses they look up at once, and come cantering towards you. Conversely one of the most frustrating scenes in the whole field of equitation is that of a lone, wet, exhausted figure trudging helplessly round a vast muddy field after a high stepping, tail flying, nostril flaring youngster who has no intention of coming in out of the rain.

If the foal is being properly handled it should not be difficult to catch. He should always be rewarded when he comes to you, and initially the reward should be a titbit, voice and touch; with time, a word of greeting and a pat on the neck will be sufficient. The very first time he is turned out the lesson should begin and obviously before he can be caught he has to be released. Correct release is as important as efficient catching. The practice of horses wheeling round and galloping off with flying heels when released should be discouraged. The habit of hurling the lead rope at their departing hindquarters to heighten the drama is to be deplored!

Walk the mare and foal into the paddock, close the gate, and turn them to face the entrance. Make them stand still, talk to them and stroke them, give them a titbit, release the lead rope and walk quietly away.

Catching is much easier if the foal slip is left on. During the first few days the natural curiosity of the foal will probably make him approach you, and he should be welcomed with something sweet to lick off your fingers. Approach him at his shoulder, rub your hand up his neck and the slip may be held and the rope attached almost without the foal knowing. On no account should you stand in front of him and make a lungeing grab for the noseband.

If, because of the various dangers involved in turning out a foal with his slip on, the foal has no slip, then catching may be more difficult. For the first few times it is better to just let the foal approach and reward him without any attempt at capture. Once the foal has learned to approach you willingly and to accept a titbit, and you can touch his shoulder and stroke his neck, he should be restrained with an arm round his neck and if necessary also round his hindquarters. The lead rope should then be placed round his neck to prevent escape, and the slip put on.

Some foals will respond more quickly than others, some will require more patience whilst others will always be eager and ready for the next lesson. By the time he is six months old and ready to be weaned, he should come when called and be easy to catch. He should lead willingly and allow his trainer to groom him and pick up his feet. He should be familiar with stable routine, and he should respond to the commands "Walk on", "Whoa", "Stand", "Steady", "Foot", "Back" and "Over". Once he has been weaned he is ready for the next stage of his basic training.

LEADING

I believe that there is no more important part of a horse's training than being led in hand. It is no exaggeration to say that successful leading is halfway to training the young horse and that the remaining half will be simple if the leading phase was effective. If the young horse meets, and becomes accustomed to, the sights, sounds and surprises of life whilst he is being led, then once he is under saddle they will be second nature to him. Educational leading demands the time of a skilful trainer, and in these days of financial constraints neither time nor staff may be available. However if competent staff and time are available, use them to the full, for the effort and sacrifices will be well rewarded.

Although the young foal has been taught to walk quietly both behind his mother, and a little way from her, serious leading only begins once he has been weaned. To start with he should be led from his headcollar, but later on when he has been introduced to a bit he may be led using a bridle. The headcollar rope should be of ample length, 2–3 metres (8–10 ft), and preferably of cotton, although nylon is more commonly and cheaply available today. Even though he is not yet shod it is a good idea to introduce him to exercise boots at an early stage. Initially the boots should be put on in the stable and the horse allowed to get used to their feel before he is walked out. Horses nearly always walk with a high "stringhalt" action when boots are first put on and this is perfectly normal. The boots should be lightweight, easy to clean and simple to fasten. Modern synthetic materials provide a smooth strong outside surface, a soft wool-like lining, and self adhesive contact fastenings. These boots may be hosed down after work, or put in the washing machine, they dry quickly and remain soft and supple.

Although a whip was not carried when leading the foal, the weaned horse should not be led without one. A long polo whip enables us to reach his hindquarters, where if necessary he should be touched, but not hit. The whip should be carried in the outside hand, and when changing hands to lead from the opposite side care must be taken not to wave the whip about in front of the horse. The whip should only be used when voice and lead hand have failed, it should only be used on the hindquarters and it should never be used as a weapon. Henry Wynmalen said in his classic book *Dressage*: "It is never necessary to hit a horse; it is a sin to do so in his stable; and it is quite unpardonable to do so at his head".

When leading I always carry a small canvas bag to hold a plentiful supply of nuts. The horse is still very young and must be rewarded when each new hurdle has been crossed. The use of a special bag means that one's pockets do not get soiled with food, and perhaps more importantly the horse does not learn to investigate pockets and clothing in search of food; this practice often leads to him becoming a nibbler or biter. If he does start to search for food he must be reprimanded immediately by gently but firmly pushing him away with a stern word of discouragement.

The trainer should wear comfortable shoes, waterproof if he is going into wet fields or through water, and if possible with strong uppers to prevent injury to the foot should the horse tread or jump on the trainer. The short daylight hours of winter sometimes mean that young horses have to be led on the road after dark. If this is unavoidable then reflective tape should be put over the horse's exercise boots, the handler should also wear a reflective jacket and if possible carry a lamp in the whip hand.

The aids for this educational leading are the voice and the whip. Sometimes it may be useful to use a reliable old schoolmaster as an aid to convey to the youngster what is required of him, and to instil confidence and trust by leading him past more terrifying obstacles such as combine harvesters, and noisy pigstyes.

As was explained in the section on leading foals the use of the voice is important and all instructions must be given clearly so the horse does not become confused. It is no good saying a crisp "Walk on" one day, and a long drawn out "Waaalk" the next, or "Walk on" one minute and clicking the tongue the next. If you want the horse to move off from a standstill into a walk then the command, which he has already learned as a baby, is "Walk on". In later life if you want him to walk forward up the ramp of a horsebox, into the starting stalls at a racemeeting, or over some narrow bridge, the command is "Walk on" and the response should be immediate.

The only other word of command at this stage, in addition to the "Walk on", "Whoa", "Stand", "Steady", "Over", "Back" and "Foot" of infancy is the word "Trot". It is usually spoken with a rolling drawn out "Tr" and an emphasised final "t", but it does not matter how the word is pronounced as long as it is always pronounced in the same way.

The pace at which one leads will naturally depend on the age, weight and fitness of the trainer, but as the majority of horsemen seem to remain remarkably slim and fit whatever their age (or sex), one can assume that in addition to walking, the trainer will be able to do some trotting, and a little jumping.

What one is trying to achieve in this phase of leading is as much variation as possible. This maintains the horse's interest in his work and introduces him to as many different situations as possible. Obviously some facilities will not be available to every trainer but careful study of one's local neighbourhood should enable one to get the best out of whatever facilities are available. The route should always be planned and the lesson thought out prior to moving off — it is no benefit just sticking the boots and headcollar on and wandering aimlessly about looking for challenges.

Initially the walking should be just around the stable area. Get him used to parked cars and dogs, to dustbins and machinery. Make sure that you lead him out to get used to the dustman, skip remover or manure disposal vehicle and lead him close to the forage merchant's lorry whenever it delivers. Calm him, reassure him and reward him when he stands still and square.

The young horse will be used to walking through gateways since the days when he was led to the paddock behind his mother. Now he should not only go safely through the gate, but he must turn back to stand quietly whilst it is closed. He must be taught to stand with his head close to the latch and to move with the gate as it opens and closes. He must learn to walk through calmly and not to rush. If possible he must learn to stand by the gate whilst other horses go through, and then to stand quietly whilst the gate is closed, despite the fact that the other horses have moved away. A few minutes gate work during every lesson will save frustration, embarrassment and even accidents in later life.

Roadwork should be carefully planned so that verges can be used whilst the horse remains unshod. If busy intersections, roundabouts or flyovers can be observed from the safety of neighbouring fields it will make the eventual progress onto those roads safe and uneventful. Roadwork through the town or village should be started on quiet days, such as early closing day, or Sundays and perhaps in the company of another horse. Gradually build up his confidence so that he can eventually be led without incident through the bustle of market day or the excitement of a fair or parade.

In the fields he should be led up and down steep slopes; through narrow gaps, into and out of woods, and into as much water as possible. It is not a good idea to lead him under low branches or you may regret it when you eventually mount him! There is not so much to frighten him in the fields so this is where his basic schooling can be started. Teach him to walk forward freely and to maintain a steady rhythm when going up and down hill. Teach him to trot on command and try to maintain a speed which enables him to travel straight and not with his head bent round to the side on which you are leading. Always do half the leading from the left side and half from the right to prevent the horse becoming over developed or stiff on one side.

Once he is walking and trotting on any surface, and he will go past the unexpected and through water, we can progress to walking and trotting over simple obstacles. A small natural log pulled across a familiar path, on the way back to the stable after a simple exercise lesson will probably almost go unnoticed, especially in the company of a lead horse. As time goes on the size and number of the logs can be increased until he will willingly pop over obstacles that are 0.5 to 0.6 m (18 in to 2 ft) high (anything higher than that tends to give me problems, and I don't think it is necessary even for the more athletic of you to go higher than 0.75 m (2 ft 6 in)).

The best introduction to coloured poles is also in the country and not in the covered school or manège. Again use the familiar path, head for home and confront him with a brightly coloured pole in place of a natural log; he probably won't even notice it if you walk on with confidence.

Having learned to tackle small obstacles across paths and in hedges where it is not easy to run out, the young horse should now be introduced to small poles set at right angles to a hedge or fence. The handler should walk on the outside (away from the fence) carrying the whip as usual in the outside hand. Once he will walk and trot over these obstacles in both directions, with a hedge to prevent him running out, the poles should be put in the middle of a field or school and the lesson repeated.

The horse should now be ready to learn to move backwards. He has been taught in the stable to move over and to move back and should be familiar with the words "Over" and "Back". Using the word of command "Back" and with gentle pressure on his nose and chest move him back one pace and then reward. This lesson can be carried out daily either before or after his normal lesson, and only when you are confident in his ability should it be performed in the school or paddock. Two or three steps is all that is required, then walk him forward and make much of him.

Rewards by now should consist mainly of words of praise and patting on the neck. The titbit of youth should only be retained for success with difficult problems, or in cases where the training programme has been interrupted or hampered by the slowness of the pupil.

Once the young horse has settled to being led by his trainer on foot, most of the lessons can also be repeated leading him from another horse. This can save manpower, and it accustoms the youngster to having a person above and behind him, but I do not recommend that it is a substitute for proper leading in hand. Initially the young horse should only be led from another horse in the confines of a school or paddock, and not until he is used to traffic in hand should he be led from another horse on the roads.

Leading exercises may be carried out with varying frequency over weeks, months or even years. The end result should be a well mannered animal that responds willingly to simple vocal aids, he should be familiar with traffic, water, natural and coloured obstacles. He should be quiet to handle in the stable and easy to box, clip and shoe, he should load and travel quietly. In fact he should be what so many advertisements describe in the "For Sale" columns of equine journals. He is now ready to be loose schooled, long reined, lunged and prepared for being ridden.

LOOSE SCHOOLING

Loose schooling is not often practised today but it is a valuable method of exercising and balancing both young and mature horses. Trainers wrongly imagine that it is necessary to have an expensive purpose built school, and many have not been taught the principles and consequently do not realise what a beneficial practice it can be. The main purpose of loose schooling is to further educate and exercise the young horse prior to his being ridden. It develops his muscles and gets him fit, it teaches him natural balance, and it gives him confidence in himself without any interference from handler or rider. In principle the trainer always remains behind the horse and encourages him to go forward.

A purpose built school is a large circular or oval arena, with a 3 m (10 ft) high perimeter wall, a 3.5 m (12 ft) wide track, and a 1 m (3 ft) high central platform. There should be a close fitting door in the perimeter wall, and provision for both upright and spread jumps. In the United Kingdom the whole structure should be roofed and provided with illumination.

If a proper loose school is not available then it is easy to improvise in an ordinary indoor school, a fenced manège, or a small well fenced paddock. Plastic military mine tape, or the tape used by highway repair teams to cordon off roadworks are ideal forms of cheap, safe material to mark out the school. The tape should be used to fence off the corners so that the horse is encouraged to move on round the school and not to stop or turn. The inside of the school can be bounded with tape strung between weighted bending poles or simple jump stands.

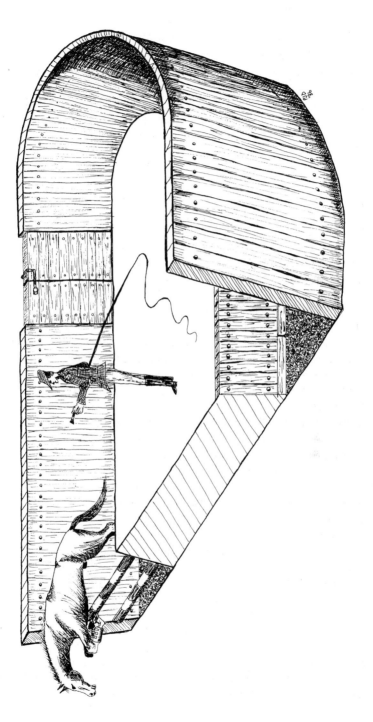

Figure 8.2. Design for a loose school.

The horse needs no tack at all for loose schooling but it is usual to put on exercise boots, and the headcollar is often left on. The trainer requires a lunge whip and a supply of nuts to reward the horse on the successful completion of each exercise.

During the first lesson in the loose school it is an advantage to have one or two assistants positioned round the school to keep the horse moving. The horse should be led round initially and allowed plenty of time to inspect and adjust to the strange surroundings. He should be led in both directions with the handler on the inside. When he is calm and settled the lead rope should be detached and the horse encouraged to walk on by himself. Trotting seems to be the easiest and most positive pace for loose schooling and should be encouraged at once. The trainer should be competent to judge the speed and impulsion of the horse and to send him on if he is slowing, and slow him down if he is rushing. There should be a calm and relaxed atmosphere with no rushing about, waving of arms, throwing of wood chips, or cracking of the lungeing whip. Use only the correct vocal aids: "Walk on", "Trot", "Steady", "Whoa" and "Stand".

Horses are surprisingly easy to loose school and after a few lessons only one person will be required for the horse will need little encouragement to maintain an even pace. The trainer should remain just behind the horse and move towards him should he lose impulsion. To slow him down the trainer should move in front of him and calmly bring him back through walk to halt. The horse should then be approached and rewarded. He can then be turned round and the exercise repeated in the other direction.

Once the horse has learned to go round at an even pace and to respond to the voice, he may be introduced to poles and small fences. For jumping it is essential to put on exercise boots. If it is possible the jumps should be placed on the long sides rather than on tight corners of the school. The great advantage of the loose school over most jumping lanes is that the distances between the fences may be varied to teach precision and to avoid boredom. In most jumping lanes or couloirs the fences and banks are permanent obstacles with fixed distances which do not suit the stride of every horse.

Horses must not be allowed to get bored in the loose school, and the pace, direction and fences should be changed frequently. The horse should be made to stand still between direction changes by offering a reward before moving off on the other rein. Do not always stop or turn him at the same place. The lessons should be kept short, and three lessons of three minutes a day will be of far more benefit than one long session.

LUNGEING

Of all the methods of training a horse on the ground lungeing is the most controversial. Even the experts cannot agree on the basic principles. The great

Henry Wynmalen insisted in his book on equitation that the lead rein should only be attached to the back of the noseband, whilst Sylvia Stanier in her treatise on the subject insists that the rein be attached to the front! Nearly every aspect of lungeing is subject to varying definitions and interpretation, to argument and debate, but one thing remains certain; it is easy to lunge a horse badly and difficult to lunge him well, and for this reason lungeing is often considered to be an art.

The horse, like the cheetah and the gazehound, was designed to travel quickly in straight lines. Anatomically these animals are not able to pronate or supinate their limbs, which means the legs will only flex and extend in one plane and there is no facility to twist like a human wrist. Most fast galloping animals have a spine which flexes dorso-ventrally, but which has little lateral movement; the horse is no exception. This anatomical inflexibility means that when these animals turn quickly they do so in a series of leaps and bounds. To trot horses in small circles is a totally unnatural action and in skeletally immature horses puts great strain on the joints and ligaments of the spine and limbs.

I believe that because skilful lungeing is such a difficult talent to acquire, and because of the fact that young joints may so easily be damaged, the immature horse should never be lunged as part of the basic training. I feel even more strongly (and this is probably the veterinary surgeon in me speaking out) that the last thing on earth you want to do to a horse convalescing from injury or lameness involving the joints, is to make him trot round in tight circles on the end of a short rein. Take him in straight lines, take him up and down hills, take him on hard and soft surfaces and take him through water, but don't lunge a horse recovering from limb injuries.

Despite the foregoing remarks lungeing is commonly used to exercise horses and to school them both on the flat and over jumps. Sadly it is all too easy to attach a 30 foot length of rope to an unbacked youngster and wind him round to the music of a cracking lunge whip. This type of forced exercise teaches the horse nothing except to fear the whip, and it can seriously damage his legs and his vertebrae. If lungeing is to be used at the start of formal education then it should be carried out professionally, with proper tack and in a calm environment.

Lungeing should be carried out on an area of flat ground large enough to allow a 20 yard circle. It can be done in the indoor school or outside, however it is better to have an area enclosed by fencing or hurdles and away from distractions so that the horse is able to concentrate on his task.

The horse should always wear exercise boots when being lunged. The young horse especially, which as has already been explained is not anatomically designed to travel in small circles, does not have effective control of his limbs and is liable to knock his legs and cause permanent damage and scarring.

Splints are often caused by physical trauma during lungeing rather than sprain of the interosseous ligaments during concussive exercise. This type of injury is known as an "acquired splint".

Lungeing is usually performed with a cavesson which must be correctly fitted. The noseband should be well padded and be fitted high so that it does not interfere with the horse's breathing by compressing the soft tissues above his nostrils. The noseband must be tight enough to prevent it flapping, and the jowl strap tight enough to prevent the cavesson being pulled off should the horse run backwards. A tight jowl strap also prevents the cheek straps from rising up and damaging the eyes. The lunge rein should not be less than 6 metres (20 ft) in length, and although most are still made of canvas webbing, modern nylon ones are much lighter and cause less jarring on the noseband. As has already been pointed out, even the experts cannot agree on where to attach the rein. I prefer to strap the rein to the back of the cavesson underneath the horse's chin where it seems to be more comfortable and less liable to damage the nose.

At a later stage the horse may be lunged in a snaffle bit, initially attached to the cavesson but later to a bridle. The rein should only be attached to the inside snaffle ring; not through the inside ring to the outside one which would tend to squeeze the bit and bruise the hard palate; and not from the outside ring over the poll and through the inside ring to the trainer which would also cause the snaffle to fold and bruise the hard palate, tongue and bars of the mouth. A surcingle or lunge roller should be used, and these can eventually be replaced by a saddle.

Side reins, like lungeing itself, are cause for debate. Many people use side reins to put the horse's head into the correct position which is bad training. Loose side reins do no good at all and tight ones cause the horse to be overbent, and interfere with his lateral bend and with the free use of the head and neck to balance himself. Side reins should only be used to encourage the young horse to seek contact with the bit, and they should have an elastic insert so that they give in the same way as sympathetic hands.

A whip is essential, and if the horse has already been loose schooled he will be used to the notion that it is only used as an aid to indicate that some response is required of him. Using a lunge whip is not easy. The novice trainer should spend time practising the use of the whip in both hands, and become proficient at changing the whip from one hand to the other.

The trainer should never wear spurs when lungeing as there is a possibility that they could become entangled in the rein or the lash of the whip.

As with many other aspects of basic training it is better for an assistant to help with the first lunge lessons. Horses usually lunge best on the left rein, so the assistant should be on the horse's right hand side and lead him from the outside in an anti-clockwise direction. The trainer should remain in the centre of the circle with the rein in his left hand and the whip in his right. If the horse has been properly led and loose schooled he will understand the commands and should quickly settle to the more restrictive movements of lungeing.

It is important for the trainer to anticipate, and by use of voice and whip to correct any faults as, or even before, they occur. As with leading and loose schooling all movements must be free and forward and the lessons should be kept short to prevent tiredness or boredom. Work equally in both directions and vary the direction in which he starts from day to day. Do not always halt in the same place, and do not always increase or decrease the pace at the same spot. If possible get him used to being lunged in different places, so that in later life he will lunge willingly at shows or events.

Jumping on the lunge is not to be encouraged in young horses. Let them jump free in a jumping lane or on the long side of a loose school, but do not teach them to jump out of a tight turn. Jumping out of a circle, after a sharp turn, or at an angle is a skill which comes with advanced training prior to speed events and it is not a skill for the unbacked novice. It is impossible for all but the most highly schooled and athletic horses to get both hocks underneath them for take off at an angle or on a turn. The novice horse will invariably leave his outside back leg behind thus putting more strain at take off on the inner stifle, hock and fetlock. He will also tend to land more heavily on his inner front leg. Immature limbs will not accept this type of insult.

The Dutch showjumping trainer Anthony Paalman, who achieved international success with his Natural Training Method, considers that the Chambon is the only equipment that should ever be used for lungeing. The Chambon was designed by a French cavalry officer and is used to stop the horse raising his head by putting pressure on the poll. It does not put pressure on the mouth, and it enables the horse to balance itself and to move freely forward. It will calm a nervous horse and even appears to give impulsion to a lazy one. The Chambon consists of a special poll piece with two rings. A length of cord passes through the poll piece rings to the snaffle rings, and the free loop is attached to the girth by an elasticated side rein between the horse's front legs (Fig. 8.3).

Only if the horse tries to lift his head does the Chambon cord tighten and exert pressure on the poll; it is a gentle action which automatically forces him to lower his head again. Mr Paalman's view is that because there can never be any backward pressure on the mouth a horse that has been lunged with a Chambon will never require a martingale, it will be well muscled and balanced, and it will have an easy mouth. We can only benefit by considering the teaching of such an expert, and the Chambon is certainly more forgiving than side reins, running reins, or even a Market Harborough, the Chambon should only be used for dismounted lungeing.

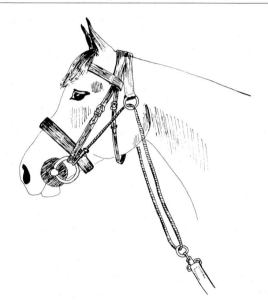

Figure 8.3. The Chambon

LONG REINING

Long reining, like lungeing, is an art, and it is often neglected because it requires considerable skill on the part of the trainer. The time and patience taken to learn the skill are well worth the effort, as it is the only form of training from the ground where the horse is controlled in a similar way to that of being ridden, with a rein on either side. Although used a great deal in the past in the basic training of driven harness horses, long reining introduces the riding horse to the idea of aids being given by the reins of a rider. Initially the reins are attached to the side rings of a cavesson, but later when the horse is going freely and obediently forward they may be attached to the snaffle rings. The two reins touching the horse on his flanks also gets him used to the contact that he will experience from the rider's legs once he has been backed.

The equipment required for long reining is a cavesson, or snaffle bridle without reins, depending on the stage of training of the horse. The two lunge reins are attached to the cavesson or snaffle with a buckle, or preferably a clip and swivel, the other end of the reins should have a loop or knot, and on no account should the two reins be stitched together. I prefer each rein to have a simple figure of eight knot which prevents the rein slipping through the hand. With a loop there is a tendency for the trainer to slip his hands through and have the loops around his wrists which could be dangerous if the horse should bolt.

A lungeing roller or saddle can be used depending on the stage of training. The saddle should have its stirrup irons secured by a rope or spare leather under the belly so that the irons hang well above the point of the elbow. The lunge whip is the only other item of equipment required, but although exercise boots are not essential I advise that they are used.

Before the horse is introduced to long reins he should have reached a stage of training where he is confident, responsive, and has trust in his trainer. It is essential to have an assistant to help familiarise the horse with the feel of the rein against his hindquarters, tail and back legs, and also to help when changing the rein. The outside rein should pass through the outside roller ring or stirrup iron and be positioned just above the horse's hock. The inside rein should come direct from the cavesson or snaffle to the trainer. With the help of the assistant leading on the outside, walk the horse round the school. The assistant should be ready for the two common reactions to the feel of the rein on the back leg which are either kicking back at the outside rein, or shooting forward with his tail clamped down.

After two or three circuits the horse should be used to the feel of the rein above his hocks and he can then be turned round with the help of the assistant, and the rein threaded through the opposite ring or stirrup. Once the horse reponds to the aids and is no longer upset by the outside rein touching his hindquarters and hocks, both reins may be passed through the rings and training can continue without an assistant. The trainer now has complete control, but for the first time in the whole training process the trainer is often positioned directly behind the horse in the only place where the horse cannot see him. The trainer must make special use of his voice, and keep talking to the horse so that the horse is always fully aware of the trainer's position.

As long reining progresses the trainer can do everything with the young horse that he will be doing mounted in a few weeks time. In fact a skilled trainer can even teach an advanced horse to do lateral work. There is little point however, if the horse is to be ridden, in teaching him on foot what can be taught with proper weight distribution and balance from the saddle.

No hard and fast fules can be laid down as to how long the various stages of training should last, or at what age they should begin. Handling should certainly start on day one and leading can be done at any age. The more leading in hand, the easier the subsequent lessons and challenges will be. Loose schooling and long reining are lessons for the more mentally mature horse and as I have already stated I do not consider lungeing an ideal training method for the skeletally immature. Thoroughbreds are raced at 2 years of age and yet I would not advise serious basic training and backing until the riding horse is in his third or fourth year.

One of the main dangers of any type of training is boredom. Work must be varied, leading one day, loose schooling the next, long reining the next, or short combinations of each. Exercise should not be carried out in the same place every day and it is good practice to box up the youngster with a schoolmaster, and take him to a completely different environment for the odd day if possible. Lessons should be kept short and rewards should not be forgotten, although at this stage a good pat and a word of encouragement can replace the nuts and carrots of youth.

BACKING

Before backing it is a good idea to remind ourselves of the horse's origins and his psychological outlook. In the wild he is almost totally defenceless and relies on his acute senses and fleetness of foot to escape from his predators. He has a blind spot behind him and a natural fear of anything landing on his back, for it was on his back that the sabre-toothed tigers and lions of old would leap. This is one of the reasons why horses that are "broken" by cowboys are so terrified, and put up such a fight to dislodge their riders. The reason why the majority of horses in harness wear blinkers is to prevent them being frightened by what is going on above and behind them.

Although the horse will probably be more afraid of the height and position of the rider, the increase in weight may also be alarming. The combined weight of the rider and saddle may well be nearly 20% of the horse's own weight and this is a considerable bulk to have placed on his back. It will take time for the horse to adapt to this increase in weight and develop his muscles and it will also take time for him to regain his balance as most of the new weight falls on his forehand.

As with all aspects of basic training, calmness and patience are essential. If there has been a gradual introduction to increased weights on the horse's back, and a progressive tightening of the stable roller, lunge pad or girth, and if the horse has been led from a schoolmaster or has been accustomed to the height of a rider standing on a box or bale at his shoulder, then when the time eventually comes to mount there should be no problems. A horse that panics or reacts wildly to being backed has not been properly prepared.

Before putting on a saddle it is best to introduce him to a folded blanket. If he has been stabled during a European winter he will already be used to a blanket, and this blanket will smell warm and familiar to him. It can be folded and placed under a surcingle, and can gradually be applied more vigorously, and eventually put on together with a saddle. It is best to use an old saddle, not one that has a broken tree or lacks stuffing, but one where the leather is worn and the stitching unsafe for a rider. If the flaps are stitched down and the saddle well oiled there will be no flapping or squeaking when the horse walks and trots.

Care must be taken when handling the girths as the buckles, if allowed to swing, can frighten the horse or even damage his legs. If an assistant is available let him stand on the other side to lower the girths and pass them under the horse's belly. Keep one hand on the saddle to prevent it slipping and tighten the girths very gently. Many horses which blow out when being girthed up have developed the habit because of unsympathetic handling at this stage of training. An elasticated race girth may avoid the danger of over tightening on a young horse, but it should only be used in conjunction with a breast plate to ensure the saddle does not slip back.

There are many other ways of introducing the horse to weight on his back. An empty sack may be attached to a surcingle and filled with increasing amounts of hay to provide bulk and weight. Some trainers prefer to physically lean on the horse, increasing the pressure each day by leaning across the shoulder to pat the far side.

Once the horse is used to being saddled in the box it should be left for half an hour with the saddle on to get used to the feel. Naturally one would not do this with a new saddle, but the old one described above will not come to harm if rubbed on the walls or even nibbled by the inquisitive youngster. The horse should then be led out and walked with the saddle on for several days. During this time stirrup leathers and irons can be introduced and the leathers lengthened so that the horse becomes used to them hanging by his sides. The dangling irons should never fall below the level of the point of the elbow as they might bruise the elbow. A fractious horse might even catch his hind hoof in the iron if the leathers are too long, with serious consequences.

When the horse has become used to the saddle and is no longer intimidated by the rider standing over him, the rider should start leaning across the saddle. An assistant should steady the horse's head to prevent him walking forward, for the horse must learn to stand when mounted from the very start. If the horse remains relaxed and does not object to the weight then the assistant can lead him round the box. At the first sign of tension the rider can slip gently to the ground. The horse must be frequently reassured during this phase of training with constant use of the voice.

So far all the tacking up and preparations for backing have been carried out in the horse's box where it is in a friendly and familiar environment. As long as the box is large enough and does not have low beams the final backing should also be performed where he feels secure and relaxed. Once the horse has accepted the rider getting up and lying across the saddle it is one small step for the rider to swing his leg carefully across without touching the hindquarters, and settle quietly into the saddle. The rider should remain crouched which makes for a quicker exit if necessary and is not so alarming for the horse. If the horse remains calm then the command "Walk on" (to which he has been responding since he was a tiny foal) is given; the assistant moves forward at the horse's head, and the rider slowly sits upright.

For the next week the mounting process should be repeated several times a day and the rider should always be given a leg up. Only when the horse is thoroughly used to being mounted in this way and to being ridden, should the rider mount by using the stirrup. Mounting with a stirrup introduces a considerable lateral strain which may unbalance a young horse and the girth must be much tighter to prevent the saddle from slipping round. To help overcome these problems the assistant should lean on the opposite iron gradually reducing the pressure he applies as the horse becomes accustomed to the rider's weight.

It is well worth spending a lot of time teaching the horse to stand still whilst being mounted. He should never move off until asked to do so. A horse will usually only move off, turn round, run back, or collapse his back if his early training was rushed or clumsy. The horse should be stood square with his weight evenly distributed on all four feet. The reins should be held in the left hand (for mounting from the left) and a gentle contact made. At no time during mounting should the horse's mouth be jabbed or he will learn to fidget or move backwards. The left foot should be placed in the stirrup taking care not to touch the horse's ribs with the toe of the rider's boot, for this will make him swing about, or move off. Do not grip the back of the saddle (the cantle) when mounting as this tends to pivot the saddle laterally and cause the underblanket or skin to wrinkle and fold. Put the right hand across the middle of the seat, and swing the right leg over the quarters without touching them. Finally the rider should stand in both stirrups, gripping with the calf, knee and thigh, before lowering himself gently into the saddle.

Dismounting correctly is as important as mounting properly. The horse should be stood square and with the reins in the left hand, the right hand grips the front of the saddle (the pommel) and takes most of the weight of the rider. The rider removes both feet from the stirrups and swings the right leg over the quarters, before dropping gently to the ground. Again, care must be taken not to jab the mouth or touch the hindquarters when dismounting. The rider should never swing his right leg over the horse's neck, for if the horse were to raise his head, it would tip the rider off backwards. Similarly, especially in young horses, both feet should be removed from the irons prior to dismounting; the habit of stepping down with the left foot in the iron could lead to the rider being dragged if the horse were to move unexpectedly. Riding convention decrees that it is correct to both mount and dismount from the left side of the horse, but it is important that the horse will allow the rider to mount from both sides, and mounting and dismounting on the right hand side should be practised frequently.

The young horse has now been backed and led at both the walk and trot with a rider in the saddle. He may even have done a little walking and trotting on his own, or in the company of a schoolmaster. If possible these early experiences under saddle should not have been confined to the school, nor should they have been limited to level ground. A varied environment, and gentle hill work are essential to break the monotony of early lessons under saddle. We are now ready to teach the horse to be ridden quietly, to respond to basic aids and to become physically fit.

INTRODUCTION TO SCHOOLING

Schooling is the stage of training concerned with teaching the horse to respond to the basic aids. He must become confident and obedient, but blind obedience without calmness and confidence must be avoided. Schooling is also the time when we start to develop and control the horse's natural ability and balance, and it is the stage where we start to get him fit.

There can be no short cuts in schooling and the use of restraining gadgets should be avoided as they only lead to resistance. As with every other phase of basic training, reward and reprimand should be the substitutes for force. The horse should never be allowed to get his own way, for if the command and respect for the rider become eroded then authority will be lost and the basic principle of training will be destroyed.

During this stage of training the rider should adopt the classic "remount seat". This seat was specially developed in European army remount centres and cavalry schools for the training of young horses. In the remount seat the rider remains well down in the saddle but leans the upper part of his body slightly forward so that weight is transferred from the seat to the thighs. The rider's weight is closer to the withers which prevents strain on the back and loins of the young horse. The hands are carried low, and away from the horse on either side of the withers. Because of the forward position of the rider the reins are slightly shorter than in the classic dressage seat. In the remount seat the stirrups should be short enough to allow a firm grip with the knees, rather than gripping with the thighs which are supporting the rider's weight, or the calves which remain free to give very positive aids.

Throughout training the practice of using a schoolmaster has been encouraged to make use of the gregarious nature of the youngster. A schoolmaster will instil confidence and enable lessons to proceed without delay. Do not however teach the young horse to rely entirely on a companion so that he will not perform on his own. During every lesson make sure that the old horse can leave the youngster and vice versa without panic or playing up. Every lesson should have a large proportion of solo work, and the schoolmaster should only be used for demonstration, example and to instil confidence.

So far in his training the young horse has responded to voice, hands and whip. He must now be taught to respond to the legs and seat of the rider. The legs are used to indicate forward movement by activation of the hindquarters, and to indicate lateral direction. To indicate forward movement the legs, that is the inside of the calves, are applied just behind the girth. They are applied simultaneously, lightly and rhythmically rather than with a constant squeezing pressure. To indicate lateral movement the legs are used independently. If the left leg is applied just behind the girth it will encourage the horse to bring his left back leg forward so that his hindquarters may be moved away from the contact. The degree of lateral movement is then controlled by applying the opposite leg slightly further back behind the girth. Inside leg for impulsion and outside leg for control.

To use the seat as an aid the rider must be both supple and experienced. At this stage of training the seat should only be deepened to make closer contact and so influence impulsion; and the weight transferred slightly onto one or other seatbone or stirrup to indicate the direction of movement.

The use of the schooling whip is controversial. The whip is intended to reinforce the leg aids, but how can we reinforce the action of two legs when we only have one whip? A great deal of time is spent teaching novice riders how to change the whip quickly and quietly from one hand to the other which is useful later on in schooling, but is nearly always too late in the very early stages. My own opinion is that a competent rider should be able to handle two whips just as easily as one. For the first few weeks of schooling a 3'6" schooling whip in each hand applied just behind the rider's legs will emphasise the aid and call for the attention of the horse in a totally symmetrical and balanced way.

All the aids must be used in harmony with each other which calls for thought and coordination on the part of the rider. The combined use of the aids should dictate both forward movement and direction without exaggerated gesticulation on the part of the rider, or excessive reaction on the part of the horse. The correct application and harmony of the aids is vital if we are to achieve absolute obedience. From an early age the horse has been taught to move off on the command "Walk on". Even with the distraction of a rider on his back the newly mounted horse will probably walk forward on command. To make sure that he does respond we have an assistant to lead him whilst we are introducing the two additional aids of leg and whip. To start with the three aids are applied simultaneously, but over the next few weeks the voice and whip will be phased out, so that ultimately the horse will respond willingly and confidently to the contact of the legs and allowing of the hands.

It is far easier for the young horse to concentrate on his lesson if he is taught in the confines of the indoor school or a small paddock. One must be careful however not to let him become bored by a repetitive routine in the same place and on the same level surface. It is a good idea to start off his daily lesson in the school and then after a short time, but only when the lesson objective has been achieved, move outside which in itself will be an ample reward. A sloping field provides dozens of permutations for a 20 minute lesson. The horse can be worked uphill or downhill, it can be worked diagonally or along the contours, it can be worked at different paces, and the speed of the paces can be varied. To help the horse learn to balance, and to teach him to stand square whilst being mounted, the rider can stand the horse facing up, down or parallel with the slope, and then mount and dismount from both sides. Gymkhana bending poles are another idea to provide interest; if put in different places they provide an interesting guide to help the horse to turn accurately in response to the aids. Introduce both coloured and natural poles on the ground at all stages of training, but be careful to peg them down to prevent them from rolling when used on a slope. A variation in the schooling surface also prevents boredom and helps the young horse to find his balance. If possible, vary the texture, from road to short grass, to long grass, to broken ground or plough, and to sand, and allow him a free rein so that he extends his neck as he explores each surface.

In the wild the horse was a creature of habit and did not have a day of rest. Every day he walked in search of grazing, he ate, drank and rested. Giving a day of rest, because it is unnatural, can cause upsets in the metabolism and lead to colic or musculoskeletal damage. Although the exact causes of azoturia, tying up, and Monday morning disease are still being unravelled, the latter name indicates that the syndrome often occurs after a day of rest. I believe that the young horse should be worked for six days and then given a light hack out on the seventh and not have a complete rest. At this stage of his training the weekly programme could consist of three days where the majority of the work is outside, two days where the majority is inside, one day of dismounted long reining or loose jumping, and one day where he is hacked out to beach or river, woodland or park.

Hacking out itself can be the basis for useful instruction and all the lessons of the riding school can be put into practice. In advanced training the paces may be varied from collected to extended, the horse made to move laterally across roads and paths, and gates opened by correct turns on the forehand and hocks. During basic training it is important not to develop bad habits such as always trotting up hills, or always cantering along some inviting grass verge. Many horses become excited and on their toes whenever they arrive at the foot of a hill, and this undesirable trait has developed entirely because during early training they were encouraged to canter up such slopes. Although the hack can be used as an extension of the riding school it should be reserved as a pleasant reward at the end of the week where the young horse can relax on a loose rein.

During basic training the young horse develops physically and becomes fit. The change from dismounted work, to carrying a rider puts a considerable strain on the limbs and the back, as well as a totally different weight distribution and balance. The muscles of the back have to develop and the muscles and tendons of the legs, together with the ligaments of the joints, must be strengthened by steady exercise. New exercises should be introduced slowly, and any change from hard to soft going should be undertaken gradually. Thoroughbred trainers who gallop on the same stretch of sand every day will always check first that the tide has not left a soft patch which could sprain joints or break down tendons.

Fitness also means an increase in stamina, and the ability of the horse to gradually do more work for a longer time over a greater distance. Stamina depends not only on the muscles of locomotion, but also on a more efficient heart and lungs which, through the blood, supply those muscles with nutrients and remove the waste products of energy production.

Once a horse has reached a satisfactory standard of fitness, and his rations have been balanced in quantity and quality to cope with his new exercise level, it will be relatively easy to keep him fit, and the fitter the horse, the far less liable he will be to injury or disease.

THE HALT

The importance of the young horse learning to stand still whilst being mounted, and not moving off until asked to do so has already been emphasised. He must also learn to stand equally still when halted from the walk.

At the halt the horse should stand attentive and motionless, he should be straight and not turn off the track, or deviate from the direction in which he was moving prior to halting. His weight should be evenly distributed on all four legs, and his feet should be square, in pairs abreast with each other. It is difficult for the young horse to stand square and not too much emphasis should be placed on this until his training has reached a standard whereby he is able to get both back legs balanced beneath him. It is important however even at this early stage not to rest a back leg, but to keep an even distribution of weight on all four legs even though one may be left slightly behind. The head and neck should be held so that the poll is the highest point and the head is slightly in front of the vertical.

The aids for the halt are the same as for a downward transition from canter to trot, and from trot to walk. During basic training the halt should be a progressive transition from the walk taking several paces to completely come to a standstill. The rider must make sure that the horse walks forward into the halt and does not, as one so often sees, allow the horse to stumble to a standstill with considerable resistance. If the horse has been well taught prior to being ridden he will be used to stopping at the word of command "Whoa", and he should continue to do so when mounted. At the same time as giving the vocal command the rider should use his seat, and both legs evenly behind the girth, to maintain impulsion, and he must apply gentle restraint with the hands to accept and control this impulsion. The hands must allow as well as restrain until the horse has come to a standstill. In the trained horse the whole process may take less than one stride, and the horse will halt as soon as he feels the rider's legs and seat driving him into restraining hands.

If the rider yields with his hands just before the horse actually stops, and continues impulsion with the seat and legs, the horse maintains his forward momentum with an increase in attention and balance. This is known as a half-halt and is used to generate impulsion, to rebalance the horse and to warn him to be ready for some new instruction. When prepared by a half-halt the horse will not be taken by surprise when the aids for the next movement are applied, and the new movement will take place smoothly and without abruptness. Half-halts are used mainly at the trot and canter and should not be confined to the riding school. The half-halt is not used in the early stages of schooling and it is described here only because the initial aids are exactly the same as for the halt.

The halt should be achieved without any resistance or throwing up of the head during the downward transition from the walk. If there is a tendency for the horse to raise his head then the rider's hands must be more yielding and the transition may thus take several more strides. With patience the horse will learn to come to a standstill without resistance as soon as he feels the correct aids.

It is not always possible to feel if the young horse has completed his halt correctly and is standing square; the use of a mirror in the school, or his shadow on the ground may help. Rather than lean over to look and so upset the horse's balance which may make him move his feet, have an assistant check that he is square.

Some young horses find it difficult to halt straight. As soon as they hear the word "Whoa" they stop, but in doing so move their hindquarters out of the track. If the horse still fails to halt straight after two or three weeks then the leg aids must be applied unilaterally to straighten him, and it may be necessary to reinforce the leg with the whip.

Always start his halts along a wall, fence or the school kicking boards so that he has something to guide him. Only when he goes forward willingly into a straight halt against the boards should halts be attempted in the centre of the school. Practise halts when hacking, and not always in the same place.

THE WALK

The walk is a slow marching pace of four time, with two or three legs always in contact with the ground and no period of suspension. The four beats of the hooves making up each stride should be even and regular producing a steady rhythm. If the rhythm is uneven caused by an irregular tempo between footfalls then the walk is described as "broken" or "disunited". The tempo of the rhythm is the number of four beat cycles per minute, in this case tempo does not mean the speed at which the horse is travelling. A seventeen hand high Hanoverian and a twelve hand high child's pony may both be walking side by side at 5 mph with a rhythmic walk; the tempo of the Hanoverian rhythm will be less than 40 per minute, whereas the little pony will be scurrying along with a nonetheless regular rhythm of more than 80 per minute.

It is relatively easy to maintain the rhythm of this pace when walking on a hard surface in a straight line, but it becomes more difficult when turning and on a soft surface. It may be a help to both horse and rider if the rider actually calls out the time like a metronome when the hoofbeats are not clear. By calling out the time one can vary the tempo of the rhythm which confirms that it is the rider who is controlling the pace, and not the horse.

The Fédération Equestre Internationale (FEI) describes four different walks but the collected and extended walks are not introduced during basic training. Of the remaining two, the free walk and the medium or ordinary walk, it is the latter which concerns us most at this stage of training. The medium walk is a free regular unconstrained walk of moderate extension. The horse remains on the bit, walks calmly but energetically with even determined steps, the back feet coming to the ground in front of the hoofprints of the front feet. The rider maintains a light but steady contact with the mouth.

The aids to start walking are an even pressure with both legs just behind the girth. The pressure should be symmetrical but not constant, and it is better to give a series of even squeezes, rather than a constant numbing pressure. It is impossible for this even pressure to be reinforced by a single schooling whip which would have to be applied unilaterally, so either do not use the whip at all or, as suggested earlier, use a whip in each hand. As the legs are applied the rider also deepens his seat slightly, without leaning backwards and without putting uneven pressure on his seatbones. The hands maintain contact with the mouth, but allow the horse to move forward freely into a straight and energetic pace.

Once the horse is moving forward at the walk, the leg aids are changed from simultaneous pressure on both sides, to alternate pressure with every other step. If the leg pressure was maintained evenly on both sides then the pace would become irregular and disunited. During the walk the rider must use his legs to maintain impulsion and a free forward movement by applying them alternately. As each back leg, which is the driving force, is about to come to the ground, so the pressure is applied on that side to make him push himself forward more energetically and to make the leg move higher and further forward during the next step.

It is easier for the rider to learn to use his legs alternately at the walk if he practises without stirrups. As the horse walks the rider will feel his right hip moving forward just as the right front leg reaches the ground. At the same instant as the right hip moves forward the left hip moves backwards pushing the lower leg closer to the horse and by an extension of this movement the rider applies pressure to the left side behind the girth just before the left leg comes to the ground. Apply the right leg when the horse's left foot comes to the ground, and apply the left leg when the horse's right front comes to the ground; in this way the pressure is applied just before the back leg on each side comes to the ground and that leg is then activated immediately.

Once the horse walks with energy and impulsion from his hindquarters, in straight lines and in circles, the rider should improve the horse's balance by getting his hocks further under him, by slightly raising his head and neck, and bringing his nose more vertical. This is achieved by a slightly more positive use of the aids, by driving the horse forward with the legs into slightly less yielding hands. There should be no resistance from the horse and if he throws his head about then the aids, especially the hands, are being applied more strongly than is expedient for his stage of training.

The fourth type of walk is the free walk, this is a pace of relaxation and should be ridden on a loose rein to allow the horse complete freedom to lower and stretch his head and neck. Impulsion is maintained by continued alternate pressure of the rider's legs to prevent a lazy or stumbling gait.

During turns at the walk, as with trot and canter, the direction of turn is indicated by contact with the rein on that side whilst the outside rein allows, and so controls, the bend. Similarly the outside leg behind the girth controls the hindquarters whilst the inside leg maintains impulsion. The horse's level of intelligence is such that he quickly learns to alternate between a series of bending poles. Initially they should be 5 or 6 metres apart and in a straight line on level ground, but the combinations for the more advanced horse are endless and poles may be placed in circles, on slopes and with varied distances between them. They are a great asset in teaching the horse the aids to turn and in helping him to bend correctly in the direction he is turning. The use of a line of trees has the same effect when out hacking.

In the school, make sure that the horse maintains an active walk during all his lessons, vary his routine as much as possible by introducing frequent changes of rein, circles, loops and serpentines. Halt him at different places around the school and make sure the transitions from walk to halt and halt to walk are smooth and straight. When he can perform well at the walk in the school then move the lessons to the paddock, or continue them when out hacking where he is no longer confined and guided by the kicking boards. Maintain a calm but firm authority at all times, and reward frequently by making.much with hand and voice.

THE TROT

The trot is a pace of two time on alternate diagonals separated by a moment of suspension. The FEI describes four trots, and as with the walk, the collected and extended trots do not form a part of basic training. The two remaining trots are the working trot and the medium trot. Ten years ago it was common only to describe three trots, the old "ordinary" trot having since been split into working and medium. During basic training one still in fact teaches the horse an "ordinary" trot where the strides are of medium height and length, and the period of suspension is neither too long or too short. The neck should be carried fairly long but contact should be maintained, and the nose should remain in front of the vertical.

As training progresses the "ordinary" trot can be split into working trot and medium trot. The working trot is closer to the "collected" trot and involves an even elastic step, and a fairly high hock action indicating impulsion from active hindquarters. The medium trot is closer to the extended trot and involves a more lengthened step and a slightly more relaxed head carriage, with a longer neck and the head more in front of the vertical. As with all paces the steps should be even, there should be a regular rhythm and the horse should be balanced and unconstrained.

The diagonals are called left or right depending on which front leg is on the ground; the right diagonal is when the right front and left back leg are on the ground and, after a period of suspension, the left diagonal is when the left front and right back legs are on the ground. Most horses, like people, are asymmetrical and prefer the rider to ride on the left diagonal, which means the rider sits in the saddle when the left diagonal touches the ground. To prevent this one sidedness the rider should practise changing diagonals frequently and should concentrate more on the one which is least comfortable. To develop good balance and a correct bend on circles it is best to ride on the outside diagonal, sitting down when the right fore hits the ground on a left hand circle and vice versa.

The debate as to which diagonal should be used when circling to the left or right during dressage tests is obviated by remaining in sitting trot and not rising at all! However if the test requires a rising trot, in Europe the rule varies depending on which side of the Channel you were trained. I prefer to ride on the outside diagonal, sitting when the outside front leg comes to the ground as this puts less strain on the inside leg and prevents the horse from falling in on his circles. However, there is no hard and fast rule and as long as you are consistent and change the diagonal when you change the rein you will not be penalised.

Changing diagonals in rising trot may be done by keeping the seat in the saddle for an extra step or, and this is a smoother movement for the young horse with an underdeveloped back, by keeping the seat out of the saddle for one more step.

The aids for the trot are firstly the voice with which the young horse should be familiar. At the same time the legs are applied firmly just behind the girth and the hands yield to accept the increased impulsion.

The sitting trot puts too much weight onto the immature spine and the young horse will evade by dropping his back, raising his head, and trailing his hindquarters. All trotting in the young horse should be with the rider rising on alternate diagonals. This takes weight off the horse's back, makes it easier to establish a regular rhythm, and is generally more comfortable for both horse and rider. Sitting trot should be introduced when the horse is well schooled, and only then for short periods after having warmed up in rising trot.

During the walk, impulsion was maintained by the application of alternate leg pressure, the rider's right leg being applied as the horse's left front leg came to the ground and vice versa. In the rising trot the legs should be applied simultaneously when the seat comes into the saddle. If trotting on the right diagonal the legs should be applied together when the right fore comes to the ground and the seat is in the saddle. If the rider tries to apply his legs when the seat is out of the saddle he will unbalance himself and upset the rhythm of the pace.

The hands act in the trot as they did in the walk. By allowing and yielding with the hands, the impulsion created by the legs is controlled and balance and rhythm are established. Care must be taken to maintain a steady contact, and not to let the hands rise and fall in time with the rider's seat rising from the saddle. In the turns the inside hand indicates and the outside hand allows, whilst the legs are applied as before, but only for that fraction of a second whilst the rider's seat is in the saddle.

During initial schooling lessons the periods of trotting should be kept short and turns and circles restricted until the horse is used to the weight and rising action of the rider. Once the horse is moving forward well and the trot is developing from an "ordinary" trot to a medium trot, then the aids can be increased gently to raise the head and to bring the hocks under the horse, to slow the pace and produce more cadence, which is the development of the working trot.

During trotting lessons make use of both the indoor school and the open fields and again use hills and bending poles to vary the programme, to increase the general fitness and suppling process and to avoid boredom.

Most exercises involving turns, circles, loops, serpentines and changes of rein will help to supple the horse and develop a lateral bend in his spine. It is important to remember that this lateral bend of the spine is not natural and that too much bending, like too much lungeing, in the skeletally immature horse, will cause irreparable damage.

Longitudinal bending of the vertebral column is a more natural movement in the horse. To graze and then raise the head to look for danger bends the spine longitudinally. To rear, buck and kick also bend the spine longitudinally. In order to supple the horse longitudinally during training he must do plenty of hill work and plenty of transitions. In the early stages of schooling all transitions should be carried out in straight lines, and they should be smooth, progressive and accurate. Frequent changes of pace by progressive upward and downward transitions will develop the muscles of the neck and hindquarters, and contribute to longitudinal suppleness.

THE CANTER

The canter is a pace of three time with a moment of suspension between each stride. If cantering to the right the footfall sequence is as follows; left back, left diagonal (left front and right back together) right front followed by suspension when all four feet are off the ground. The FEI describes four canters; the collected canter, an advanced movement with a short stride, active hindquarters, and a raised neck; the extended canter, where the horse covers as much ground as possible whilst remaining in three time; and the two canters which concern the young horse, the working and medium canters.

As with the working and medium trot, the working and medium canter have developed from the "ordinary" canter. The working canter is closer to the collected canter and the medium canter is closer to the extended canter. Again as with the trot, during early canter schooling the young horse performs neither working canter nor medium canter, but starts off in "ordinary" canter. "Ordinary" canter is a pace between extended and collected canter where the horse should move freely with long strides, the rider maintaining the rhythm and a certain amount of cadence with his seat and legs. Early canter work should only be on a loose rein to allow the back muscles to develop without restraint, and any tendency to buck should be controlled by the rider moving forward off the horse's back, and not to sit back as he would do in a mature horse with well developed back muscles.

When a horse canters on a left hand circle it should "lead" with the left front leg and vice versa on a right hand circle. This is known as a true canter. The "false" or "counter" canter is when the horse is cantering on a left hand circle with the right front leg leading, or vice versa. In advanced schooling the counter canter is used as a suppling exercise; it should not be encouraged in the young horse. The disunited canter is when the horse leads with a diagonal rather than an individual front leg, it is unnatural and uncomfortable for both horse and rider and should be corrected at once.

The change from right to left canter and vice versa is carried out on the young horse by bringing him back to a trot and then asking for a transition to canter on the opposite lead. This movement is known as a "simple change of leg". Correctly the simple change of leg should involve a few steps in walk but in the young horse at this stage of schooling a few steps of trot are less confusing. The flying change which occurs in the moment of suspension between one canter stride and the next has no place in basic training.

The transitions from halt to walk, and from walk to trot involved increased impulsion by the legs, and containment and control of this impulsion with the hands. To move from trot into canter is similar but one should from the very start insist that the horse leads with the leg that is on the same side as you are turning or are about to turn. The asymmetry of the pace means that the aids are also applied asymmetrically. To go forward into left canter, the left rein indicates the direction of bend, and the right or outside rein controls the bend and prevents the horse from running on at a faster trot. The left leg maintains impulsion at the girth and the right leg is applied further back to control the hindquarters. In canter to the left it is the right back leg which is the first limb to strike off, and so the rider's outside leg is also in this case applied to provide impulsion. The same aids are applied to maintain impulsion with every stride during canter at the moment when the leading front leg comes to the ground.

Breaking into the correct canter lead is complex and involves precise application of the aids. It is difficult for the young horse to understand the aids and he must be given every assistance by transference of the rider's weight, and use of the corners of the school.

If the horse breaks with the wrong lead then bring him back into trot and start again. Only when he has learned to break correctly, freely, regularly, and without resistance can the rider stop shifting his weight to the inside, and start to increase the bend to the inside. If the latter bend is applied too strongly in the early stages of training it restricts the free forward movement of the inside shoulder and prevents the horse striking off with the correct leg.

The downward transition from canter to trot is much easier because the rider no longer has to worry about the correct lead. To prevent any resistance and head turning, use only the inside rein to control the impulsion. Do not actively pull on the rein, just stop the hand moving with the nodding of the horse's head, and at the same time use the word of command "Whoa" that he has responded to since he was a few days old.

There is always a strong temptation for the young horse to want to take off and have a good gallop when he first canters under saddle in the open.

To prevent this the rider must use plenty of common sense and not canter towards the stable, or past other loose horses in a neighbouring field. He should only canter towards the corner of a field and not towards open ground, and he should not habitually always start to canter at the same spot. If the horse does start to increase his speed he should be steered into a large circle, the circumference of which is gradually reduced until the horse is settled, then very gently increase the circumference of the circles again and continue with the exercise.

Once the horse is cantering freely forward and leading with the correct leg we can then start to split our "ordinary" canter into the working canter and the medium canter required by FEI definition. The working canter is a pace between collected and medium canter in which the horse, not yet trained and ready for collected movements shows himself properly balanced and remaining "on the bit" goes forward with even, light and cadenced strides and "good hock action". "Good hock action" implies impulsion originated from the activity of the hindquarters.

Medium canter is a pace between working canter and extended canter. The horse goes forward with free, balanced and moderately extended strides and an obvious impulsion from the hindquarters. The rider allows the horse, remaining "on the bit", to carry his head a little more in front of the vertical than at collected and working canter, and allows him at the same time to lower his head and neck slightly. The strides should be long and even, and the whole movement balanced and unconstrained.

THE REIN BACK

From an early age we have been teaching our youngster to move sideways and backwards by use of voice saying "Over" and "Back" respectively, in conjunction with gentle pushing of his flanks or chest. During the leading phase and when being schooled with long reins, the exercise of a few steps backwards should have been continued, so that when we are at last mounted it should not be difficult to make him move back in a controlled and relaxed manner without fight or resistance.

To rein back means to walk straight backwards, however the FEI defines the rein back as an equilateral backwards movement in which the feet are raised and set down almost simultaneously by diagonal pairs, each front foot being raised and set down an instant before the diagonal back foot so that, on hard ground, as a rule, four separate beats are clearly audible. The feet should be well raised and the hind feet remain well in line.

The rein back appears to be contrary to all we have been trying to teach during basic training. In every movement and pace we have been encouraging free forward movement, and now we are asking the horse to go backwards.

The rein back is not a natural movement for the horse, although all horses can move backwards if they have to. In the wild they would prefer to swing round or rear up to escape, rather than to step back out of a tight corner. If they do move back of their own accord it is more commonly a true two time diagonal movement like a trot without a period of suspension.

If the horse has been properly educated and taught to step back in hand it will be easy to teach him under saddle. Lead him out with saddle and bridle into the school, halt him, stand in front of him and ask him to move back, pushing his nose and chest gently if necessary. When he responds reward him well. Repeat two or three times and then with the help of an assistant repeat the exercise when mounted. Do not use either leg or hand aids from the saddle, just the voice and a little physical encouragement from the assistant on the ground if necessary. When he responds reward him and move on to another exercise. Repeat the above exercise without mounted aids for several days and then gradually, and still using an assistant, introduce the correct aids for the rein back.

The aids for the rein back are applied with the horse halted and standing square. Apart from the voice, which at this stage is the most important aid, we apply light even symmetrical pressure with the legs as if to move forward from the halt, but instead of an allowing hand, the hands do not yield and remain fixed. There is no pulling back on the mouth or tugging on the reins, just a restraining hand which is relaxed as soon as the horse steps back. After taking a step or two back the horse must be rewarded and asked to walk forward.

After a few days the exercise should be repeated without the assistant and then over the next few weeks the rein back should be carried out in different places and at different times during the daily lessons. Eventually the horse will be able to walk back straight and without resistance from a halt, in the middle of a field with no boards to guide him. To teach him to pick his legs up higher, halt him facing down hill and then ask him to back up the slope for one or two steps. Always reward by making much and asking him to move forward again.

JUMPING

During basic training if the horse has been led in hand as described earlier in this chapter, he will have become used to simple obstacles such as natural and coloured poles, ditches and water. One of the real values of a comprehensive and well planned leading phase, is that when it comes to jumping under saddle the horse is already fully experienced in going over all the obstacles he is likely to meet during the next few months. Also if the horse has been loose schooled either in a loose school or down a fixed jumping lane, he will again have had experience of going over both natural and coloured obstacles. The muscles used in jumping will also have been developed by loose school jumping but the horse will now have to rebalance himself and develop different muscles to carry both himself and his rider safely over the same obstacles.

I strongly believe that the best introduction to jumping when mounted is exactly the same way as his introduction to jumping when being led. The horse should continue, during the time he is being backed and schooled at the walk, trot and canter, to spend some time each week being walked in hand on the long rein or loose schooled. During these periods of relative relaxation he should pop over a few of his old familiar obstacles to keep himself both fit for and in the habit of jumping. His first jump when mounted should be to walk over a simple, natural log across the path home at the end of the daily lesson. Invariably he will not even notice the obstacle especially if he has been led over it many times before.

The horse will require no aids other than the normal asymmetrical pressure of the legs to maintain his free forward movement. The reins should be allowed to go loose so that he can lower his head to get a proper look at the obstacle but the rider must be ready for the massive leap which some young horses feel is necessary in order to lift horse plus rider over a one foot high log!

During the first week introduce several different logs and poles at various places and walk and trot the horse over them initially when going home but eventually when going away from the stable. Also if you are fortunate enough to have schooled him on foot over small ditches and through water, then carry out the same exercise when mounted. The use of a schoolmaster may be a useful adjunct if your young horse shows any tendency to be unwilling, for the herd instinct will invariably make him follow the older horse.

Once the horse is going freely forward over the obstacles which have been placed across the paths on his rides it is time to introduce him to the more formal poles on the ground in the school which are used to teach him to be accurate before he carries his rider down the jumping lane.

The initial lesson should consist of three or four cavaletti placed around the school on their lowest level. The horse should find these no problem if the build up has been thorough. Once he has walked and trotted over all of them they should be arranged in a row 1 m to 1.06 m (3 ft 3 in to 3 ft 6 in) apart for walking and 1.30 m (4 ft 3 in) apart for trotting. Not all horses have the same stride and at this stage of their training they cannot be expected to adjust their stride, so if the distance between the poles is obviously wrong for your horse then adjust the poles accordingly.

When the horse is trotting quietly over the cavaletti on the ground at their lowest level, the last one should be turned up to its highest level. If the horse remains relaxed then the second, third and eventually first one may also be raised. Even if the horse appears totally confident, this cavaletti work should be spread over several weeks and should be repeated frequently. To avoid boredom the poles should not always be put in the same place. It is important for the horse to spend time gaining confidence, building up his muscles and learning to adjust his stride so that he takes off at the correct point. The stride adjustment at this stage of training should always be to lengthen rather than shorten in order to meet the pole correctly. Any tendency to shorten his stride, or worse still to put in an extra stride, should be counteracted by more impulsion (but not more speed) when approaching the pole.

The next stage in cavaletti work is the introduction of both higher and wider final obstacles so that eventually after two or three months the horse will be able to negotiate calmly and without any increase in speed a combination of several poles. The combination should consist of three or four introductory cavaletti 1.3 m (4 ft 3 in) apart, followed by two higher fences 3.5 m (11 ft 6 in) apart and a final fence 6.4 m (21 ft) further away. The spacing of the final fence allows one stride of canter before the jump which should by now be 1.22 m (4 ft) wide but only 0.75 m to 0.90 m (2 ft 6 in to 3 ft) high.

The horse is now ready to take his rider down the jumping lane and over natural obstacles in the fields. He will have built up his muscles and confidence naturally and steadily over a three or four month period without resistance, refusal or boredom.

At no stage should there be any rush, and any overkeenness must be curbed by arranging the fences in a curve which reduces the horse's natural impulsion. Rushing at fences can be avoided by circling in front of the fence until such time as the horse is completely relaxed and only then putting him at the obstacle.

The landing after the fence is just as important as the take off. After landing the horse must be taught to continue in a straight line and he must not rush away from his fences but should come calmly back to a trot. Young horses may require ten or twelve strides of canter to become balanced again after a fence, although as their back muscles and hindquarters develop they will require fewer strides to recover. Do not try to slow him down during this rebalancing phase or you will teach him to pull and once he learns to run on after a fence it will be difficult to retrain him. Always vary the jumping lessons so that he does not become bored. Teach him to jump in company and on his own (although one should never jump a young horse out of sight of an assistant in case an accident should happen). Intersperse formal school work with lessons in the country over natural fences.

Most fences should be jumped from the trot rather than the canter at this stage of training although the young horse will often canter on for a few strides after landing. Trotting over fences teaches the horse to judge his take off properly and to use his hocks. At the canter his speed will provide sufficient impulsion to carry him over these low fences, but at the trot he must obtain his jumping impulsion from his hindquarters. It is always better to increase the spread of the fence rather than the height; in jumping the increased spread the horse will automatically jump higher. Continue to reward by making much after success.

I have deliberately omitted any reference to seat, hands and aids in this section on jumping for as long as the horse is moving freely forward over his fences and the rider is going with him without any interference from either hands or seat we are achieving our aim. It is only later in the more advanced training for showjumpers, cross country, or steeplechasing that a specialised seat is required, and special aids become necessary; both seat and aids are very different for each discipline.

CONCLUSIONS

It is appreciated that not every trainer will be fortunate enough to be presented with a young foal and asked to train it as described in this chapter. Many trainers will have to tackle unhandled yearlings or wild two-year-olds, and the principles of basic training will have to be modified. I cannot over-emphasise however the importance of gaining confidence by leading in hand at whatever age. I am convinced that every hour spent leading the young horse in an intelligent and educational manner will save many hours of frustration in later schooling under saddle.

In reminding the reader that the horse is not designed anatomically to trot in small circles, and that lungeing, especially the skeletally immature youngster, can damage the joints of the limbs and the spine, I would emphasise that lungeing is as much an art as is riding.

Not only are lungeing and riding arts in themselves, but the whole subject of training is an art. Individual descriptions of aids, movements and methods is scientific, the art lies in blending them all together in the harmony of the aids and the fluency of the movements.

Many aspects of equitation are controversial and no attempt has been made in this chapter to unravel the complexities that better men than I have argued about over the centuries. Does one fix the lunge rein on the front of the cavesson or the back, or does one not lunge the youngster at all? Only the individual trainer can answer the questions for himself, perhaps basing his conclusion on advice from others. Not many people will ride a young horse with a schooling whip in each hand, yet how else can a symmetrical reinforcement of the leg aids be achieved?

No effort has been made to describe lateral work or collection. Both of these subjects are outside the scope of a chapter which concerns itself with the basic education of the young horse.

The complexities of equitation are often far more difficult to describe in writing than to put into practice. What may have been formidable to comprehend in this chapter will I hope be easier in the stable or the manège.

Many horses will not have the temperament or the conformation to progress beyond the stage that has been reached at the end of basic training, but he should be a pleasant companion who is well mannered at all times, who can walk, trot, canter and jump, and who is fit enough to work willingly for up to two hours. For those horses that are able to progress one has a solid foundation on which the advanced skills of equitation can be built.

REFERENCES

Chamberlin, H. D. (1934). *Riding and Schooling Horses*. Derrydale Press.
Chamberlin, H. D. (1952). *Training Jumpers, Hunters and Hacks*. D. Van Nostrand.
d'Endrödy, A. L. (1967). *Give Your Horse a Chance*. Allen.
Jackson, G. N. (1967). *Effective Horsemanship*. Compton Press.
Littaur, V. S. (1963). *Common Sense Horsemanship*. D. Van Nostrand.
Müseler, W. (1937). *Riding Logic*. Methuen.
Paalman, A. (1978). *Training Showjumpers*. Allen.
Phillips, G. M. (1966). *You . . . and the Horse*. Allen.
Rees, L. (1984). *The Horse's Mind*. Stanley Paul.
Rodzianko, P. (1950). *Modern Horsemanship*. Selley.
Rose, M. (1977). *Training Your Own Horse*. Harrap.
Stanier, S. (1972). *The Art of Lungeing*. Allen.
Wheatley, G. (1968). *Schooling a Young Horse*. Cassell.
Wynmalen, H. (1938). *Equitation*. Allen.
Wynmalen, H. (1962). *Dressage*. Allen.
Xenophon. *The Art of Horsemanship*. Allen.

9 Exercise and Training

D. H. SNOW

INTRODUCTION

The horse has evolved as an athlete par excellence and this, together with its ease of domestication, has led to its important role in the development of mankind over the last 5000 years. Initially, man required the major role of the horse to be as a means of military transport and for various agricultural activities. The advent of the internal combustion engine, however, limited the power-based function of horses and led to a worldwide decline in the horse population during the first half of this century. However, recently there has been a reverse in this decline due to a marked increase in the number of horses for both leisure and entertainment. For example, the horse population in the USA in 1985 was estimated to be 8.5 million.

This increased popularity has resulted in vast rises in prices which purchasers are prepared to pay for elite animals, particularly in the racing and showing worlds. The early necessity for horses to be able to adapt to different climatic conditions, together with more recent requirements for specific tasks, has led to the development of numerous breeds with special characteristics. For example, for very heavy work the draught breeds evolved, whereas for speed, the warm-blooded breeds have developed.

Until recently the thoroughbred was considered to be the superior breed for speed. However, recent developments have seen the arrival of the American Racing Quarterhorse, the equine equivalent of the greyhound. It should be realised that this breed has largely developed from the thoroughbred rather than from the stock quarterhorse. In many cases, racing animals are up to 15:16 parts thoroughbred.

Although breeds have been selected to satisfy the varying requirements of man, this is not enough. In order to carry out these requirements to best advantage, the full genetic potential has to be obtained. This can only be

Horse Management 2nd edition
ISBN: 0–12–347218–0 case

done by appreciating the importance of the interaction of numerous factors such as correct nutrition, the proficiency of the rider and, of course, proper training to bring the animal to peak fitness. The attainment of fitness has generally been subjective as there are few criteria which allow its complete determination. In the past when horses were often an integral part of the daily life of man, continuous work per se often resulted in a high degree of fitness. There are many records of horses being ridden for long distances at relatively high speeds and covering 80–100 miles per day for several consecutive days. However, nowadays, the horse is principally used in leisure sports and economic considerations demand a considerable reduction in the time available to achieve optimal fitness. To try and compensate for this reduction in time, new training methods have to be developed to improve on the largely empirical historically-based fitness development programmes.

The development of new training methods depends on a knowledge of the metabolic demands of specific tasks and how these can be best harnessed to enable the body to adapt (in fitness) to meet them. To obtain this basic information the scientist has to measure the normal physiological and biochemical changes that occur during exercise and training. Selected training programmes can be studied to assess the best methods to meet the animal's requirements.

Although research into the demands of exercise and the selection and development of human athletes has progressed during the 20th century and there has been an increased momentum in the last 25 years, until recently similar work in the equine has been negligible. Perhaps this can be paralleled by the vast overall improvements in world record times in man, but the considerably smaller improvements recorded in both standardbred and thoroughbred racing. Since 1900, running speed in the Epsom Derby has improved only 2% although the German trotting Derby speed has improved by 16%. It is interesting to note that the greater improvement has taken place in the standardbred since it is in this racing discipline that greater research effort has been directed. More scientifically based training programmes have been developed in the standardbred, whilst thoroughbred training follows the more traditional methods. In addition to improved training methods, better performance times in the standardbred can also be attributed to lighter racing sulkies and better track designs.

Since the 1970s more and more research has been undertaken into the physiological and biochemical changes occurring in the horse at exercise. This work has been undertaken in a number of centres including those at Uppsala, Sweden; Glasgow, Scotland; Washington and Kansas State Universities; the University of Sydney and the Equine Research Institute in Japan. Recently an Exercise Physiology Unit has been established at the Animal Health Trust, Newmarket, where unique facilities are being developed to examine the integrated responses of the horse to exercise. That such facilities are under development at Newmarket is only fitting, as it has been considered the home of modern racing since the reign of King Charles I. It is hoped this research effort will be continued and if possible expanded, and the knowledge gained will not only lead to possible improvements in racing times, but more importantly will benefit the welfare of the horse. More scientifically based training methods may reduce the need to resort to various pharmacological aids in the maintenance of performance in certain equestrian events. This is especially important as international legislation on drug abuse becomes stricter and the drug testing laboratories become increasingly proficient at detecting the numerous compounds that have been used to improve or impair performance.

Obviously racing and other equestrian disciplines are dependent on efficient muscular contractions reflected in movement, whether it is the precise disciplined movement required in dressage, the short explosive propulsion seen in jumping, the continual maximal effort in racing, or the sustained effort in eventing or long distance rides. Effective and maintained movement is dependent upon integration between many of the body's systems. These include the muscular and nervous (central and peripheral) systems, which are essential for effective movement, and the cardiovascular and respiratory systems which are of vital importance in maintaining the movement by assisting in the fuelling of working muscle, the disposal of waste products and the dissipation of heat generated during intense muscular activity. Furthermore, the optimum utilisation of these systems is dependent on the mental attitude of the animal to its given task and its relationship with driver or rider.

Thus, proper training can and will lead to the attainment of optimal performance. Nevertheless, it is the genetic potential which largely determines whether a horse will be an outstanding or an elite athlete. Just as in human sports, the trainer plays a vital role in helping the athlete reach his optimum output and it is also the trainer who can often either make or break his pupil.

It is not the purpose of this chapter to provide easy answers on how best to train a horse, but rather to review the available scientific data on the physiological and biochemical demands of various types of exercise and leave it to the reader to use these as a guide to devise his own training programme. It should always be remembered that, as in man, whilst certain criteria and guidelines can be laid down, each animal is an individual for which a training programme should be tailor-made.

Within a chapter the diverse subject of exercise physiology can only be very superficially covered, and for further general information there are presently three excellent books available on exercise physiology in man (Lamb, 1984; Astrand and Rodahl, 1977; Fox, 1984). Also, since the first edition, two International Symposia on Equine Exercise Physiology have been held, which have resulted in the production of Proceedings. In addition a number of books have been produced on equine exercise physiology. Suggested references for further reading are provided at the end of this chapter.

ENERGY SOURCE FOR WORK

Production of adenosine triphosphate

In man it has been calculated that there is well over a 100-fold increase in energy needs of muscle from rest to exercise. In the horse, the total energy requirements are even greater as reflected by VO_{2max} (maximal oxygen consumption) being nearly twice that in man. The energy used by the cells in the body to perform work is chemical and the immediate energy source for muscle contraction is a "high energy phosphate molecule", adenosine triphospate (ATP).

When the last phosphate group is split off this molecule, by the catalytic action of an enzyme called ATPase, a large amount of chemical potential energy is released.

$$ATP \xrightarrow{ATPase} ADP + P, + energy$$

The ATPase enzymes responsible for the breakdown of ATP are located throughout cells, wherever processes requiring energy are underway, e.g. for the active transport of ions. The specific ATPase that splits ATP to provide energy for muscle contraction is located on one of the major contractile proteins, myosin, and therefore allows the production of energy directly to the molecules that generate the contractile force.

Although ATP is the immediate source of chemical energy, it is only stored in small amounts within cells and therefore it has to be continually synthesised from its precursor adenosine diphosphate (ADP) and P in increasing amounts as the energy requirements of the cell are increased. This is seen to the highest degree in maximally contracting muscle.

The energy sources for rephosphorylation can be derived from several reactions which can be considered as (i) those requiring oxygen (aerobic processes), and (ii) those that can continue in the absence of oxygen (anaerobic processes). The aerobic generation of ATP occurs within the cell mitochondria. The capacity for this process thus largely depends on the size and number of mitochondria within the cell and the uptake of oxygen.

The basic fuels that supply the substrates for oxidation and generation of ATP within the mitochondria are (i) glucose, which can enter the cell from the blood or from the intracellular stores of glycogen and (ii) free fatty acids which can enter from the blood or to a limited extent from triglyceride depots within some cells including the myofibres. Both these substrates are metabolised within the cytoplasm to smaller sub-units which can then enter the tricarboxylic acid (TCA, Krebs' citric acid) cycle within the mitochondria (Fig. 9.1). The production of CO_2 and hydrogen ions in the TCA also results in the formation of the reduced coenzymes $NADH_2$ and $FADH_2$. In order to maintain this cycle it is necessary for these reduced coenzymes to be oxidised. This occurs in the respiratory or electron-transport chain in the presence of O_2. In addition, $NADH_2$ formed in glycolysis also enters the mitochondrion for reoxidation (Fig. 9.1).

When oxygen is insufficient, as in the early stages of exercise, before full readjustments in blood flow have been made, or in very high intensity exercise, generation of ATP can occur by anaerobic means, for which there are essentially two pathways. High energy phosphate is stored bound to creatine, as creatine phosphate (CP) and in the initial stages of exercise, this is broken down by the enzyme creatine kinase to yield high energy P, which rephosphorylates ADP. This anaerobic process has been shown to provide sufficient energy for the first ten seconds of exercise in man.

$$ADP + CP \rightleftharpoons Creatine + ATP$$

Once this source has been depleted, regeneration of ATP is dependent on aerobic generation or by a second anaerobic pathway in which glucose or glycogen is broken down by glycolysis or glycogenolysis to yield two molecules of pyruvate, which would then normally enter the TCA cycle for oxidation. However, when energy demands are severe, insufficient ATP is generated via the TCA cycle, and the pyruvate is instead further degraded to lactate and the formation of $NADH_2$, itself important for continued glycolysis. Thus, lactate production allows glycolysis to continue when anaerobic conditions in the cell retard $NADH_2$ oxidation in the mitochondria. As is shown below this method of utilisation of glucose is very much less efficient in the immediate generation of energy. However, since it appears that energy availability is the prime factor in obtaining maximum muscle force, any extra supply is likely to be beneficial.

Although it might be thought that in the absence of oxygen, lactate production could proceed indefinitely, this is not the case *in vivo*, as the process is self-limiting. Associated with the production of lactate are hydrogen ions (H^+) which, due to a slow removal from the myofibres, accumulate within the cell, and result in a considerable lowering of the intracellular pH. Once the pH falls below certain levels, inhibition of metabolic pathways occurs, and it has been suggested that this is responsible for the onset of fatigue seen during high intensity exercise. In other words, the cell is being "poisoned" by its own metabolic products. To some extent these detrimental

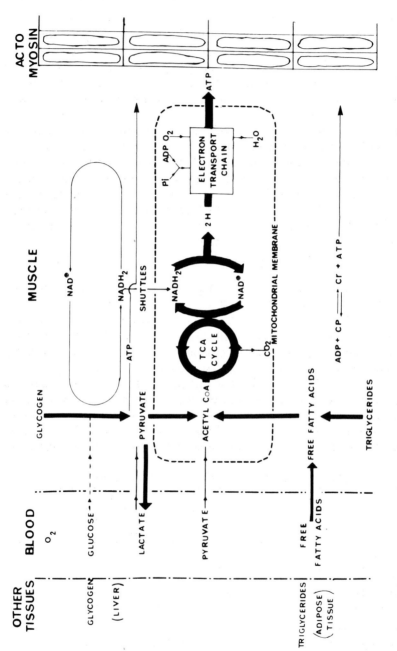

Figure 9.1. Substrates and metabolic pathways important for the generation of ATP in muscle fibres (Snow and Guy, 1977).

effects of increased acidity are counteracted by two mechanisms: the intracellular buffering capacity, and the diffusion of lactate out of the cells. Once lactate is removed from the cells producing it, it enters other less active muscle cells where it can be converted back to pyruvate and then either enter the TCA cycle for oxidation, or in inactive muscle or liver undergo gluconeogenesis whereby it is reconverted via pyruvate to glucose and glycogen via the Cori cycle.

The number of ATP molecules produced from the metabolism of different substrates is shown in Table 9.1. On the basis of carbon atoms about 30% more ATP can be produced from FFAs than from carbohydrate. However, when energy production is considered in terms of oxygen requirement, glucose oxidation is about 12% more efficient. This may account for the metabolism of carbohydrate rather than FFAs in high intensity exercise. In the oxidation of either FFA or glucose about 40% of the energy within the molecule is generated as ATP, the rest being lost as heat.

Regulation of substrate utilisation

The utilisation of different substrates to generate high energy phosphate depends on the intensity of the exercise and the fitness of the animal. During the rest state, muscle utilises FFAs, but when work commences, the contribution of glucose becomes greater as the intensity increases until at higher intensities, glucose (or glycogen) becomes the major substrate. It is possible, however, that during very heavy workloads in the horse, FFAs may contribute up to 15% of energy expenditure. The preference for FFAs at lower workloads is obvious when considered in terms of total energy supplies in the body, as energy stores of FFAs are of the order of 40 times greater than those of glucose. The extent of FFA utilisation at any workload is not fixed, but varies influenced by the concentration

Table 9.1. Energy metabolism in skeletal muscle

	ATP yield			
	per atom of carbon	per mole of substrate	per mole of O_2	RQ
Anaerobic				
creatine phosphate	—	1	—	—
Glycolysis				
from glycogen	0.5	3	—	—
from glucose	0.33	2	—	—
Aerobic				
complete oxidation of				
glycogen (glucosyl unit)		37–39	6.2–6.5	1.0
glucose		36–38	6.0–6.3	1.0
fatty acid		138	5.6	0.7

of circulating FFAs as uptake is related to concentration and also by the fact that training will result in adaptations within muscle which will increase its utilisation.

As the workload rises so the contribution of glycolysis increases resulting in the respiratory quotient (RQ) approaching unity. The pyruvate formed from glycolysis can enter the Krebs cycle for complete oxidation or can be converted to lactate anaerobically. Anaerobosis increases as the workload becomes greater, as a result both of the initial utilisation of creatine phosphate and the production of lactate. At workloads greater than that providing $VO_{2\,max}$ the RQ becomes greater than 1.0 due to the large quantities of lactate formed. The relative contribution of aerobic and anaerobic metabolism during different running times for elite human athletes is shown in Table 9.2. This relationship has not been studied closely in horses. However, following thoroughbred races over both the flat and jumps, blood lactate concentrations in excess of 30 mmol/1 have been recorded, whilst in standardbreds over 1900–2500 m (mean speed 12 m/sec) concentrations were only slightly lower indicating that such activity is highly anaerobic.

Generally, anaerobic metabolism is determined by an increase in blood lactate concentrations over resting levels. The original concept was that the anaerobic formation occurred due to a lack of oxygen supply to the muscle. However, recent studies have led to a modification of this concept. It is now considered that lactate formation occurs in three phases, and it is only in the last phase that large scale production occurs which causes fatigue. In phase I, exercise at relatively low intensities is almost purely aerobic, but although it would appear that sufficient oxygen is available, some production of lactate occurs. However, little or no change in lactate levels occurs in either muscle or blood as its metabolism keeps pace with its formation. In phase II, at moderate intensities of exercise, there is an initial rise of lactate production to blood levels of about 2 mmol/1, and this may subsequently increase to

Table 9.2. The contribution of aerobic and anaerobic metabolism for running at maximal speeds for different durations

Time	Aerobic (%)	Anaerobic (%)
10 s	15	85
1 min	65–70	30–35
2 min	50	50
4 min	70	30
10 min	80–85	10–15
30 min	95	5
60 min	98	2
120 min	99	1

From Strauss (1979)

about 4 mmol/1 depending on the intensity of the workload. During this stage no accumulation of lactate occurs within muscle fibres as diffusion into the circulation equals the rate of formation. No alteration in blood pH occurs, as blood buffering compensates, and there is an increase in ventilation. At this workload it is thought that lactate production occurs, not due to inadequate oxygenation, but as a result of increased energy demands. The pyruvate formed from increased glycolysis is unable to enter the Krebs cycle due to preferential entrance by FFA sub-units.

Onset of phase III occurs at the greatest workloads, as anaerobosis increases to meet the increased metabolic demands. Lactate production increases markedly and leads to a rapid rise in blood lactate concentration. Possibly more importantly, lactate accumulates within the muscle fibres as the diffusion from muscle cannot keep up with production. Accumulation at these higher workloads involving maximal contractions may not result from inadequate blood oxygenation, but rather from (i) maximal contractions resulting in occlusion of vessels which impede blood flow to working muscle fibres, and (ii) the recruitment of additional muscle fibres of lower oxidative capacity.

The transition between phases I and II has been referred to as the aerobic threshold, whilst that between phases II and III, as the anaerobic threshold. Today, many workers believe that such threshold terms are misleading since some lactate formation occurs during all intensities of exercise.

The role of amino acids produced by protein catabolism during exercise as an energy substrate is controversial. Recently, detailed studies in man have shown that during moderate long-term exercise there is a decrease in whole-body protein synthesis and a rise in whole-body protein breakdown. The source of the degraded protein is not known but is not thought to come from myofibrillar protein degradation. It was calculated that during the exercise studied the amino acids may have contributed 4–8% of energy expenditure. Their major role may be as gluconeogenic precursors, and be important in the maintenance of glucose blood concentrations during the latter stages of exercise.

SKELETAL MUSCLE

All alterations during exercise are related to movement and therefore are either directly or indirectly related to the requirements of the involved skeletal musculature. Therefore, in the determination of speed of movement characteristics of skeletal muscle are important in influencing both the frequency of stride and stride length (Fig. 9.2). Furthermore, an understanding of the demands of various types of exercise requires a knowledge of the specific changes that occur within working muscle and the modifications

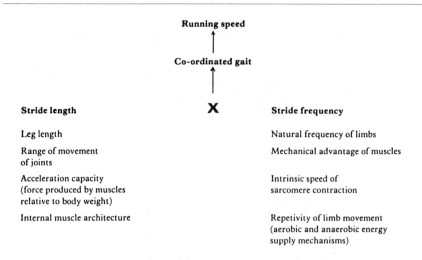

Figure 9.2. How muscle influences the speed of running in horses (from Gunn, 1983).

that result from training. During any movement coordination of contraction and relaxation of large groups of muscle is necessary. The way in which the different muscles are involved in a particular motion can be studied by measuring activity within them using electromyography (EMG). Unfortunately, few such studies have been carried out in the horse, so much information on the utilisation of different muscles at various intensities of exercise is lacking.

Over 90% of muscle consists of the muscle fibres (cells), with the rest comprising nerves, blood vessels and the connective tissue which separates individual fibres (endomysium) and bundles (perimysium). The connective tissue merges with the muscle origin and its tendon of insertion (Fig. 9.3).

The muscle cell is a multinucleated elongated cell, which is generally referred to as a fibre. Its diameter varies both within and between muscles, the size being affected by growth and activity. It is generally accepted that increased muscle mass is brought about by hypertrophy (i.e. increase in size of individual fibres) rather than hyperplasia (increased number).

Myofibrils

Within the muscle fibre the contractile element is the myofibril and within each fibre there can be hundreds of these myofibrils which are aligned parallel to each other, each consisting of repeated units called sarcomeres. Each individual sarcomere is able to develop tension and when all the sarcomeres contract together, a measureable pulling force is generated by the fibre.

MUSCLE

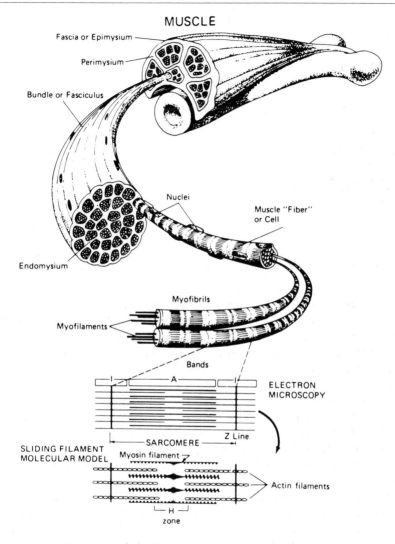

Figure 9.3. Skeletal muscle structure (from Lamb, 1984).

Increase in muscle size is brought about by the increased number of myofibrils, whilst increase in length during growth occurs due to additional sarcomeres developing at the ends of fibres. Each sarcomere consists of proteins the arrangement of which gives rise to the characteristic cross-striations of skeletal muscle. The two major muscle proteins are actin and myosin. It is the actin which slides between the myosin filaments by the interaction with its crossbridges that produces contractions.

Cytoplasm

Although the myofibrils constitute 50–75% of the volume of the myofibres the other organelles are vital for the myofibrils to carry out their function effectively. As a continuation of the plasma membrane (sarcolemma) muscle fibres have a unique structure, the T-tubules, which conduct an action potential into the interior of the fibre. These T-tubules lie adjacent to parts of the sarcoplasmic reticulum, which contain large concentrations of calcium ions. The transmission of the action potential down the T-tubules results in Ca^{2+} being released from the sarcoplasmic reticulum (SR) and the initiation of myofibrillar contraction.

The mitochondria are the "power-houses" of the cells and maintain the large requirements of ATP for myofibrillar contraction, the re-uptake of calcium by SR, changes in membrane permeability and other cellular processes. The mitochondria are located under the sarcolemma and between the fibrils and vary in size and number in different fibres. The source of fuel for these "power-houses" are derived from either the breakdown of glucose or free fatty acids obtained from either the circulation or from stores within the myofibres. Numerous glycogen granules can be seen under the sarcolemma and between the myofibrils. In association with these glycogen granules are the enzymes of the glycolytic pathway. Free fatty acids are stored, to a much lesser extent than glycogen, as triglycerides in lipid droplets that are found in close proximity to the mitochondria (Fig. 9.4).

Neuronal component

Muscle tissue is supplied by both sensory and motor nerves which course through the interfascicular space. Each muscle fibre (or extrafusal fibre) is supplied by one α-motorneurone, which originates in the spinal cord. However, any individual motorneurone is able to supply more than one fibre, the number varying from ten to several hundred, so that an impulse passing down one motorneurone will result in contraction of all fibres innervated by it. The motorneurone and all the fibres it innervates are classified as a "motor unit". The force generated by a muscle depends on the number and type of motor units recruited. For the formation of low tensions, small numbers of motor units are activated, whilst increased requirements can be achieved by recruiting large numbers. Muscle tensions can also be controlled by altering the frequency of stimuli since by increasing the frequency of action potentials elevated tensions within the whole muscle results.

The pattern of α-motorneurone activity is controlled by the complex interaction of both excitatory and inhibitory stimuli that come from the brain, as well as from peripheral sensors in the muscle, joint and tendons. These peripheral sensors are responsible for modifying muscular contractions so that rapid corrections can be made during movement.

Figure 9.4. Ultrastructure of a high oxidative muscle fibre with numerous mitochondria (M), lipid droplets (L) and glycogen granules (G). (Photograph courtesy of I. Montgomery.)

Vascular supply

Muscle receives a continuous supply of blood, the flow being increased by up to 60-fold during strenuous exercise. Arteries enter through the epimysium, and branches form which enter the fasicles. From these arterioles extends a network of capillaries surrounding the individual muscle cells. These then join veins which then leave the muscle. In both man and horse it has been demonstrated that the number of capillaries surrounding a fibre is greatest for the oxidative fibres.

Muscle biopsy

Over the last decade, there has been a remarkable increase in the knowledge of the composition of muscle and how the various fibres are utilised during exercise in both man and horse. This has been attained by the technique of percutaneous needle muscle biopsy which enables, using special needles, small amounts of muscle (100–200 mg) to be sampled. Following local anaesthetic infiltration of the skin, but not the muscle, a biopsy sample can be obtained atraumatically and safely. This technique is now being used by a number of research workers in experimental horses and in field studies often involving very expensive competition horses. Sampling is painless as is confirmed by the fact that biopsies can be obtained in between bouts of exercise. Regeneration of muscle fibre rather than formation of connective (scar) tissue occurs following biopsy. Although many muscles can be biopsied in the horse, the one most commonly used is the middle gluteal, as it has been found to be very important during locomotion.

The small amount of muscle obtained is sufficient for histochemical, biochemical and ultrastructural studies and most of the information given below resulted from analysis of biopsy specimens.

Skeletal muscle fibre types

It has long been known that not all skeletal muscles are similar in composition. The earliest differentiation between muscles was based on colour (a reflection of the myoglobin content) resulting in the two categories of red and white muscle. Red muscle became synonymous for high oxidative muscle and white for low oxidative muscle. Today the application of physiological, histochemical, biochemical and ultrastructural techniques have shown that these early observations on differences between muscles resulted from the presence of fibres having different contractile and metabolic properties.

On the basis of contractile properties, two distinct types of fibre exist. These are the slow twitch (ST or Type I) fibres which have low myosin ATPase activity at a pH of 9.4, and a relatively slow contraction time and slow relaxation; and the fast twitch (FT or Type II) fibres which have a relatively faster contraction time, and faster relaxation and a high myosin ATPase activity at pH 9.4 (Table 9.3). More recent work indicates that the fast twitch fibres can be further sub-divided according to the lability of their myosin ATPase at acidic and high alkaline pHs. This has resulted in a subdivision of the type II fibres into IIA and IIB fibres (Fig. 9.5). A further type IIC fibre can also be identified. It is found in large numbers during development and occasionally in mature animals as a transitional fibre, when transformation from type I to type II or vice versa may be taking place. The differences between the myosin ATPase activity of these different fibre types may be explained by the recent demonstration of minor variations in the composition of the myosin in these fibres. The differentiation of fibre types according to their myosin ATPase activity after acid preincubation has been used by a large number of workers in the classification of fibre populations present within muscle.

In addition to this classification, divisions according to the metabolic properties of the fibres, together with a division into slow and fast fibres, has been carried out. This has led to a plethora of classification systems,

Table 9.3. Characteristics of muscle fibre types in the horse

	Type I	Type II	
	ST	FTH	FT
Speed of contraction	Slow	Fast	Fast
Max tension developed	Low	High	High
Myosin ATPase activity pH 9.4	Low	High	High
Myosin ATPase activity pH 9.4	High	Low (IIA)	Intermediate
(After pre-incubation pH 4.35)		Intermediate (IIB)	(IIB)
Oxidative capacity	High	Intermediate to High	Low
Capillary density	High	Intermediate	Low -
Glycolytic capacity	Intermediate	High	High
Lipid content	High	Intermediate	Negligible
Glycogen content	Intermediate	High	High
Muscle fibres per motor unit	Low	High	High
Fatiguability	Low	Intermediate	High

ST: slow twitch fibre.
FTH: fast twitch high oxidative fibre.
FT: fast twitch low oxidative fibre.

groups and sub-groups, according to the number of metabolic markers examined, and the degrees of activity into which each marker is sub-divided. The two most popular metabolic markers are the measurement of an enzyme representative of oxidative capacity (either succinic dehydrogenase (Fig. 9.5) or NADH diaphorase) and glycolytic capacity (phosphorylase or α-glycerol phosphate dehydrogenase). Using these metabolic markers in combination with a myosin ATPase stain differentiating fibres into slow and fast types, the following three basic fibre types are obtained: slow twitch high oxidative (ST or SO), fast twitch high oxidative, high glycolytic (FTH or FOG) and fast twitch low oxidative high glycolytic (FT or FG).

In the young or untrained horse it has been found that a fairly clear distinction can be made between high and low oxidative fibres, but as training proceeds, sub-division becomes more difficult as there is a general increase in oxidative ability, and a broad spectrum of activity occurs through the fibres. This is one reason why many workers have now resorted to using a system based solely on myosin ATPase activity, although more recent studies have shown that using sensitive techniques, a spectrum of type II fibres can also be found.

As well as being differentiated physiologically, histochemically and biochemically, ultrastructural differences can also be seen. In association with aerobic capacity, high oxidative fibres have many more subsarcolemmal and intermyofibrillar mitochondria, as well as having varying amounts of lipid droplets when compared to the low oxidative fibres.

Figure 9.5. Illustration of histochemical stains used in the identification of fibre types: (a) myofibrillar myosin ATPase pH 9.4 showing slow (Type I) and fast (Type II) fibres; (b) myofibrillar myosin ATPase pH 9.4 after preincubation at pH 4.35 showing Type I, Type IIA, and Type IIB; (c) succinic dehydrogenase activity showing Type I high oxidative fibres (ST). Type II high oxidative fibres (FTH), and Type II low oxidative fibres (FT) (Snow, 1983).

With respect to their contractile properties, fast twitch fibres with their greater requirement for rapid Ca^{2+} turnover, possess a more prominent sarcoplasmic reticulum.

The speed and frequency of contraction together with the metabolic properties, determines the fatigueability of the fibre types as shown in Table 9.3. Differences in fibre types are also outlined in Table 9.3.

Muscle fibre recruitment

During most muscular contractions it is not necessary for the muscle to generate maximum tension, and therefore not all fibres have to be stimulated. The orderly selection of the required pool of muscle fibres for either maintenance of posture or movement is regulated by the pattern of impulses coming from the α-motorneurones. The smallest diameter motorneurones, which have the lowest threshold, innervate the slow twitch fibres whilst the largest innervate the fast twitch IIB fibres. In order to maintain posture it is only necessary to recruit type I fibres but, as the speed of movement increases, the development of more and more tension to generate the required torque is necessary, and the type IIA fibres are also recruited. The very forceful contractions required for rapid acceleration or jumping result in the type IIB fibres being recruited. Because the frequency of impulses coming down the neurones varies, asynchronous contraction of different motor units will occur. However, this does not lead to jerky contractions as the various speeds of contraction of the different fibre types results in a smooth contraction generating the required tension. Therefore, this system of selective recruitment results in controlled development of different degrees of tension to suit the movement. A schematic representation of fibre recruitment is shown in Fig. 9.6.

Fibre type populations within muscle

In the horse as in man most muscles contain a mosaic of the three major fibre types. This mosaic is not uniform throughout the muscle. In general the proportion of type I fibres increases within the depth of a muscle. Also, with increasing depth, a higher proportion of oxidative fibres within the type II population occurs. The presence of all three fibre types within a muscle allows it to have a wide range of contractile responses and is therefore adaptable to different physiological requirements. Until recently the description of function of different muscle groups has relied on anatomical features, but electromyography has resulted in a better understanding of the sequential involvement of muscle groups during movement. The different functional requirements of muscle is responsible for variations in the proportion of the three fibre types between muscles in all species including the horse.

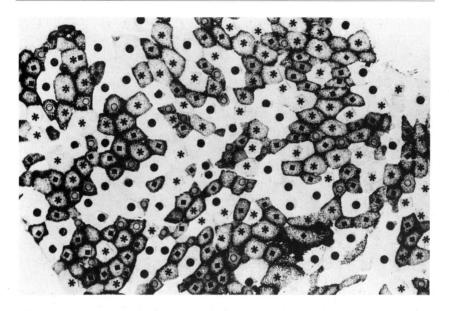

Figure 9.6. Schematic illustrations of the recruitment of muscle fibres at different intensities of exercise. O = fibres recruited at walk, O + + recruited at moderate trot, O + + * = at fast canter, O + + * + = at a gallop. Note that as speed increases more and more fibres contract so that sufficient power can be generated. In this not fully trained animal, increases in speed result in low oxidative fibres being recruited at the canter and trot. These fibres will rapidly build up lactate and become fatigued.

Even within one specific muscle, marked variations in composition between mature individuals have been described in both man and horse. In man the most frequently studied muscle is the thigh muscle, the vastus lateralis, which has been found between individuals to have a slow twitch proportion ranging from 10 to 100%, with a mean of approximately 50%. This is an important muscle in running and it has been shown that elite marathon runners have a very high proportion of slow twitch fibres, whereas elite sprinters in contrast have a very low proportion of slow twitch fibres. These differences are not surprising when considered in terms of both rate and duration of differing energy requirements for competition over these varying distances.

In examining a number of equine limb muscles it has been found that in the horse a relationship exists between performance ability and the proportion of the fibre types in the middle gluteal muscle. Because of this it has been suggested that taking a biopsy of the middle gluteal and determining fibre type composition may be of assistance in predicting performance, even to the extent of predicting an ideal racing distance for thoroughbreds and in selecting animals for breeding. However, the small difference in fibre composition between individuals, and the variation in fibre composition within the middle gluteal makes reliance on a single muscle biopsy unwise. Although the proportion of types I and II fibres are genetically controlled, studies in the soleus muscle of the mouse have indicated that this involved a polygenic mode of inheritance and breeding predictions are, therefore, risky. Of greater benefit may be the use of a biopsy sample to estimate metabolic characteristics. Undoubtedly a young untrained horse will have a much higher proportion of low oxidative fibres than a well trained animal.

In addition to the importance of the proportion of fast and slow twitch fibres in assessing athletic ability, the cross-sectional area of individual fibres is also significant as this influences force output. It has been shown that the greater the cross-sectional area of fast twitch fibres the greater the force output. Hence the importance of hypertrophy of fast twitch fibres, especially the low oxidative fibres in competitors involved in events requiring explosive movements. Even in the untrained horse, the American Racing Quarterhorse, for example, has a greater proportion of the middle gluteal muscle occupied by large fast twitch low oxidative fibres than in the thoroughbred or standardbred. On the other hand, in those horses well adapted for endurance work, very large diameter fibres are lacking, all fibres being of similar size and highly oxidative.

Muscle fatigue

The period at which work at any given load can continue is limited mainly by the onset of muscle fatigue. For a long time scientists have argued whether or not this fatigue is central or peripheral in origin. If fatigue is peripheral it could originate either within the transmission system, i.e. at the neuromuscular junction, or be associated with the contractile mechanism by interfering with the supply of ATP through depletion of substrates or interference with metabolic pathways. In determining the causes of the onset of fatigue, the type of exercise has to be considered, and for the horse fast sprinting has to be distinguished from endurance exercise requiring moderate to high workloads.

During short-term exercise, such as flat out galloping at greater than 100% $VO_{2\,max}$, anaerobic metabolism becomes an important source of ATP. As a result of this degree of anaerobosis, lactate accumulation occurs within the muscle as its diffusion rate into the circulation is below the rate of formation. Moreover, lactate levels in the blood also build up, so that blood concentrations in excess of 24 mmol/1 following galloping are not uncommon. In association with lactate accumulation is a build up of hydrogen ions (protons) which result in a decrease in intracellular pH. Recent studies in the horse indicate that when the lactate build up or pH drop reaches a critical level, there is also a gradual decrease in the amount of ATP within

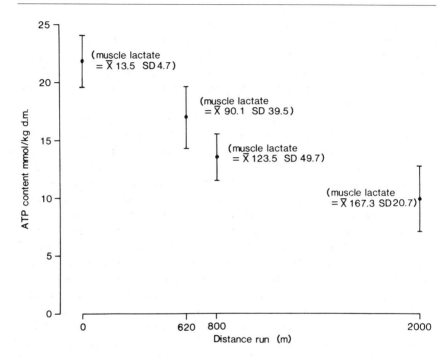

Figure 9.7. The relationship between muscle ATP and lactate concentration to the distance galloped maximally (Harris, Marlin and Snow, unpublished data).

the muscle. After very strenuous exercise, levels may be only 50% of normal. This decrease has been seen after racing in both thoroughbreds and standardbreds. As shown in Fig. 9.7 the extent of this decrease would appear to be dependent on the amount of lactate formed, and the distance covered. The decrease in ATP is probably a result of an interference in the metabolic pathways; the decreased pH is also thought to cause impairment of the contractile elements. It is this build up of lactate or decrease in ATP within muscles that is probably responsible for the "fading" of many horses during the last furlong of a race. This fading represents a decreased force output from the muscles and may arise either from a complete or partial failure of certain muscle fibres, probably the less oxidative and those only recruited at the fastest speeds. To try and delay the deleterious decrease in intracellular pH, the administration of sodium bicarbonate prior to strenuous maximal exercise has been advocated. However, whether this has a beneficial effect is unproven with varying results being reported from studies carried out in man. It would appear that the administration of bicarbonate has little intracellular effect, but rather enhances the rate of removal of lactate from the cells and is therefore of some advantage only if a number of bouts of maximal exercise are to be undertaken in rapid succession. Bicarbonate has also been administered to horses by many racehorse trainers, but it is not known whether it has any beneficial effects. In a recent investigation 0.4 g/kg was given orally 2 hours before maximal galloping to study a possible benefit in increased speed, but results were equivocal.

Another possible contributing factor to fatigue is a change in ionic balance between intracellular and extracellular compartments, as well as alterations in specific electrolytes. Recent work has indicated that with maximal exercise, plasma potassium increases dramatically, the rate of rise being related to the intensity of exercise. At very high intensities peak plasma values in the range 9–10 mmol/1 have been recorded. Further work has indicated that maximal plasma potassium concentrations are seen at any intensity of exercise after 2 minutes and if this exercise is continued for a longer period, the electrolyte concentration plateaus. The exact significance of the early increase is unknown, but despite these high circulating levels of potassium, no cardiac abnormalities have been detected. Following the completion of exercise the plasma potassium concentrations return to normal within 2 to 3 minutes.

Low glycogen content resulting from maximal effort has also been suggested as a limiting factor to performance. However, studies carried out indicate that under racing conditions, at least in thoroughbreds, glycogen content is only reduced by about one third and is unlikely to be important.

Once the accumulated lactate is removed from the muscle fibres and ATP supplies restored, exercise can then be continued at a similar workload although after a hard race, a longer recovery period is required than would be predicted from biochemical alterations. The removal of lactate from muscle is aided, however, if mild submaximal exercise is continued (Fig. 9.8). This is because a better blood supply is maintained to the muscles than if the horse is kept stationary. In addition the muscle at low workloads is able to metabolise much of the lactate. The restoration of ATP will take approximately one hour after a hard race, whilst complete restoration of glycogen may take several days.

Fatigue from exercise at submaximal workloads below those which result in lactate accumulation, occurs when supplies of glycolytic pathway precursors, i.e. either intramuscular glycogen or circulating glucose, become depleted. Energy sources from fat, and catabolism of protein, are still abundant. In accordance with findings in man it has been shown in the horse that after prolonged exercise the different muscle fibre types become preferentially and progressively depleted of glycogen. The type I fibres are depleted first, probably as a result of being recruited in the initial stages of exercise and also having a lower glycogen concentration than the type II fibres. During prolonged exercise the type IIA fibres are also recruited, although

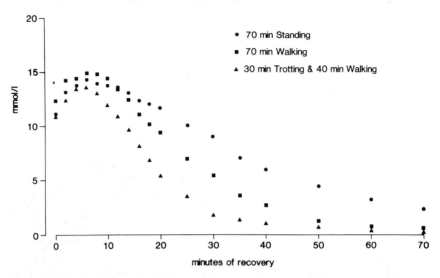

Figure 9.8. The influence of different modes of activity on blood lactate concentrations following 2 min at 12 msec with stand, walk and trot/walk recovery (Marlin, Harris and Snow, unpublished data).

not all motor units are in use at any one time, and therefore they gradually become depleted at different rates. Although as outlined previously, type IIB fibres are only normally recruited at very high intensity workloads, it appears that once the other fibres are glycogen depleted, mechanisms operate by which the motor neurones innervating IIB motor units are stimulated so that exercise can continue until all fibres are depleted. An example of the glycogen depletion pattern seen over an 80 km ride is shown in Fig. 9.9.

As with maximal exercise the onset of fatigue during prolonged exercise can be increased by various factors including the degree of training and the increase of fuel availability. As glycogen depletion is the primary cause of fatigue, factors that can either slow down its rate of utilisation or increase its concentration within muscle will prolong exercise time. In addition to glycogen, FFAs are a very important alternative substrate for the aerobic generation of ATP. The extent that FFAs can be utilised in preference to glycogen during exercise is dependent both on their plasma concentration and their ability to be incorporated into the TCA cycle. This latter factor may depend on both the ability of muscle to take up circulating FFAs, and the concentration of enzymes responsible for its degradation into subunits for incorporation into the TCA cycle. The uptake of plasma FFAs has been shown to be dependent on their concentration with a greater uptake of higher concentrations. The raised circulating FFA concentrations brought about by increased lipolysis or the administration of certain FFAs has been shown in

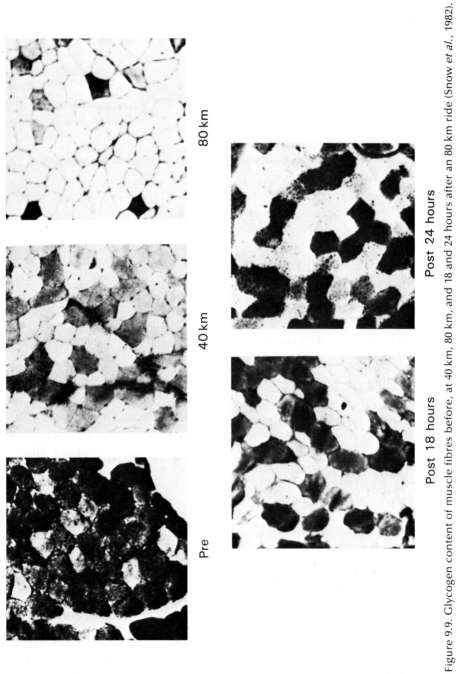

Pre   40 km   80 km

Post 18 hours   Post 24 hours

Figure 9.9. Glycogen content of muscle fibres before, at 40 km, 80 km, and 18 and 24 hours after an 80 km ride (Snow *et al.*, 1982).

man to prolong exercise time. Training also increases the utilisation of FFAs by muscle and creates a glycogen sparing effect.

In addition, the increase of glycogen concentrations within muscle also prolongs time to fatigue. In marathon runners various dietary manipulations are used the week before a marathon to produce very high muscle glycogen concentrations. Whether similar manipulations would be effective in the horse is not known, but it is unlikely that many of these dietary regimes could be successfully imposed on the horse. Equine muscle normally has a high glycogen concentration although training has been shown to increase this further.

Following complete depletion of glycogen it takes at least several days for complete replacement. The repletion within fibre types is in the reverse order to that seen with depletion (Fig. 9.9).

Fatigue may also have a central component, but its contribution is difficult to assess in man, and probably virtually impossible in the horse. Asmussen (1979) concluded that "central fatigue is caused by an inhibition elicited by nervous impulses from receptors in the fatigued muscles". This inhibition is thought to arise from the reticular formation and can be overridden by other signals of peripheral or central origin which may explain the important role of motivation and other signals, e.g. the whip, and certain stimulant drugs, in extending the onset of fatigue through the development of central fatigue.

Adaptations with training

Muscle is a tissue which can display a great deal of plasticity in that it can generally adapt to the varying demands placed on it. These adaptive changes are brought about via the pattern and frequency of impulses arriving at the fibres from the α-motorneurones. Using experimental techniques, where normal patterns of stimulation are markedly altered, very large changes, such as reversal of fibre types from I to II and vice versa, can be brought about. However, in normal training it is considered unlikely that any major shifts in the proportion of these two basic fibre types occur. The main effects of training are on the metabolic rather than contractile elements of the fibres. It is axiomatic that a muscle fibre can only adapt to its requirements in competition if it has received stimuli during training. In other words one can only get a favourable effect from training if it entails the same type and similar intensity of exercise as that encountered during competition. For instance, the largest IIB fibres are only recruited during high intensity exercise or after very prolonged aerobic work. Obviously if development for sprinting is required then sufficient sprinting should be included in the programme. If the competition requires exercise below or just above the anaerobic threshold, prolonged exercise at these levels is needed to increase oxidative capacity in order to increase both the time till onset of fatigue and the rate at which anaerobic metabolism proceeds.

It has been shown in numerous studies in man and laboratory animals that with endurance training a marked increase in oxidative capacity and the enzymes responsible for FFA degradation occur. Similar findings have now been made in studies using various breeds of horse, involved in both racing and endurance events. This has been demonstrated in a number of ways: by an increase in both the size and number of mitochondria, oxidative enzyme activities, and the number of capillaries around the fibres. The extent to which this occurs is best seen histochemically in endurance ride horses, in which all muscle fibres react intensely for SDH activity. In studies involving purely endurance training programmes, there is generally no alteration in glycolytic enzyme activity. However, where a sprint component has been introduced into training the results have been equivocal with respect to this pathway. Recently alterations with training in activity of enzymes responsible for purine degradation have been reported.

As well as metabolic adaptations within the muscle fibres, it has been found that in parallel with these changes a proportion of the type IIB fibres are transformed to IIA fibres, possibly via an intermediate IIAB fibre. This indicates alterations in the light chain sub-units of fast myosin. How this alters the contractile characteristics of the fibres is not fully understood.

CARDIOVASCULAR SYSTEM

The cardiovascular system (CVS) can be considered as the plumbing system of the body. It has the vital role of delivering fuels, including the oxygen required for the generation of energy, and removing metabolic waste products thereby preventing accumulation and poisoning of cells. The heart is the CVS pump that ensures that blood is able to flow to and from the active tissues and at the required rate through the vessels supplying them. The horse's CVS can adapt much better to meet the increased demands of exercise than that of any other species. During muscular activity metabolism increases enormously and the circulation must adjust rapidly to meet the demands for activity to continue. Several factors influence the capacity for aerobic muscular activity including (i) pulmonary ventilation, (ii) oxygen diffusion from lung alveoli into blood, (iii) cardiac output the oxygen transporting capacity of the blood, and (iv) the diffusion of oxygen from the blood into the muscle.

The following section will concentrate mainly on the third of these factors.

Oxygen uptake (consumption)

Maximal oxygen uptake ($VO_{2\,max}$) is the term used to define maximum oxygen uptake from air during physical work at sea level. Other synonymous terms include maximal oxygen consumption, maximal oxygen intake and maximal aerobic power. At a particular stage of fitness the greater the $VO_{2\ max}$ the greater the workload that can be performed before anaerobic metabolism becomes a limiting factor. In man it is well established that elite athletes have a higher $VO_{2\ max}$ than the general population. This higher capacity is, however, largely genetically determined; training only results in a 10–20% increase in $VO_{2\ max}$. Difficulties in carrying out measurements in horses accounts for the availability of limited data on oxygen consumption during exercise in this species. Nevertheless, oxygen uptake can be estimated using the following equation:

$$O_2 \text{ uptake} = (\text{Cardiac output}) \times (\text{Arteriovenous } O_2 \text{ difference})$$

It will, therefore, be apparent that increasing factors contributing to the right side of the equation will lead to an elevated O_2 uptake. In practice whole body O_2 uptake is usually determined by measuring the difference between the rate at which inspired oxygen enters and the rate expired oxygen leaves the lungs during standardised exercise tests (SET). In man, $VO_{2\ max}$ is determined by increasing workloads (e.g. on a treadmill or bicycle ergometer) until workloads are attained which result in a plateau of O_2 consumption. Unfortunately, because of the obvious difficulties in exercising thoroughbred horses at high intensities (e.g. galloping) on a treadmill, and the problems of collecting expired air, data on O_2 consumption in horses has until recently been lacking. However, there appears to be in the horse a reasonable linear relationship between workload and both heart rate (HR) and oxygen consumption. Thus, $VO_{2\ max}$ can be extrapolated after increasing submaximal workloads. In man, extrapolation from heart rates and $VO_{2\ max}$ at submaximal loads results in an error of about 10% in VO_2 estimation. Hopefully, it is likely that O_2 consumption in horses exercising on the track will soon be measurable following the recent development of a telemetric method for assessing O_2 uptake using flow and PO_2 sensors incorporated in a face mask.

There have already been several reports of the $VO_{2\ max}$ measurement in standardbred and thoroughbred horses and it appears that values of about 140–150 ml/kg/min are not unusual. This suggests that a horse can increase its $VO_{2\ max}$ over resting O_2 consumption by 50-fold, roughly twice the 80 ml/kg/min recorded in top-class athletes. Such figures highlight the elite nature of the equine athlete.

Heart rate

The normal resting HR is between 30–40 beats/min (bpm). However, in well trained horses, rates as low as 25 bpm are frequently recorded and may result from training, or be genetically determined. HR during and following exercise is usually measured using a harness or adhesive electrodes. The signals from the electrodes are either collected by radiotelemetry or on a small portable tape recorder carried on the saddle or by the rider. Recently commercial cardiotachometers have become available. These give a digital readout of heart rate and can be easily attached to the rider, saddle or neck (Fig. 9.10). One such model allows the reading to be stored on a memory chip for later direct playback or computer printout (Fig. 9.11). This allows mean HR to be recorded every 5, 15 or 60 seconds and can be used by the rider to monitor workloads during training, as well as HR recovery following exercise. Care has to be taken in the selection of a suitable cardiotachometer, as some models on the market are less reliable and can give rise to misleading readings. Two independent tests suggested that the Hippocard (Leuenberger Medizin Technik AG, Switzerland) and Equistat (EQB, USA) give good accuracy and have easily placed electrodes.

When submaximal exercise starts there is a rapid increase in HR followed after 2 to 3 minutes of steady work, by a drop which soon plateaus. At high intensity workloads this initial "overshoot" does not occur and the plateau occurs more rapidly. It is possible that the initial overshoot in the horse arises

Figure 9.10. On board heart-rate meter (Equistat — EQB, USA). The meter can be attached around the riders thigh or arm or on the horses neck.

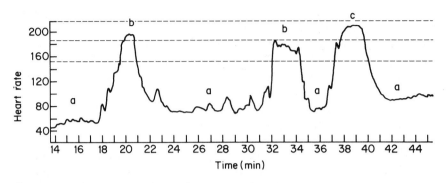

Figure 9.11. Computer print out from an on board heart rate meter (Hippocard PEH 200, Leuenberger Medizin Technik AG, Switzerland): (a) walk, (b) canter, (c) gallop.

because of a delay in the splenic mobilisation of erythrocytes, since the adjustment of circulating blood volume to meet the requirements of the workload takes about 3 minutes. As in man, a linear relationship between HR and workload exists in the horse, although this tails off at the very highest workloads. During exercise the HR escapes from vagal inhibition and progressively comes under control of the sympathetic system whose activity increases proportionally.

As well as measuring HR at varying workloads, some investigators have also monitored rates during competitive events. In the thoroughbred, a pre-race HR of 132 bpm has been recorded when the horse is in the starting stalls. The rate then increased to a mean maximum of 223 bpm at the end of the race (Fig. 9.12). Studies on fast and slow working days on a track, revealed a linear relationship between HR and speed between 6.7 to 13.4 m/sec. However, a levelling off occurred at higher speeds and at racing speed of 16.7 m/sec this relationship did not apply. Maximum HR of the order of 200–220 bpm were recorded in these studies but other workers have reported rates in excess of 220 and even up to 250 bpm. As in man, it would appear that maximal HR is age dependent; 2-year-old horses have maximal HR of 220–250, but older horses, e.g. those over 10 years of age, may only reach 200 bpm. Similar findings also occur in the standardbred.

Often as part of a training programme or during recovery from lower limb injuries swimming exercise has been used to try and maintain aerobic fitness. It has been found that the HR of a swimming horse will increase to between 140 and 200 bpm — higher values being recorded when the horse swims against a weight or induced current. This type of exercise is good training for the oxygen transport system.

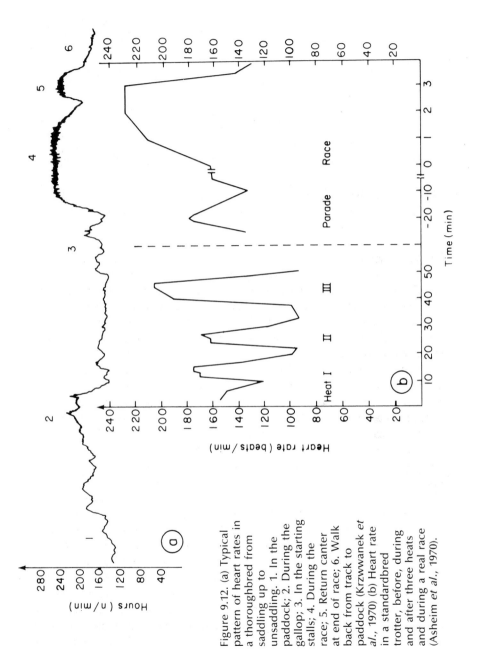

Figure 9.12. (a) Typical pattern of heart rates in a thoroughbred from saddling up to unsaddling. 1. In the paddock; 2. During the gallop; 3. In the starting stalls; 4. During the race; 5. Return canter at end of race; 6. Walk back from track to paddock (Krzwwanek *et al.*, 1970) (b) Heart rate in a standardbred trotter, before, during and after three heats and during a real race (Asheim *et al.*, 1970).

Post exercise recovery   After the completion of maximal exercise effort there is a rapid fall in HR during the first 30–60 seconds rest. This is followed over the next few minutes by a more gradual decline to around 100 bpm after which a plateau is reached. This very rapid decline in the first minute explains why monitoring post exercise heart rates to determine how hard the horse worked is of little value. As rest continues there is a much slower decline and HR following maximal effort may not return to normal for at least an hour or more. This long recovery time after maximal effort can be attributed to the body overcoming the oxygen debt incurred during exercise. In the post exercise period additional oxygen is required to break down the vast amounts of lactic acid produced. Following submaximal exercise, as is seen in endurance rides, the third portion of the recovery phase is more rapid. Longer recovery periods in these circumstances are often due to fatigue of the horse, or thermoregulatory stress, requiring maintenance of high blood flow to the skin.

In man, the time for the heart rate to return to resting levels is related to the intensity of exercise and fitness. Whether a similar situation is seen in the horse is unclear. Some studies have reported no alteration and others more rapid returns with training. This discrepancy may be due to the failure to use uniform workloads in "in field" trials, and psychogenic factors producing sudden shifts in heart rates. However, despite these various external factors, HR have been and are being monitored both during and after endurance rides to try and ascertain the degree of fatigue/fitness of the animal. Such work has shown that there is a good relationship between HR recovery measured at 30 minutes, and other parameters considered indicative of fatigue in an endurance ride. It is generally agreed that careful monitoring of HR, together with clinical examination, has greatly reduced the incidence of fatigue and other problems in endurance rides.

Effect of respiratory and cardiovascular disease on exercise

Heart rate   The effects of disease on HR have been studied by a number of workers during a standardised exercise test comparing results with those from normal healthy animals. Horses suffering from chronic obstructive pulmonary disease (COPD) were found to have heart rates 20 bpm higher than in normal horses during exercise; they also had a diminished working capacity. The more severe the COPD the greater the rise in HR. Horses with valvular disease and atrial fibrillation have been generally reported to have an increased exercise heart rate as well as impaired performance. The electrocardiogram (ECG) has been found to be a useful diagnostic aid in examining horses with a history of poor performance. One group of workers reported a 93% incidence of ECG abnormalities in poor performing horses compared to 21% in a similar sized group of normal animals. Changes in the T wave are quite commonly seen but the significance of this is debatable. It has been suggested that abnormal T waves *per se* do not reflect decreased performance, but rather may be secondary to other problems associated for example, with overtraining and/or with red-cell hypervolaemia.

Heart score

Of importance in the equation determining the increase in cardiac output during exercise is the ability for stroke volume, i.e. the volume of blood ejected from the heart per beat, to increase. As well as being dependent on venous return, stroke volume is also regulated by cardiac mass, which is often related to ventricular volume. In man, it has been reported that a large heart mass is commonly found in elite athletes, although the extent to which genetics and training determines their larger heart mass/body weight remains controversial. A larger heart mass, by increasing stroke volume, is however, likely to improve the aerobic capacity of an individual.

A method was described by Steel (1963) whereby the approximate heart size of a horse can be determined from ECG readings. This value is called heart score and is defined as the arithmetic mean of the QRS intervals in each of the three standard limb leads. Each QRS interval is measured to the nearest 10 milliseconds. Subsequent studies have shown that there is a good correlation between heart score and heart mass. It has been found that heart scores of 100 and 130 are associated with heart weight of 3.0 and 5.4 kg respectively. Fillies and mares generally have a slightly lower heart score than stallions or geldings, but whether this is solely due to a lower body weight has not been determined.

The non-invasive technique involved in the determination of heart scores makes it an attractive aid in assessing potential. In a study of the major "staying" handicap races over 1.5–2 miles in Australia, it was found that the winners generally had a heart score greater than 120. Similarly, in standardbreds, a high correlation between heart score and both stakes, winnings and kilometre times have also been reported. In studying horses competing in an endurance ride the better performers had the higher heart scores. In contrast, however, a study in British thoroughbreds revealed a poor relationship between performance and heart score but this may reflect the fact that the researchers equated performance to a Timeform rating, which does not distinguish between sprinters and stayers and although a highly efficient cardiovascular system is known to be necessary for staying racers, the importance for sprinting has yet to be determined. One of the important factors determining heart size is genetic and so measurements in yearlings may give an indication of potential both for racing and the selection of breeding stock. It has been shown that heart scores of progeny reflect those of the dams more closely than those of the sires but since the heritability estimate is low (of the order of 0.4) the usefulness of this finding has been questioned.

There undoubtedly exists a relationship between heart score and performance in aerobically demanding events, and there is a genetic component in its determination. Unfortunately, little work has been carried out to determine the influence of other factors such as age, growth and training. All these factors are likely to have an influence on the heart score.

There appears to be two schools of thought with regard to the usefulness of heart score. One group believes that despite limitations scoring has been extremely useful in selecting racehorses even at the yearling age. The second group, however, feels that because of the numerous variables determining performance, some horses have exceptional performances despite low heart scores, and therefore selection on a heart score evaluation would eliminate potentially valuable horses. At the end of the day it is up to the individual racehorse owner to decide whether the selection of yearlings on the basis of heart score for later training is useful or economically justified. The costs saved from those eliminated for low heart scores will probably outweigh the chances of missing top-class performers despite low heart scores.

Another non-invasive method of assessing cardiac dimensions is by the use of ultrasound techniques, or echocardiography. Over the past four years, echocardiography has been used to investigate not only cardiac abnormalities, but also variations in cardiac dimensions, such as ventricular wall thickness and ventricular volume both between individuals and within an animal during training. Early reports have indicated a good correlation between heart score and interventricular septal thickness. Using this and other techniques in human athletes it has been shown that left ventricular hypertrophy occurs to a greater extent in endurance runners than in sprinters who have a smaller increase in ventricle wall thickness and mass. As has been shown in a longitudinal study on the effects of training, increased cardiac dilation, and increased cardiac mass, result in an increase in maximal stroke volume and hence cardiac output during maximal exercise.

Cardiac output and stroke volume

Cardiac output is the product of HR and stroke volume, and is an important determinant of oxygen consumption. Generally cardiac output and HR are determined experimentally and then stroke volume calculated from these data. Cardiac output in the resting horse is usually measured using a dye dilution technique but thermodilution techniques are now also being used. Recently the use of an electromagnetic flow probe has also been described. This system has the advantage that cardiac output can be measured during non-steady state conditions. However, because of the major surgery involved in the placement of the probe, it will probably remain basically as a research technique. In the resting state, cardiac output in the horse is about 80 ml/min/kg body weight, and stroke volume is 1.8 ml/kg. Little is known of changes occurring during exercise, especially at the higher workloads, but the main factor increasing cardiac output is known to be the elevated HR. One study has shown that at low, submaximal workloads an elevation of HR to 135 bpm was accompanied by a cardiac output increase of 3.5 times and a 15% increase in stroke volume. At higher workloads producing a HR of between 190–200 bpm, a 7-fold increase in cardiac output to 523 ml/min/kg was recorded with increases in stroke volume up to 41% (2.7 ml/kg). In man training increases maximal cardiac output by increasing stroke volume, but it has been reported in standardbreds that no change in cardiac output or stroke volume occurs with training. Testing in this study, however, used a standardised exercise test which would represent a lowered relative workload after training.

Blood flow

With the onset of exercise there is an almost simultaneous increase in blood flow to the active muscles. This is brought about by a local vasodilator mechanism which is not neurogenic, but possibly due to the release of certain substances, e.g. potassium ions. In the resting state only about 15% of the cardiac output goes to striated muscle tissues although these comprise approximately 40% of body mass. During strenuous exercise in ponies, however, it has been found using radionuclide-labelled microspheres that blood flow increases over 70-fold to the muscles of the forearm and thigh. This is in contrast to the much smaller 20-fold increase in blood flow to similar muscles during maximal exercise in man. Furthermore, to aid in heat dissipation there is an increased blood flow to the skin, and to account for the increased blood flow to muscle and skin, there is a redistribution of blood from other tissues. Thus the blood flow to the kidneys, although unaltered during moderate exercise, is decreased at maximal exercise to 19% of the level seen at rest. The increased blood flow to muscle results in an increased oxygen extraction compared to that seen in the resting state. During strenuous exercise, oxygen extraction may rise to 90–95%, but at near maximal workloads, the forceful contractions constricting arterioles, may impede blood flow to the actively contracting muscle fibres and establish a transient hypoxia.

Blood pressure

Mean systemic arterial pressure (MAP) depends on cardiac output and on peripheral resistance. Therefore, as both these parameters alter during exercise it would be expected that MAP would also change. In the horse, MAP is generally measured from the carotid artery or aortic arch. Studies have shown that there is a rise in MAP with increasing workloads, although variation in the sites of measurement has produced some difference between data. At all workloads, however, there appears to be an increase in systolic arterial pressure. Diastolic arterial pressure falls slightly at low workloads and then slightly increases above resting values at higher workloads. The increase in blood pressure occurs within the first one or two minutes of initiation of exercise. Swimming horses show higher peak systolic pressures. Pulmonary arterial pressure also increases markedly during exercise and this may be relevant in the aetiology of exercise induced pulmonary haemorrhage.

There are virtually no reports on the effects of training on blood pressure although one study has shown a reduced blood pressure following training at the same workload in two horses. At maximal workloads, it is unlikely that a reduction in blood pressure would occur.

Blood volume, haematocrit and haemoglobin

Blood or circulatory volume is the volume occupied by the plasma and cells. In man, a slight decrease in circulatory volume occurs with exercise due to fluid shifts which result in a decreased plasma volume. The decrease is greater as work intensity increases; more marked decreases occur when the anaerobic threshold is exceeded. This also occurs in the horse as indicated by an increase in total plasma protein concentration, but blood volume increases markedly due to the increase in circulating red blood cells. It has been well documented that the additional red blood cells are released due to contraction of the spleen following sympathetic stimulation. The increased proportion of circulating red blood cells results in an increased capacity for oxygen to be transported to working muscles. In man it has been found that total haemoglobin (THb) is related to the working capacity, or $VO_{2\,max}$, and as man has no splenic reserve, resting Hb can be used in the calculation of THb. In the horse, however, this can only be determined after the spleen has been emptied after which there is a direct correlation between work performance and THb and total blood volume (TBV). The importance of determining total values rather than evaluations from resting Hb levels was well established in a series of studies using standardbreds by Persson (1967, 1968, 1969) in Sweden. This work further showed that splenectomy decreases work performance in horses.

To empty the spleen requires contraction of the smooth muscle within the splenic capsule which is controlled by α-adrenoceptors. It follows that injections of adrenaline and near to or maximal exercise will bring about complete splenic contraction. Although there are slight differences, both methods can be used to determine TBV and THb and both are highly reproducible. With exercise the extent of splenic contraction is dependent on the workload, as this determines the increase in sympathetic activity. At low workloads only partial emptying occurs, and this is seen with horses

Table 9.4. Mean relative values ± standard deviations for plasma volume (PV/BW), total blood volume (TBV/BW), and total haemoglobin (THb/BW) in four breeds of horses (one cold blooded and three warm blooded)

Breed or service purpose	(n)	PV/BW (ml)	TBV/BW (ml)	THb/BW (g)
North Swedish horses	10	45.5 ± 5.3	87.3 ± 9.3	13.6 ± 1.6
Saddle horses	18	53.1 ± 4.9	116.0 ± 9.4	20.3 ± 2.3
Trotters	159	53.1 ± 4.9	120.9 ± 14.1	22.6 ± 3.9
Thoroughbred racehorses	19	53.1 ± 4.9	142.1 ± 13.9	29.2 ± 3.0

From Persson (1968).

competing in endurance rides, where haematocrits little different from resting values are generally seen.

Several factors may affect TBV and THb including (i) breed, and as would be expected these values are lower in cold blooded than warm blooded horses (Table 9.4); (ii) age; (iii) body weight; (iv) training; and (v) sex. Persson however, found sex made no difference to the values, and found that TBV was best correlated to body weight, than with degree of training and age, whilst THb was most dependent on training and age. Therefore, training results in an increased oxygen carrying capacity. Furthermore, Persson concluded that state of training could be estimated in normal racehorses from either TBV or THb provided that variations due to body size, age, sex and breed are considered. In some groups of animals studied a relationship between performance and TBV/BW and THb/BW existed. In contrast a group of workers in Germany found that although THb may be helpful in assessing training progress, it was not reliable enough for differentiation of performance.

Both groups of workers demonstrated that there is no relationship of resting Hb to THb or to performance. The reason why resting measurements are of little use is that great variability, which is largely dependent on splenic tone, occurs both between and within horses. Horses having different resting haematocrits have been shown to have similar haematocrits when the spleen empties. As splenic tone is under the control of the sympathetic system any factor that can alter this may alter the resting haematocrit. Obviously exercise will cause a marked increase in packed cell volume and a return to true basal state may take up to 48 hours. Any excitement (which may not be visually obvious) causes an elevation in haematocrit in the horse. Certain drugs, e.g. some tranquillisers, produce a decrease in resting haematocrit and should never be used to collect samples for haematological examination. Conditions resulting in a decreased plasma volume will also cause false elevations in Hb. It has been shown that blood sampling after feeding hay results in an elevation of up to 15% in haematocrit as well as total plasma protein concentration. This effect is most marked within the first hour after feeding.

Table 9.5. Normal range of erythrocytic parameters in thoroughbreds in work in Victoria

	Mean −2 SD	Suggested lower limit	Mean-precision estimate	Mean	Suggested upper limit	Mean +2 SD
RBC ($\times 10^6$/mm^3)	7	7.5	9	9.5	11	12
Hb (g/100 mg/100 ml)	11	12.5	14.5	15	17.5	19
PCV (%)	29	33	38	40	48	51

OPTIMAL

← — — — — →

DIAGNOSIS   ANAEMIC   SUB-OPTIMAL   POLY-
                    ←————————— — — —→   CYTHAE-
                                              DOUBTFUL   MIC

From Stewart and Steel (1974).

Despite all these variables, attempts continue to be made by numerous veterinarians to assess the state of fitness and likely racing performance of horses from resting blood samples. It has been generally thought that an increase in resting haematocrit with training is indicative of improving fitness, although studies investigating this aspect have produced equivocal results. The Australian workers, Stewart and Steel (1974) have suggested that if true basal samples can be obtained the measurement of such parameters may be of value in determining the likelihood of a horse racing poorly, rather than predicting a good performance. They considered that values below 2 SD of the population mean were indicative of an anaemic condition, whilst those higher than 2 SD suggested polycythaemia. The range of the values they found is shown in Table 9.5. There are a number of physiological explanations why horses with resting haematocrits at either end of the normal scale are likely to be poor performers. In horses with a low resting haematocrit, the increase brought about by splenic contraction during maximal exercise will not be as great as that seen in horses having a value within the normal range, and this will result in a markedly reduced oxygen carrying capacity. In polycythaemic animals complete splenic contraction will result in haematocrits in excess of 0.70 l/l. It has been found in standardbreds that some cases of poor performance can be attributed to over training resulting in a red cell hypervolaemia, in conjunction with an adrenocortical insufficiency. Whether such effects occur in the thoroughbred is unknown. The extremely high blood viscosity resulting from haematocrits in this range will place a considerable load on the circulatory system and possibly on oxygen transport. However, the exact effects of high blood viscosity on performance awaits investigation, although work currently being undertaken at the Animal Health Trust,

Newmarket, indicates that the measurement of blood viscosity in the horse is considerably more difficult than in man.

The increase in TBV and THb during exercise is of physiological importance. Increased TBV will allow for the maintenance of adequate blood flow to more tissues. The increase in the THb allows a much greater increase in oxygen delivery to working muscle and therefore permits a greater uptake. This is one of the reasons why $VO_{2\ max}$ is 2–3-fold greater in the horse than in man. As the mean oxygen carrying capacity of Hb is 1.34×10^{-3} litres/g, at a resting Hb of 130 g/l (haematocrit 0.41 l/l) the oxygen capacity is 0.174 l/l blood, whilst at maximal exercise with a Hb of 210 g/l (haematocrit 0.65 l/l) capacity is increased to 0.281 l/l or by 62%. However, because of the effects of a decreasing pH and elevated temperature during maximal exercise, the Bohr effect results in a lowered oxygen saturation. This effect is not disadvantageous as within the musculature more oxyhaemoglobin is reduced, contributing to greater uptake by muscle. As discussed later the theoretical oxygen carrying capacity is also not attained throughout maximal exercise because of incomplete saturation occurring during passage of erythrocytes through the lungs.

Blood doping (blood boosting)   Induced polycythaemia has been used by a number of human athletes to improve their aerobic endurance performance. Increased haematocrits are induced shortly before competition by auto-transfusion of red blood cells removed several weeks earlier and either refrigerated or frozen. In a review of the effects of blood doping in man, it was concluded that as long as sufficient blood volume is reinfused an increase in $VO_{2\ max}$ will occur. The use of blood doping in the horse has been discussed but, through the horse's "self-blood boosting capacity" the spleen can regulate its output according to the exercise requirements, and such attempts would then appear to be completely unjustified.

White blood cells   Resting red blood cell parameters are used by many trainers and veterinarians in assessing fitness. Others place reliance on white blood cell counts and, particuarly, the ratio of neutrophils to lymphocytes. Undoubtedly leucocyte counts and differentials are important in the assessment of the presence of clinical and possibly sub-clinical diseases. However, as physiological as well as pathological conditions can alter the normal resting leucocyte count, great care has to be taken in appraising these values. The collection of a blood sample may *per se* markedly alter the white cell picture. Prolonged exercise causes a leucocytosis due to a neutrophilia, with return to normal values taking more than 24 hours. This alteration in the leucocyte picture is similar to that reported following ACTH administration, and results from the prolonged release of cortisol during such exercise. Following maximal exercise, there is also an increase in the circulating leucocytes, but because of the increased total blood volume due to the concomitant release of red blood cells, there is little, if any, change in the overall leucocyte count per litre. This increase in circulating white blood cells does not arise from release from the bone marrow, but rather the mobilisation of cells sequestered in capillary beds and the spleen. Generally more lymphocytes than neutrophils are added to the circulation resulting in an alteration, and often a reversal of the normal neutrophil to lymphocyte ratio. In the normal racing thoroughbred a neutrophil:lymphocyte ratio of approximately 60:40 exists. Changes in this ratio are considered to be undesirable as they may indicate an underlying disease, such as viral infection or training stress. Nevertheless, in the author's experience, although such changes may often be associated with below par performances, there are some animals that can race successfully with a "reversed" ratio.

Buffering capacity of blood   Arterial blood pH in the resting state is normally approximately 7.4, with venous pH being slightly lower. Under most circumstances the pH remains within narrow normal limits, although metabolic waste products, e.g. chiefly carbon dioxide are continually entering the system. Constancy is maintained by the blood's own buffering capacity, which involves blood proteins and the bicarbonate systems. The blood proteins comprise plasma proteins and haemoglobin, which act as H^+ acceptors. The buffering potency of the plasma proteins in man is about one ninth of the Hb, and probably even less in the horse, where the concentration of Hb per ml is even greater than in man during exercise. When oxygen dissociates from haemoglobin, the proteins become weaker acids and act as H^+ acceptors. With the bicarbonate system carbon dioxide released in cells during metabolic activity is readily diffused into plasma and then into the erythrocytes where it is catalysed by carbonic anhydrase:

$$CO_2 + H_2O \rightleftharpoons H_2O_3 \rightleftharpoons H^+ + HCO_3^-$$

The equilibrium is determined by the concentration of the various molecules and ions. If the H^+ is removed, the reaction will go to the right. For example when H^+ binds to dissociated Hb, the CO_2 produced will then shift the reaction to the right, to maintain H^+ and pH will be maintained. The HCO_3^- is freely diffusible into the plasma and can be replaced by Cl^-, so that about 70% of HCO_3^- is carried within the plasma. In addition to the released CO_2 entering this HCO_3^- cycle, some is also transported to the lungs combined with haemoglobin as carbaminohaemoglobin. This combination is favoured at the capillary level by the simultaneous reduction of HbO_2.

Normally the extra CO_2 produced by oxidative metabolism during moderate submaximal work can be handled by the buffering systems, and no change in blood pH occurs. However, when workloads are increased to the extent that lactate production commences, both PCO_2 and pH cannot be regulated at the control value. This occurs because the buffering of lactic acid depends on the following system:

$$Na^+ \; HCO_3^- + H^+ \; lactate \; \rightleftharpoons \; Na^+ \; lactate \; ^-(H_2CO_3) \; \rightleftharpoons \; CO_2 + H_2O$$

Therefore, in contrast to the CO_2 produced in aerobic metabolism, the added lactate results in a reaction in the reverse direction and the removal of HCO_3^-. When lactate production is low, this metabolic acidosis can be compensated for by an increase in ventilation, i.e. the onset of hyperventilation. Blood pH then remains at control values, although there is a reduction in both HCO_3^- (which provides approximately 90% of lactic acid buffering) and arterial PCO_2. However, at extremely high workloads, as seen during racing, hyperventilation provides incomplete compensation and there is a progressive decrease in blood pH, with a large efflux of lactate into the blood. Although the lowest blood pH compatible with life is generally stated to be 7.0, following maximal exercise blood pHs as low as 6.8 in man and 6.75 in the horse have been recorded. However, usually after maximal effort, levels between 6.9 and 7.10 are seen. Immediately after exercise stops the venous pH is about 0.1 lower than arterial pH due to a higher CO_2; in the next 5–6 minutes there is only a small increase as lactate efflux from muscle continues. Not all the lactate produced by muscle is dependent on buffering of the blood, and much of it may also be buffered by proteins in other tissues. Once blood pH decreases below about pH 6.9, there will be a very low muscle pH, interference of normal cellular functioning, and onset of fatigue.

The efflux of lactate may be dependent on the blood concentration of HCO_3^-. Therefore once large amounts of lactate are produced a vicious cycle is established: as blood $[HCO_3^-]$ decreases, slower efflux of H^+ and lactate occurs, a more rapid decrease in intracellular pH occurs and the onset of fatigue due to impairment of metabolism results.

Due to the pivotal role of blood $[HCO_3^-]$ in controlling blood pH during anaerobic metabolism, the administration of HCO_3^- prior to strenuous exercise has been investigated in man but although it has also been tried by some horse trainers, its use in both man and horse have, to date, produced equivocal results.

RESPIRATORY SYSTEM

For exercise to continue, even during the shortest of races, adequate gas transport has to occur between the muscle cells and the atmosphere. The

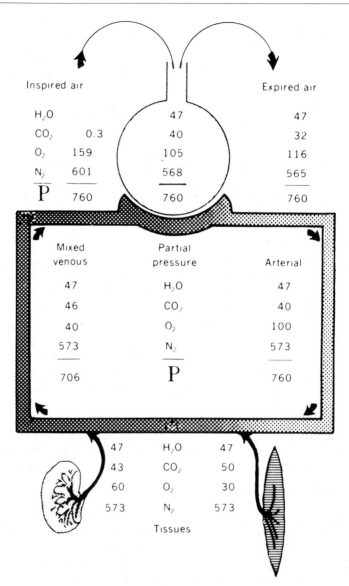

Figure 9.13. Typical values of gas tensions in inspired air, alveolar air (encircled), expired air, and blood, at rest in man. Barometric pressure 760 mmHg; for simplicity, the inspired air is considered free from water. Tension of oxygen and carbon dioxide varies markedly in venous blood from different organs (from Astrand and Rodahl, 1977).

exchange of these gases between the blood and the atmosphere is defined as pulmonary gas exchange. It was previously considered that the capacity of the lungs in the normal animal or man, was not a limiting factor to exercise capacity. However, recent evidence has confirmed that this is not the case at least for the racehorse, and some highly trained human endurance athletes. Our knowledge of the respiratory system during exercise has improved as the availability of high speed treadmills, plus suitable monitoring technology, has increased.

Respiration results not only in the uptake of oxygen and the dissipation of carbon dioxide but also in maintaining normal blood pH. Blood gaseous exchange occurs in extremely small calibre capillaries, located only a few microns (10^{-3}mm) from alveoli air, where rapid passive diffusion of gas molecules from regions of higher to lower chemical activity proceed. Examples of the partial pressures of the gases in alveoli and in the arterial and mixed venous systems are shown in Fig. 9.13.

To ensure an adequate passage of gases within the alveoli there is a regulated pulmonary ventilation. Ventilation (V_E) is a measure of the volume of gas moved in one minute (l/min) and is therefore the product of the tidal volume (l) and respiratory rate (breaths/min). In man, during exercise, ventilation increases with the workload, although at very high workloads with high anaerobosis there is a disproportionate increase in V_E as a result of hyperventilation caused by the decreasing blood pH. In the horse, however, it is unlikely that this is the main cause of the increased respiratory rate, as the increase is particularly marked as soon as galloping or cantering commences.

Studies in which the simultaneous recording of events in the respiratory and limb cycles were monitored during different gaits, yielded some interesting findings. It was found, for example, at the canter and gallop that whilst the respiratory rate was identical to the stride rate, each phase of the respiratory cycle was also constantly related (in time) to a particular event in the cycle of limb movements. At these gaits inspiration occurs while both forelimbs are protracting and this locomotory-respiratory coupling would appear to be mechanically advantageous. However, this constant relationship can be altered by pathological and by certain physiological conditions. In horses galloped from a standing start, it may take up to 20 or more strides before the "normal respiratory pattern" is established. In contrast, at the walk and trot, there is no correlation between the timing of the respiratory cycle and the stride (Fig. 9.14).

Ventilation is brought about by the coordinated movement of the thoracic and abdominal respiratory muscles during inspiration and expiration. The work carried out by the respiratory muscles chiefly performs both elastic work (compliance of the lungs), i.e. against the pulmonary tissue and chest wall

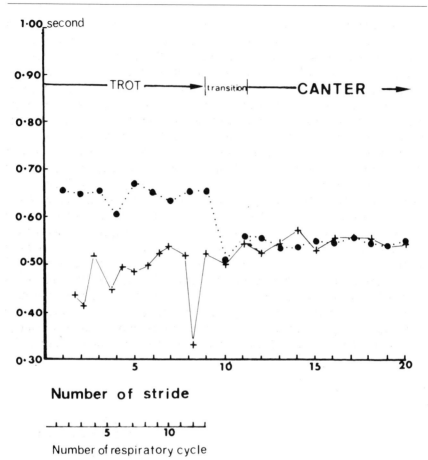

Figure 9.14. Relationship of respiratory cycle to gait. Periods of each respiratory cycle (+) and stride (●) in a sequential series at the trot and canter and during a transition from trot to canter (Attenburrow, 1982).

resistance, and non-elastic work, i.e. overcoming airway resistance. Airway resistance is variable and depends on a viscous resistance due to friction in the tissues, and a resistance to the movement of air in the passages. Bronchial smooth-muscle contraction can double the normal airway resistance and relaxation can halve it. In this respect, the increased sympathetic activity seen during exercise is extremely effective as it induces bronchodilation. Airway resistance may also be increased by mucous oedema or intraluminal secretions. At rest, the energy cost of respiratory muscular work is only a small fraction of the resting energy turnover; however, with high workloads, the energy cost of breathing becomes relatively much greater.

At rest, ventilation in the normal thoroughbred racehorse is usually between 100–150 ml/kg/min, which requires 12–15 breaths per minute. With exercise this increases, the magnitude being related to the intensity of the exercise. Hornicke *et al.* (1977) using thoroughbreds fitted with a mask and monitoring system recorded ventilations of 3.25–3.5 l/kg/min (1170–1400 l/min) at maximal speeds. At these speeds the respiratory rate was 120–140 per minute. Tidal volume increased to 10–12 litres from 7–8 litres at rest. These large increases would indicate that ventilation is unlikely to be a limiting factor to performance even at the highest workloads. However, recent studies have shown that despite high ventilation at maximal workloads, arterial hypoxaemia will develop. For example, Bryly *et al.* (1983) at Washington State University found that P_aO_2 decreased from 96 mm Hg at rest to 62 mm Hg after a 1.6 km gallop. However, this decrease is less marked when corrections are made for an increase in arterial blood temperature. A decrease in P_aO_2 is even seen in submaximal workloads and occurs in the very early stages of exercise (Fig. 9.15). It indicates a lowered oxygen saturation of blood and results in a decreased oxygen delivery. Exactly why this decrease in P_aO_2 occurs is currently being investigated but it seems that the hypoxaemia may result from suboptimal alveolar-pulmonary capillary gas diffusion, rather than from hypoventilation. The very high cardiac output during strenuous exercise results in a transit time of red blood cells in the pulmonary capillaries

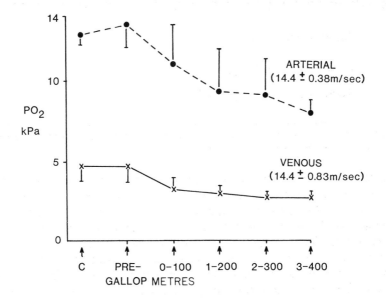

Figure 9.15. Changes in arterial and venous PO_2 during the first two furlongs of a gallop (Littlejohn and Snow, unpublished data).

being markedly reduced and this results in decreased oxygen uptake in cells which already have an extremely low oxygen content. It therefore appears that in contrast to most species studied at exercise, adaptations within the pulmonary system in the horse may be a major factor in restricting performance.

Limitations within the pulmonary system in the healthy horse make it even more evident why respiratory disease can have such a disastrous effect on racing performance. Damage to lung tissue may vary from small discrete areas, to more widespread involvement, which will even decrease exercise tolerance at low workloads. In chronic obstructive pulmonary disease (COPD, or "heaves") the early stages are considered reversible as there is reduced ventilation resulting from airway obstruction. This can be alleviated by the use of specific bronchodilators (e.g. Clenbuterol). As the disease develops, an irreversible condition is established as emphysema and tissue changes occur thus preventing adequate diffusion of gases across the alveolar wall. As hypoxaemia is already present in the resting animal, it can be appreciated why exercise tolerance is considerably reduced.

Following racing, 2.5% of horses are affected by nasal bleeding (epistaxis). In the vast majority of these cases the blood originates from the lungs, rather than from injuries to the upper airways. This bleeding from the lungs is now referred to as exercise induced pulmonary haemorrhage (EIPH). Using flexible fibroptic endoscopes, a number of studies over the last 6 years have been carried out to determine the extent and incidence of EIPH. Its presence has been shown in all warm-blooded breeds (thoroughbred, quarterhorse, standardbred, Arabian) examined after racing. An incidence of between 50–75% has been reported in thoroughbreds and 30% for standardbreds, when examined on a single occasion. However, when animals are examined on repeated occasions, as many as 80% of horses have been shown to suffer from EIPH. The extent may vary from a very small trickle of blood in the trachea, to splashing of blood on the larynx, and to visible bleeding from the nostrils.

In an extensive study undertaken on thoroughbred racehorses autopsied in Hong Kong, 96% showed evidence of old alveolar haemorrhage and bronchiolitis, with lesions most commonly found bilaterally in the dorso-caudal region of the caudal lung lobe and it was suggested that airflow to these regions had been impaired. In conjunction with the diseased airways in affected regions, a bronchial arterial neovascularisation was also found to have developed.

Precisely why EIPH occurs is still unknown. It may result from a normal physiological response to strenuous exercise, or as a result of an existing inflammatory response in the pulmonary tissues. It is possible that the increase in pulmonary arterial blood pressure during maximal exercise leads to capillary fragility and rupture. This in itself would produce an inflammatory response during healing. EIPH could even result from the establishment of pulmonary lesions following viral exposure or after the inhalation of particulate matter during exercise. It is known, however, that the occurrence of EIPH is related to the intensity of exercise, and it has not been found to be common in horses competing in endurance rides.

Whether EIPH has any effect on performance is equivocal since no detectable effects nor impaired performance have been reported following this condition. Obviously quantitative information is extremely difficult to obtain, as statistically it is very hard to detect performance effects over 1 or 2 seconds, which can be the difference between first and last horses in a sprint race. However, if a marked bronchiolitis is present it is likely that maximal performance will be compromised anyway, as the available area for effective gaseous exchange will be reduced, at a time when full capacity is required. This is also a long-held belief that the injection of the diuretic frusemide (Lasix®) several hours prior to racing reduces the severity of EIPH, and improves performance. Investigations to substantiate this have produced inconclusive findings.

THERMOREGULATION, FLUID AND ELECTROLYTE BALANCE

Heat gain by an animal arises either from internal metabolic activity or by an influx from the external environment. The normal thermoregulatory mechanisms control the effect of heat gains by maintaining core body temperature within the narrow limits necessary for life. However, if the two factors causing heat gain are operating at very high levels at the same time, the regulatory mechanisms will be overwhelmed and can be potentially life threatening. Although few studies have been carried out in the horse, this species does appear to have evolved a very efficient thermoregulatory system and is only rarely troubled by hyperthermia.

The dissipation of heat is largely dependent on the circulatory system to transfer heat by conduction and convection from the sites of production to the external surfaces. As the heat production rises, blood flow to these areas increases, and during heat stress there can be a several-fold increase in cardiac output. The heat is transferred to the skin and loss to the environment occurs. The mechanisms involved in this loss are *Conduction*, which is usually low; *Convection*, which is related to air velocity, i.e. as the wind speed increases the contribution of losses by convection rises. The greater the temperature gradient between skin and environment the greater the convection losses can be; *Radiation*, which is of minor importance during exercise, and *Evaporation*, which is by far the most important method of heat dissipation in the horse. Every kg of water evaporated removes 600 Kcal of thermal energy. The efficiency of sweating, however, depends on the external environment. In a hot dry atmosphere evaporative cooling is highly efficient, whereas in humid heat, sweating may be inefficient because up to two thirds of the sweat secreted will run off the skin without being evaporated. Coat thickness also affects the efficiency of sweating. Studies at the University of Glasgow showed that horses with winter coats are much less able to tolerate ambient conditions of 40°C and 60% relative humidity than those with a summer coat since the evaporative loss is impeded. Sweat dropping freely from an animal indicates an effective sweating mechanism but an inefficient system of heat dissipation. Horses transported from European to tropical countries may develop a condition called anhydrosis (dry coat). This is a complete or partial loss of the ability of the horse to sweat due to degeneration of the cells in the body of the sweat gland. Hyperthermia rapidly develops and the exercise tolerance of these animals is markedly reduced.

In addition to sweating which, in the horse as in man is an extremely well developed system, evaporation also takes place from the respiratory tract. In man, it is estimated that this contributes to about 5% of the heat loss during exercise, whilst in naturally panting animals such as the dog, the respiratory route provides the most important means of heat loss. Although it is generally considered that evaporative loss from the respiratory tract is of minor importance in equine thermoregulation, studies using an environment chamber have shown that when evaporative losses from the skin are insufficient, rapid breathing can occur to aid in the maintenance of body temperatures. Respiratory rates of up to 230 breaths per minute have been recorded with respiration of a markedly shallow nature. Heat loss occurs under such conditions by a process known as dead space ventilation, as air is moved rapidly within the nasal cavity and upper respiratory tract. In endurance exercise where increased alveolar ventilation is required together with heat dissipation, a respiratory alkalosis may then result.

In the resting horse muscle temperature (36.8°C) is slightly below body temperature (37.5°C). However, during exercise it will rise rapidly in working muscle and heat dissipation becomes necessary for continued effective muscular contractions. The rate of heat production within muscles is dependent on the intensity of the exercise, as at higher intensities more ATP is required. During intense exercise heat production can increase 20-fold. Although no data is available in the horse, extrapolation from man suggests that the energy required to increase by 1°C the body temperature of a 500 kg horse will be of the order of 415 Kcal (1738 KJ). If during the prolonged exercise of endurance rides where the intensity approximates 50% $VO_{2\,max}$, an oxygen consumption of 60 ml/kg/min is being utilised to metabolise a mixture of fat and glucose, an energy output of 4.8 Kcal/litre O_2 would result in the production of 142 Kcal of heat per minute. Therefore, there would be a 1°C rise in temperature approximately every 3 minutes of exercise if no temperature regulatory mechanisms were available. Heat production is usually well tolerated in endurance rides however and it is only in the presence of a high temperature and humidity that problems occur. Under particularly adverse conditions rectal temperatures as high as 47°C have been recorded following endurance rides and the cross-country stages in 3 day trials. Obviously as workload increases, heat production rises and the ability to exercise in high temperatures is reduced. In a standardbred race, rectal temperatures have been shown to rise to 39.8°C, whilst muscle temperatures reached 41.5°C.

Maintenance of body temperature within normal physiological levels is dependent on a negative feedback system between temperature sensors, the integrator (thermoregulatory) centre and the effector (sweat glands) organs. The water lost by evaporation from the skin is produced by sweat gland activity in response to stimuli coming from the thermoregulatory centre in the hypothalamus or circulating adrenaline. The rate of evaporation is determined by the rate of sweating, as well as by environmental conditions. The temperature sensors are nerve endings within the skin surface but are also found in the pre-optic area of the hypothalamus and it is these which are extremely important in detecting increased blood temperature associated with exercise.

Sweat gland function

In the horse sweat glands are found in almost all areas of skin associated with hair follicles and are therefore referred to as epitrichial. Changes within these sweat glands during prolonged sweating have been recently described. Alterations occur which may partly explain fatigue or reduction in the sweating rate after several hours of profuse sweating. The glands are innervated by the sympathetic nervous system and have β_2-adrenoceptors. Work in donkeys has demonstrated that heat-induced sweating is controlled via sympathetic nerves, whilst exercise sweating is mediated both via the sympathetic system and circulating adrenaline released by the adrenal medulla during exercise. Adrenaline is also responsible for the psychological factors which cause sweating in the horse.

In addition to the water that is secreted to maintain thermoregulation, electrolytes and other constituents are excreted in the sweating process. Although most of the physiological and biochemical changes which occur during exercise are similar in man and the horse, it is of interest to note that although sweating in both is the main thermoregulatory process, the concentration of many of the constituents are markedly different. As shown in Table 9.6 horse sweat is hypertonic: in that the major electrolytes Na^+, K^+ and Cl^- are in concentrations in excess of that in the plasma, whilst in

Table 9.6. *Electrolyte concentration in plasma, muscle and sweat (horse and man)*

	Electrolytes (mmol/litre)				
	Na$^+$	K$^+$	Mg^{2+}	Ca^{2+}	Cl$^-$
Plasma	140	3.5–4.5	0.8	2	100
Muscle	9	162	15.5		6
Horse sweat	130–190	20–50	16		160–190
Human sweat	40–60[a]	4–5	0.8–2.3[a]	2–4[a]	30–50[a]

From Costill (1977).
[a] Dependent on sweating rate.

man, it is hypotonic. The hypotonicity in man may be an evolutionary adaption to preserve electrolytes, as, during secretion, electrolytes are reabsorbed in the tubular portion of the sweat gland. With increased sweating rates, reabsorption cannot proceed fast enough with a consequent increase in electrolyte concentration in sweat. The higher electrolyte concentrations in equine sweat means that at comparative sweating rates electrolyte loss is considerably greater in horse than man.

In the initial stages of sweating, calcium and magnesium are secreted at concentrations considerably higher than those found in plasma. However the rate gradually declines to plasma levels with time. Peculiar to the horse is the finding that protein is also secreted in relatively large amounts in the sweat. It is this high protein content that causes the lather seen with sweating. It was originally thought these high amounts could contribute to a significant body protein loss during prolonged sweating. However, recent studies have shown that although initially protein concentrations are high, they decrease as sweating continues (Fig. 9.16). The protein is thought to be stored within the sweat gland cells in granules which disappear during sweating. Recent studies have identified two major proteins of relatively low molecular weight in horse sweat. Albumin, contrary to earlier suggestions, only constitutes about 1% of the total protein present. So far only one of the two major proteins has been characterised and has been named latherin. It has an exceptionally high content of hydrophobic amino acids, with leucine contributing 1 out of every 4 amino residues. Latherin adsorbs very readily to hydrophobic surfaces, making them wettable. It is possible that this property allows the sweat to form a thin film, rather than droplets on the horse's hair thereby aiding more efficient evaporation of sweat. Interestingly changes in magnesium concentrations parallel the decline in protein loss, although this ion is not protein bound.

Fluid loss

No satisfactorily controlled fluid balance studies have yet been conducted during different types of exercise in the horse. However, in a small study

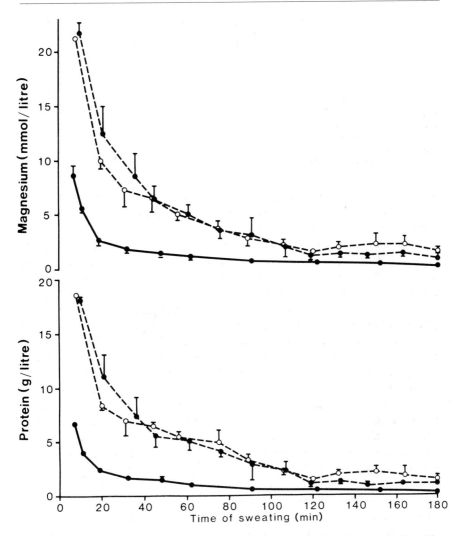

Figure 9.16. Change in magnesium and protein concentration of sweat induced by a prolonged infusion of adrenaline. ● Sweat from neck. ○ Sweat from body. Broken line indicates sweat from unwashed skin and a continuous line sweat collected following prior washing of skin. Mean ± SE (Kerr and Snow, 1983).

using four thoroughbreds ridden for 80 km at a speed of 16 km/h over easy terrain in a mild climate, we have been able to calculate sweating rates of approximately 7–8 litres/h from body weight losses. Over an exercise period of 5 hours, sweat loss can approach almost 8% of body weight which is indicative of moderate to marked dehydration. Speed is an important factor in determining sweating rates, but terrain and environmental temperature can also have marked effect. This accounts for the varying degrees of dehydration reported, based on calculations of changes in haematocrit and total plasma protein concentrations, during endurance rides. A loss of 10–15 litres/h may occur in horses exercised at a fast pace in a hot dry climate. As sweat induced dehydration progresses the circulatory volume is decreased, and when this reaches a critical level, a reduction in the sweating rate occurs. Once this happens the upward spiral of the body temperature to a life-threatening hyperthermia commences. It is presently uncertain whether decreased urinary production occurs to compensate for the fluid losses in sweat. However, with marked fluid loss it is likely that some reduction in renal blood flow will occur, leading to a decrease in urine formation.

When moderate to marked dehydration develops clinical signs are seen such as loss of skin elasticity, sunken eyes and dry mouth, reflecting that most of the fluid lost is from the interstitial spaces. To prevent large losses occurring during prolonged exercise, frequent rehydration is desirable. Horses should be encouraged to drink whenever possible during endurance rides and, to overcome natural reluctance to do this, they should be encouraged to drink during training. Although for a long time it has been thought both in man and horse that the consumption of large quantities of cold fluids was undesirable, recent studies in man do not support this view, and indeed it now appears to actively aid in heat dissipation.

Electrolyte loss

Prolonged sweating leads to a marked electrolyte loss. In the 80 km study mentioned previously it was concluded that mean sodium and chloride concentrations in the interstitial fluid decreased from 140 mmol/l and 115 mmol/l to 132 and 85 mmol/l respectively, whilst potassium loss from the intracellular compartment was of the order of 6–7%. The large loss in chloride ions which results in decreased plasma chloride concentrations is considered to be responsible for the metabolic alkalosis seen in endurance rides. To maintain ionic balance bicarbonate ions are retained. In the light of these findings, it appears remarkable that clinical symptoms such as synchronous diaphragmatic flutter (thumps), characteristic of ionic losses, do not occur more frequently. To overcome these electrolyte losses it is important to add electrolyte supplements to drinking water and train horses to drink salty-tasting water. Many electrolyte preparations are available, but few have the correct composition to replace electrolyte losses. A correct formulation should contain four times the weight of NaCl to KCl, with smaller amounts of calcium and magnesium. Preparations containing bicarbonate should not be used as these can exacerbate an alkalosis.

Table 9.7. Sodium and potassium balance in the exercising horse

	Sodium (g)	Potassium (g)
Amount in feed per day (hard feed plus hay)	36	95
Daily requirements	14	60
Excess intake per day	22	35
Loss per hour of sweating	27	15
Situation after 1 h of sweating	−5	20
Situation after 3 h of sweating	−59	−10

If a horse is exercised daily at a pace which causes sweating for prolonged periods, it is likely that the feed will not provide sufficient quantities of some of the electrolytes to replace those lost in sweat. The effect of a moderate rate of sweating on sodium and potassium balance is shown in Table 9.7. Although no studies have been carried out in the horse, it is likely that similar adaptations to that found in man occur during heat acclimatisation and with training. Even after exposure to a hot environment for a few days, acclimatisation occurs in man. This is brought about by an increase in sweat production due to an increased sensitivity of the glands. Training also enhances the sweating mechanism and for optimal heat acclimatisation, simultaneous exposure to both heat and exercise is recommended. Although these mechanisms are undoubtedly of benefit in thermoregulation, the increased fluid and electrolyte losses make the availability of replacement fluids even more important during a ride.

ENDOCRINOLOGY OF EXERCISE

The demands of exercise result in an alteration in the normal homeostasis of the body. To meet the requirements of exercise and to maintain long-term homeostasis, numerous hormonal substances are released from endocrine organs. The release of these hormones is dependent on the intensity and duration of exercise and they play an immediate role in the control or modulation of the physiological and metabolic responses. In addition, many bring about the long-term training adaptations. Unravelling the roles of the various hormones is difficult because of complex inter-relationships between many of them which are required to elicit an integrated response during exercise. It would appear that the sympathoadrenal system is pivotal in the adjustment of circulating levels of many hormones. The anterior pituitary gland also seems to have a vital regulatory role.

The hormonal responses to exercise have been extensively studied in man and laboratory animals, but few such investigations have been carried out in the horse. Most equine studies have been restricted to the measurement of alterations in circulating hormone concentrations during "one-off" exercise bouts, with little attempt to analyse the significance of the changes on the physiological and metabolic responses. Interpretation of results from the measurement of circulating hormone concentrations is difficult because plasma concentrations depend on the rate of release from the endocrine organ, its uptake or interaction with membrane receptors of the target organs, and its metabolism in non-target organs such as the liver. In addition a number of hormones circulate as both plasma-free and protein-bound forms, and it is only the free concentration that exerts biological activity. Therefore when exercise at a similar relative workload is performed after training, a decrease in circulating concentrations of a hormone does not necessarily suggest a lesser importance of that hormone, but may indicate an increased sensitivity of the target organ or indeed that more of the hormone is circulating in the free form.

Hormones can elicit specific responses from their target organs by acting on receptors. These receptors may be located on the cell membrane and may require a "second messenger", such as cyclic AMP to catalyse the production of, for example, protein kinase which then initiates the characteristic action of the hormone. On the other hand they may be transported into the cell and bring about a change in cellular protein synthesis. It is likely that this latter action results in a less immediate response to the demands of exercise. Only those hormones which have been shown to have an important regulatory role during exercise will be considered.

Sympathoadrenal activity

The release of the catecholamines, adrenaline (from the adrenal medulla) and noradrenaline (the neurotransmitter) results from the stimulation of the sympathetic nervous system. The best way of assessing sympathetic activity during exercise is to measure plasma concentrations. The presence in plasma of adrenaline results from an overflow into the circulation following release into the synapses. In the horse an indirect measurement of sympathetic activity is the increase in the resting haematocrit. It has been well documented in man that catecholamine concentrations vary exponentially with work intensity and that at lower workloads noradrenaline concentrations increase more than adrenaline. During endurance rides a small but significant increase in concentrations of plasma of both catecholamines has been reported. Maximal exercise on a treadmill has resulted in an increase of more than 50-fold in both plasma concentrations of noradrenaline and adrenaline.

To meet the increased catabolic requirements for a general acceleration of body functions, the release of catecholamines mediates a multitude of physiological and metabolic responses. Sympathetic activation has become synonymous with "fight or flight". The importance of the sympathetic system for normal performance can be seen when adrenoceptor antagonists are administered prior to exercise. In man the effects on performance of β-adrenoceptor antagonists such as propranolol, used widely in the treatment of hypertension, have been extensively studied, and it has been shown to cause reduced performance, especially at high intensity workloads. Similarly, in the horse, following administration of these drugs, both speed and endurance time declined with reductions in many of the physiological and metabolic responses associated with fast galloping.

In addition to such adjustments of the cardiovascular system and the initiation of sweating, catecholamines also enhance the availability of substrates to working muscle by rapidly activating lipolysis in adipose tissue and glycogenolysis in the liver. This results in high circulating concentrations of FFAs and glucose which favours their uptake by muscle. It is the release of catecholamines which is responsible for "sweating up" seen before racing and, although if excessive may be deleterious, moderate sweating, indicative of the horse having the "adrenaline flowing" may be useful in priming the body for subsequent increased physiological and metabolic demands. Following training it appears that a diminished sympathetic activity occurs and this is reflected by the lower plasma catecholamine concentrations for a given workload.

Insulin and glucagon

These two pancreatic hormones play a vital role in the control of blood glucose concentrations. Insulin is secreted by the β-cells of the pancreas and enhances the uptake by tissues of glucose, thereby decreasing blood levels and enhancing glycogen deposition. It also acts on adipose tissue to enhance FFA uptake and fat deposition. Insulin is secreted when blood glucose levels are high. Glucagon operates antagonistically to insulin and is secreted in response to low blood glucose levels by the α-cells of the pancreas. It stimulates hepatic glycogenolysis and gluconeogenesis. It is now considered that catecholamines stimulate the immediate release of glucose at the onset of exercise while glucagon provides a longer term stimulus that includes gluconeogenesis. It is likely that glucagon has a much more important role to play in meeting the glucose requirements during endurance rides when sympathetic drive is low, but the demands on circulating glucose are high.

During endurance rides there is a pronounced fall in plasma insulin concentrations but a very marked elevation in glucagon. With maximal exercise no change in plasma insulin concentration is found immediately post exercise, but during the first hour of recovery levels rise in response to markedly elevated blood glucose concentrations.

Adrenocorticotrophic hormone and cortisol

Adrenocorticotrophic hormone (ACTH) is released from the anterior pituitary gland under the control of the hypothalamus and stimulates the secretion of glucocorticoids from the adrenal cortex. As in man, cortisol is by far the major glucocorticoid released in horses. The cyclic fluctuations in the release of ACTH, however, results in a diurnal pattern of plasma cortisol with the highest concentrations occurring in the early morning in the normal resting horse. Alterations in this pattern of secretion occur in a wide variety of stressful situations including exercise. With normal resting concentrations, most of the cortisol is either tightly bound to a specific protein, transcortin, or loosely bound to albumin. As concentrations increase on exposure to stress, saturation of these binding proteins results, and a higher percentage is found in the free form, which greatly increases the amount available for biological activity. Although cortisol has an effect on general metabolism, its precise effects and their exact significance during exercise are still largely unknown. One of its most important actions however is to stimulate gluconeogenesis, in support of glucagon. Another known effect is the alteration in leucocyte count, and although there has been considerable speculation the significance of this phenomenon is also unknown.

During submaximal exercise of varying duration, there is a progressive increase in plasma cortisol concentrations in the horse. This progressive increase does not necessarily reflect an increasing rate of secretion of ACTH as the ride becomes more stressful, but is more related to the fact that cortisol has a relatively long plasma half-life (about 90 minutes) in the horse. During maximal exercise, marked increases in plasma cortisol concentrations are also seen.

During training, it has been reported that in man and in other species hypertrophy of the adrenal cortex occurs resulting in an increase in resting cortisol concentrations. However, a study in racehorses in Newmarket, found no significant change in resting plasma cortisol concentrations, and a further study reported no change in the cortisol response to a standardised exercise test throughout a 3-month training period. It is however possible that an over-extensive training/racing programme may lead to adrenal exhaustion (transient acute adrenocortical failure) resulting in lowered plasma cortisol and aldosterone concentrations at rest and after exercise and a number of cases of poor performance have been attributed to adrenal exhaustion. A study of poor performers in New Zealand suggested that a hyperkalaemia and a haemoconcentration resulting from dehydration was due to deficiency of aldosterone due to adrenal exhaustion. In Sweden, poor performance in standardbreds has been associated not with dehydration, but with a polycythaemia due to overproduction of erythrocytes, and adrenal cortex insufficiency.

Thyroid hormone

Thyroid hormones are important for the maintenance of normal cell metabolism in the resting state. In man it has been shown, and in the horse it has been suggested, that hypo- and hyper-thyroid states result in a reduced work performance. As with cortisol, secretion of these hormones is under the control of a feedback mechanism via the hypothalamus and release of thyroid stimulation hormone (TSH) from the anterior pituitary gland. Thyroxine (T_4) is by far the major hormone released from the thyroid. Peripherally, T_4 is taken up by tissues, and monodeiodinated to form two metabolites 3,5-3'-triodothyronine (T_3) and 3,3-5'-triodothyronine (rT_3). T_3 is believed to exert most biological activity of the group with rT_3 having little activity. It appears that the highest metabolism of T_4 occurs in muscle, as it is found in highest concentrations in these tissue, which also have the highest concentrations of the deiodinating enzymes. This supports the putative importance of T_3 in metabolic regulation in muscle.

Almost all the circulating T_3 and T_4 are protein-bound to the specific thyroid-binding globulin (TBG), prealbumin and albumin. The horse has an unusual binding protein content with free T_4 and T_3 concentrations being approximately 0.01–0.02% of total concentrations.

Although thyroid hormones are known to play an important role directly and indirectly in metabolism of nervous, cardiac and skeletal muscle, their significance in mediating metabolic adjustments during exercise, or with training, is still largely unknown. Because of their high protein binding and their extremely long plasma half lives (several days), it is considered that they may be more concerned with regulating long-term effects of exercise. There is conflicting evidence in man on whether plasma concentrations of these hormones change with exercise or training. The dominating finding is that total plasma concentrations of T_4, T_3 and rT_3 as well as the clearance of these hormones are unchanged, with exercise. In contrast a worker in New Zealand has found both in man and horse that with training an increased clearance of T_4 occurs leading to a lowered T_4 pool. However, because of a higher percentage of free T_4, the absolute circulating concentrations of free T_4 are maintained.

STANDARDISED EXERCISE TESTS AND ANALYSIS OF FITNESS

In training the horse towards its optimum performance, it is helpful to know when the horse can perform to its peak ability, i.e. whether it is fully fit and healthy, and also to have some idea of its genetic potential, i.e. does the horse have the necessary physiological attributes for the purpose envisaged. Obviously the latter is a difficult area and, as it is multifactorial, it is virtually impossible to obtain enough answers to make accurate predictions. Already in this chapter it has been shown how determination of muscle composition and heart score may be of value. Another criterion of potential value is the analysis of gait pattern obtained from high-speed cinematography. Whether we have enough basic information on gait patterns to presently make predictions on gait analysis is, perhaps, questionable. However, some experts believe that, in time, gait analysis will be shown to have considerable value.

The use of physiological tests is likely to be more useful in the elimination of horses that are shown not to have the correct "make-up" for a specific purpose, rather than in the prediction of winners. It is now well established that predictions of racing performance on breeding are often unreliable and it is increasingly being realised that scientific methods of assessment are a necessary part of training assessment. On the basis of Timeform ratings, for example, racing performance has only a heritability factor of 0.35. It must however be borne in mind that even if science aids in a higher predictability in the selection of winners, it cannot remove psychological factors, which can influence the ability to train as well as the "will to win".

Fitness tests

Generally, the assessment of fitness is subjectively evaluated by either the trainer and/or rider using various criteria. These vary from trainer to trainer, and judgement can similarly vary widely depending on experience and idiosyncracy. To overcome some of these subjective problems a number of scientific tests to assess both stamina and strength, under standardised conditions, have been devised for the human athlete and a number of research institutes are assessing the suitability of these for the horse. So far, tests in the horse involve examination of aerobic capacity, which although probably very useful for the assessment of fitness for events involving stamina, may be of less use in strength events, e.g. sprinting.

Traditional assessment has relied heavily on studying the animals both at work and within the stable. The ability to observe is one of the most important qualities in a good trainer. Whilst at exercise the rider and/or trainer will watch closely how the horse works, the day to day progress and overall performance in terms of expectations. Within most racing yards the performance of several animals of similar ability will be compared, sometimes aided by the timing of training gallops. As respiratory patterns are easily heard and observed these can be of particular use in evaluating fitness. The more unfit the horse, the longer forceful respirations will continue after exercise. Following high intensity exercise, increased respiration is necessary for the horse to repay the oxygen deficit arising from lactic acid formation.

An important but little used aid in assessing fitness is the weight of the horse. Most horses start training appreciably overweight and a reduction is anticipated as training proceeds. Although it would appear obvious that an overweight horse, just as a fat human, cannot perform at its best, it is surprising how many horses at the racecourse and at other forms of competition are overweight. It should be realised that an overweight horse is handicapped just as much as if it were carrying an overweight rider, and this applies to any event that involves maximum effort, whether in a race or an endurance ride.

Although many experienced trainers feel that they can accurately estimate a horse's weight, and small losses and gains, this confidence must, in some cases, be questioned. In stables where large numbers of horses are being trained, the installation of a weighbridge is probably justified. However, where only one or a few horses are trained, weight can be reasonably accurately determined using a good equine measuring tape. By regularly weighing horses, several leading thoroughbred racing stables have found that every horse has its own ideal racing weight, and anything above or below this weight may result in reduced performance. Perhaps surprisingly, a horse's ideal racing weight does not usually alter between its 2- and 3-year-old seasons although a slight gain may be seen when they reach 4 years of age.

Standardised exercise tests (SETs)

To enable fitness to be properly quantified in horses, a number of standardised exercise tests (SETs) are being developed. In man, such tests are readily accomplished by exercising subjects on a treadmill or by using a bicycle ergometer with increasing workloads. Treadmills for use in the horse range from relatively cheap models the speed of which is restricted to work only at a walk or slow trot, although workloads can be increased by raising the incline. Such treadmills however are most suitable for training purposes and not for SETs. More expensive high-speed treadmills are, however, commercially available and allow SETs to be carried out specifically for fitness assessment and to aid diagnosing the cause of poor performance (Fig. 9.17). Using treadmills which can attain speeds of 14 m/sec (30 mph) or more at slight inclines, the work effort experienced in both trotting and galloping races can be simulated. Experience with these treadmills has proved that horses adapt rapidly to work on them at high speeds. Benefits include the ability to control environmental conditions, the collection of blood samples during and after the SETs, and physiological monitoring.

SETs for field use are also under development. Obviously, reproducible conditions are more difficult to obtain in these due to daily variations in such factors as track conditions and weather. Constant speeds can be fairly readily maintained in standardbreds although this is considerably more difficult for thoroughbreds. As described previously, radio-controlled systems now allow for the collection of blood samples whilst horses are being exercised in the

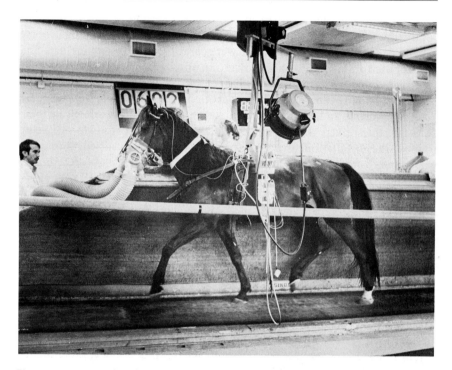

Figure 9.17. Horse fitted with mask for O_2 consumption determinations exercising on a treadmill. (Photograph courtesy of S. Persson. Upsalla, Sweden.)

field (Fig. 9.18) and monitors attached to the horse are available not only to measure heart rate, but also blood and rectal temperature, respiratory rates and oxygen consumption.

In conjunction with the SET, the next problem is to decide which parameters should be measured to give an index of the horse's potential, or actual state of fitness. Many important parameters change with increasing workload, some showing abrupt changes at certain points as the intensity increases. These all have varying technical difficulties in their measurement but recent advances including miniaturisation of recording equipment have increased the number of parameters that can be accurately measured. In man, the main fitness assessment criteria are the increase in $VO_{2\,max}$, lowered heart rates at given absolute workloads, and more recently the point of onset of the anaerobic threshold (arbitrarily set by many at 4 mmol/lactate). This latter measurement has been used because $VO_{2\,max}$ may only increase slightly with training and has not been found to be a reliable index of fitness. On the other hand, following extensive aerobic training the percentage of $VO_{2\,max}$ at which the anaerobic threshold occurs is markedly elevated. In elite marathon

Figure 9.18. Placement of catheter and blood collection tubes for radio-controlled blood sampling during exercise.

runners and cross-country skiers the threshold may not occur until 90% $VO_{2\,max}$; whilst in the untrained non-athletic individual the threshold is usually 50–60% $VO_{2\,max}$.

the determination in horses of the anaerobic threshold, also referred to as the point of onset of blood lactate accumulation (OBLA), involves finding the work intensity at which there is a sudden upsurge in the blood lactate concentration, i.e. the point at which production is not matched by the rate of removal from the blood. The anaerobic threshold is determined by working the horse on a treadmill or on the track at different speeds for a set period of time. Four or five progressively increasing work intensities, of 1 to 2 minutes duration are used. Blood samples for lactate determination are collected after each workload, the results plotted and the speed of OBLA (assumed to be 4 mmol/l) calculated by intrapolation/extrapolation. With increasing aerobic fitness there is a shift of the curve to the right as OBLA occurs at a higher speed than in the early stages of training (Fig. 9.19).

Of all the objective means of assessing fitness, monitoring the heart rate is easiest and is readily undertaken by most riders using commercially available heart rate meters. This allows accurate measurements of heart rate both during

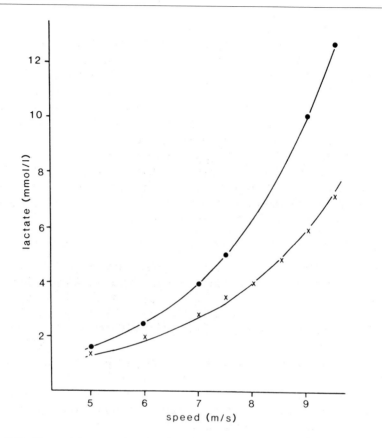

Figure 9.19. Changes in anaerobic threshold with training (modified from Thornton *et al.*, 1983).

exercise and recovery. For standardised exercise testing it has been suggested that the speed necessary to give a heart rate of 200 (i.e. V_{200}) should be obtained.

Health

To obtain optimum performance a horse has to be in prime health. Correct management procedures including vaccination, worm control programmes and foot and dental care are obviously mandatory and routine profiling of various haematological and plasma biochemical analytes may also have an important role. For these to be of maximum value, however, it is important that blood samples are collected from an individual animal at regular intervals, so that baseline values can be accurately determined. This is necessary as there is considerably less variation within an individual animal than within the general population, even for a particular breed. Small changes in some blood constituents may well indicate the onset of sub-clinical disease conditions requiring more detailed investigations. Although the measurement of red blood cell numbers and associated indices for assessing potential or fitness has been criticised, as each individual animal generally has similar values from month to month, any decreases may indicate the presence of an underlying problem. Similarly, alterations from established baseline values of total white blood cell count and neutrophil:lymphocyte ratio should be further investigated. If haematological indices are to be determined, it is very important that samples are collected at the same time. Ideally they should be collected early in the morning prior to feeding and exercise. If this is not possible, then they should be collected at evening stables prior to feeding and on a day when only light exercise has been given.

Plasma enzymes

The routine determination of certain plasma enzymes may also be useful in determining optimum health/fitness. Generally horses perform best when plasma enzyme activities remain constant. The activities of plasma creatine kinase (CK) and aspartate aminotransferase (AST or GOT) have often been used to indicate both the presence of muscular disorders and even the state of fitness. In the early stages of training, especially in 2-year-olds, when heavier workloads are first introduced, elevations in the plasma activities of CK and AST often occur due to muscle damage. With increasing fitness these elevations usually disappear following exercise until, in the fit animal, remarkably constant values are seen. When this stage is reached, even a small increase in enzyme activities should be investigated as it may well be significant. Elevated plasma enzyme activities however do not always mean that performance will be below par. For reasons as yet unknown, some horses competing successfully in both racing and endurance rides have shown quite marked elevations in CK and AST with no clinical abnormality detected. Following a race, there should only be a very small increase in these enzymes in normal horses, with a return to baseline values occurring within 24 hours. However, following endurance rides, marked increases in plasma enzyme activities has been used as an indication of lack of fitness.

Rhabdomyolysis ("tying-up", azoturia) is a well known muscular problem in horses undergoing training for competition in various events. Rhabdomyolysis is a syndrome with probably several underlying causes which may include nutritional and hormonal disturbances. Recent work at the Animal Health Trust has indicated that electrolyte imbalances may play an important role in many recurring cases. There appears to be a predisposition in young racing fillies, but in older animals there appears to be no sex predilection. Following a "tying-up" episode, plasma concentrations of CK and AST can be moderately or extremely elevated depending on the duration of the attack and the extent of the muscle mass involved. Because of the considerably longer plasma half-life of AST compared to CK, the collection of sequential plasma samples can usefully indicate whether the condition is continuing or resolving.

Recently another plasma enzyme, gamma-glutamyl transpeptidase (GGT) has been shown to have possible value in fitness/health screening. It appears that where an intensive training programme is used, some horses will have plasma GGT activities above the normal range. Often on close examination of these animals some minor injury or illness can be detected. Elevations therefore may occur due to overstress in their training programme. Why this enzyme, which normally becomes elevated in the plasma after hepatic damage, increases in training is presently unknown but, following elevation, the decline in plasma activity occurs only very slowly, as it has a plasma half-life of over two weeks.

TRAINING

To obtain the best performance from a horse requires the optimum conditioning of all of the systems involved during the specific work task. A fit horse is less likely to be completely fatigued at the end of a race. This is an important consideration as it has been suggested that fatigue is the major cause of breakdown in racehorses. Both the intensity and duration of the horse's training schedule influences the state of conditioning obtained. Obviously discretion is needed as there are limits to which different body systems can be stretched. There is no point in trying to maximise development of the muscular system if, as a result, this leads to skeletal and joint problems. The effect of age on the amount of training that can be given is also an important consideration. Just as in man, the immature, growing animal is unable to withstand the same rigorous training schedule as the mature well developed animal. This is clearly shown by the higher incidence of breakdown in racing 2-year-olds.

It was stated in the beginning of this chapter that the aim was not to describe specific training programmes, but rather to give an understanding of the basic physiological adaptations that occur with different intensities of exercise. However, a consideration of some salient points on the major features for inclusion in a training schedule may be of value. The establishment of a training programme requires specificity, individualism and common sense. Specificity implies that the type of fitness to be developed, and therefore the training programme used, must be aimed at the competition objective. Individualism refers to the fact that every animal is different so that any programme must be flexible enough to allow for continuous adjustment to meet the needs of the individual horse and its responses. Common sense implies an ability to observe how the horse is responding, and then making appropriate modifications to the training programme and diet.

When considering the design of a training programme, the trainer has to decide in which areas he wants improvement. The various improvements generally aimed for are (i) improved energy production to withstand fatigue; (ii) an improvement in skeletal structure within the animal; (iii) improved skill brought about by the development of neuromuscular coordination; and (iv) psychological familiarity leading to the horse performing in a more relaxed, energy conserving, manner.

Any training programme essentially consists of three stages, the early stage, the harder stage and the maintenance stage.

Stage 1

During this early stage, low intensity exercise of moderate duration is undertaken and the standard is generally similar for all types of eventual competition. This stage will produce the early conditioning and will improve the suppleness and mobility of joints and tendons. During this period the time spent exercising daily is gradually increased, with an increasing amount of time spent at the trot. During this stage, which will generally last at least one month, little adaptive changes in muscle and cardiovascular function are likely to occur, as only a small proportion of the total muscle fibre pool is recruited. Heart rates are unlikely to exceed 130–140 bpm. The importance of this stage is that it produces the foundation on which more intense workloads can be introduced.

Stage 2

This second more difficult stage is when real fitness starts to be developed. It is also the time when specific training schedules are introduced. For instance, horses involved in activities based largely on strength, e.g. showjumping, cutting or quarterhorse racing should have training incorporating anaerobic work and coordination in order to build up power. For the trotter, pacer and the middle distance or staying thoroughbred, the programme must consist of both power and aerobic elements. For endurance rides the objective is to place the emphasis on training which will markedly increase the aerobic capacity and thus raise the anaerobic threshold. To accomplish these objectives, a number of continuous or intermittent exercise schedules have been devised.

Continuous training programmes are generally used by horses destined for endurance activities and, throughout Stage 2, both the duration and intensity of training is increased. Horses are worked below their anaerobic threshold at heart rates between 160–180 bpm. During the training programme, some days are devoted to high intensity workouts to improve the capacity to maintain high speeds and to increase the ability to climb steep hills.

Traditional methods of training thoroughbreds have also involved continuous training programmes, generally 4 days per week of slow work, and 2 days of hard work, when the horse is given a warm-up period followed by a single bout of fast work, before being walked back to the stables. In comparison to the intensive training programmes used in human athletes, this type of training is now thought to be inadequate. As a result, more intense training is now considered by some trainers to be desirable and intermittent (interval) training programmes are increasingly being adopted. These involve the horse undergoing a series of exercise bouts interspersed with periods of relief to allow partial recovery. The rationale is that such systems allow the muscular system to build up high levels of lactate without the possible deleterious effects which would be anticipated if a similar amount was formed during a single bout of exercise. The horse therefore should be more readily able to handle the amounts of lactate generated during racing. The concepts and suggested advantages of interval training in the standardbred, thoroughbred and quarterhorse have been discussed by Tom Ivers, a leading proponent of these methods (Ivers, 1983). However, it should be realised that the use of interval training, although accepted for many years in standardbred circles, is only in its formative years for thoroughbred training. There are a number of arguments both for and against its use in this breed and it has to be borne in mind that for all horses all sorts of computations relating to speed, duration, relief intervals, and repetitions, can be devised and the optimum result will vary with the objective. On-board heart rate meters may be of value in helping to decide rest intervals and the number of repititions that can be safely undertaken.

In most interval training programmes the aim is to increase the anaerobic threshold. Although this is desirable for endurance rides, eventing and some racing distances, it may not be good for animals racing over sprint distances. This is because as muscle fibres increase their oxidative capacity, the larger IIB fibres also decrease in size. This reduces the muscle mass, which is so important in generating power. Therefore, it is not possible to develop peak aerobic capacity and peak strength at the same time. Training programmes for sprinting thoroughbreds and quarterhorses must utilise schedules that place the emphasis on power. For this it may be possible to use a version of interval training involving repeated short bursts of maximum speed, with each burst having a period during which complete recovery occurs. Recovery can be assessed by a return of heart rates to around 70 bpm. Using this technique, neuromuscular coordination is improved as well as developing power.

Furthermore, for horses involved in eventing, or even to give variety to racing animals, less structured programmes such as the Fartlek system of

warm - up

5 min trot or smooth
canter

muscle strength and
velocity training

fast gallop 120-150m
uphill, after that 4 min
walk; repetition 5-10 times

endurance training

3-4 min gallop appr. 80%
of max. speed, after that
3 min walk; repetition
3-4 times

Figure 9.20. Proposed scheme for a combined intermittent training programme for riding horses (Engelhardt, 1977).

training can be used and a programme for riding horses is shown in Fig. 9.20. As can be seen this involves working up hills which is extremely useful for building up strength.

Stage 3

This is the maintenance stage. Once a horse has attained the desired level of fitness, care then has to be taken to avoid over-training which results in a decreased performance. In Stage 3 the amount of strenuous training being undertaken can be reduced, especially if the horse is competing at frequent intervals. Interestingly there is some evidence that short periods of relative inactivity may not result in any loss of the beneficial effects of training.

To help reduce the risk of injuries, during Stages 2 and 3 when hard days of training are undertaken, it is extremely important that sufficient time is devoted to warming up and cooling down.

Training aids

To assist in the training of some animals suffering from minor injuries, to introduce change to a training programme and in some circumstances to reduce costs, a variety of ancillary equipment is often used in training establishments. Such equipment must however remain an accessory to normal training methods and obviously can never replace track work. Examples include the horse (hot) walkers which have found wide popularity in the USA and Australia, but are much less common in the United Kingdom. They are probably of little use in cardiovascular or muscular development, but can provide the light exercise necessary during warming up and cooling down. Their main benefit is economic as they are undoubtedly labour saving.

Treadmills are used for standardised exercise testing to aid in assessment of fitness and clinical diagnosis. In addition, the cheaper, slower speed models

Figure 9.21. Swimming in a circular pool. (Photograph courtesy D. Marlin.)

have, like the horse walkers, found some popularity in the USA and Australia for giving moderate workouts to horses when for some reason they cannot be ridden, due to, for example, back problems, or bad weather conditions. They also may have a place in the training of yearlings.

Swimming (Fig. 9.21) has been used for a long time as a training aid, although its usefulness has long been debated by trainers. Most horses take to swimming and some trainers believe that swimming does a lot of good, but others think it is of little benefit in the majority of animals. Swimming results in increased heart rates to 140–170 bpm with greater work being achieved by making the horse swim against a current or pull weights. It is extremely useful in the maintenance of fitness in horses suffering from joint and tendon injuries as it allows weight to be kept off these structures. It is also of use in horses that may have become stale or bored with training programmes.

Recently, a device has been introduced which is a cross between a flat treadmill and a pool of water called a "water treadmill". Rather than using swimming (which may result in unnatural muscle development) the horse actually walks and trots in water with the water level only up to its belly. The water introduces an extra workload which the horse has to overcome, but it also acts as a cushion against the effects of concussion when landing. From heart rate monitoring the workload is not very great, as heart rates rarely exceed 100 bpm. Again, the main usefulness of this machine may be in rehabilitation of horses after injury and possibly in early training of young horses.

REFERENCES

Asheim, A., Knudsen, O., Lindholm, A., Ruelcker, C. and Saltin, B. (1970). Heart rates and blood lactate concentrations of standardbred horses during training and racing. *J. Am. Vet. Med. Assoc.* **157**, 304–312.

Asmussen, E. (1979). Muscle fatigue. *Med. Sci. Sports.* **11**, 313–321.

Astrand, P.-O. and Rodahl, K. (1977). *Textbook of Work Physiology.* McGraw Hill, New York.

Attenburrow, D. P. (1982). Time relationship between the respiratory cycle and limb cycle in the horse. *Equine Vet. J.* **14**, 69–72.

Bryly, W. M., Grant, B. D., Bruze, R. C. and Kramer, J. W. (1983). The effects of maximal exercise on acid-base balance and arterial blood gas tension in thoroughbred racehorses. In: *Equine Exercise Physiology* (Snow, Persson and Rose, eds), p. 400. Granta Editions, Cambridge.

Engelhardt, Wv, (1977). Cardiovascular effects of exercise and training in horses. In: *Advances in Veterinary Science and Comparative Medicine* Vol. 21. (C. A. Bradley and C. A. Cornelius, eds), pp. 173–205. Academic Press, New York and London.

Exercise Physiology (1985). Veterinary Clinics of North America, Equine Practice, Vol. 1(3). W. B. Saunders, Philadelphia.

Fox, E. L. (1984). *Sports Physiology*, 2nd ed. CBS College Publishing, New York.

Gunn, H. M. (1983). How muscle influences the speed of running in horses. In: *Equine Exercise Physiology* (Snow, Persson and Rose, eds), p. 274. Granta Editions, Cambridge.

Hornicke, H., Engelhardt, V. W. and Ehrlein, H. J. (1977). Effect of exercise on systemic blood pressure and heart rate in horses. *Pflugers Arch.* **372**, 95–99.

Ivers, T. (1983). *The Fit Racehorse* Esprit Racing Team Ltd., Cincinnati, Ohio.

Kerr, M. G. and Snow, D. H. (1983). The composition of equine sweat during adrenaline infusion, prolonged heat exposure and exercise *Am. J. Vet. Res.* **44**, 1571–1577.

Lamb, D. R. (1984). *Physiology of Exercise*, 2nd ed. Macmillan Publishing, New York and London.

Persson, S. G. B. (1967). On blood volume and working capacity in horses. *Acta Vet. Scand.* **8**, (Suppl. 19).

Persson, S. G. B. (1968). Blood volume, state of training and working capacity of racehorses. *Equine Vet. J.* **1**, 52–64.

Persson, S. G. B. (1969). Value of haemaglobin determination in the horse. *Nord. Veterinaermed.* **21**, 513–523.

Physick-Sheard, P. W. (1985). Cardiovascular response to exercise and training in the horse. In: *Veterinary Clinics of North America, Equine Practice*, Vol. 1(2), pp. 383–417. W. B. Saunders, Philadelphia.

Snow, D. H. (1983). Skeletal muscle adaptions, a review. In: *Equine Exercise Physiology*. Eds D. H. Snow, S. G. B. Persson, R. J. Rose, pp. 160–168. Granta Editions, Cambridge.

Snow, D. H. and Guy, P. S. (1977). The structure and biochemistry of equine muscle. *Proc. Am. Ass. Equine Pract.* **22**, 199–210.

Snow D. H. and Vogel, C. J. (1987). *Equine Fitness*. David and Charles, Newton Abbot.

Snow, D. H., Kerr, M. G., Nimmo, M. A. and Abbott, E. M. (1982). Alterations in blood, sweat, urine and muscle composition during prolonged exercise in the horse. *Vet. Rec.* **110**, 377–384.

Steel, J. D.(1963). *Studies on the Electrocardiogram of the Racehorse*. Australasian Medical Publishing Co. Sydney.

Stewart, G. A. and Steel, J. D. (1974). Haematology of the fit racehorse. *J. S. Afr. Vet. Med. Assoc.* **45**, 287–291.

Strauss, R. H. (1979). *Sports Medicine and Physiology*. W. B. Saunders and Co., Philadelphia.

Thornton *et al.* (1983). Changes in anaerobic threshold with training. In: *Equine Exercise Physiology* (Snow, Persson and Rose, eds), p. 481. Granta Editions, Cambridge.

SUBJECT INDEX